The Galileo Affair

The Galileo Affair

A Documentary History

Edited and translated
with an Introduction and Notes by
Maurice A. Finocchiaro

UNIVERSITY OF CALIFORNIA PRESS
Berkeley · Los Angeles · London

University of California Press
Berkeley and Los Angeles, California

University of California Press, Ltd.
London, England

© 1989 by
The Regents of the University of California

Library of Congress Cataloging-in-Publication Data

The Galileo affair : a documentary history / edited and translated with an introduction
and notes by Maurice A. Finocchiaro.
    p.   cm. — (California studies in the history of science)
    Bibliography: p.
    Includes index.
    ISBN 0-520-06360-0 (alk. paper)
    ISBN 0-520-06662-6 (pbk.: alk. paper)
    1. Galilei, Galileo 1564–1642.   2. Religion and science—History of controversy.
3. Inquisition.   I. Finocchiaro, Maurice A., 1942–   . II. Galilei, Galileo, 1564–
1642. Correspondence. English. Selections. 1989.   III. Galilei, Galileo, 1564–1642.
Essays. English. Selections. 1989.   IV. Series.
QB36.G2G319   1989
621.5'5—dc19   88-31503
                                                    CIP

Printed in the United States of America
 2 3 4 5 6 7 8 9

Contents

Preface and Acknowledgments

In 1633 the Roman Inquisition concluded the trial of Galileo Galilei with a condemnation for heresy. The trial was itself the climax of a series of events which began two decades earlier (in 1613) and included another series of Inquisition proceedings in 1615–1616. Besides marking the end of the controversy that defines the original episode, the condemnation of 1633 also marks the beginning of another classic controversy—*about* the Galileo affair, its causes, its implications, and its lessons; about whether, for example, John Milton was right when in the *Areopagitica* he commented on his visit to Galileo in Florence by saying: "There it was that I found and visited the famous Galileo, grown old a prisoner to the Inquisition, for thinking in astronomy otherwise than the Franciscan and Dominican licensers thought." I happen to be extremely interested in this second story and second controversy, and a critical interpretation of the affair remains one of my ultimate goals. But that is not the subject of the present work, which is rather concerned with something more fundamental, namely with the documentation of the original episode.

To be more exact, the aim of this book is to provide a documentary history of the series of developments which began in 1613 and culminated in 1633 with the trial and condemnation of Galileo. That is, it aims to provide a collection of the essential texts and documents containing information about both the key events and the key issues. The documents have been translated into English from the original languages, primarily Italian and partly Latin; they have been selected, arranged, annotated, introduced, and otherwise edited with the following guiding principles in mind: to make the book as self-contained as possible and to minimize contentious interpretation and evaluation. The Galileo affair is such a controversial and important topic that one needs a sourcebook from which to learn firsthand about the events and the issues; since no adequate volume of the kind exists, this work attempts to fill the lacuna.

The originals of the documents translated and collected here can all be found in printed sources. In fact, with one exception they are all contained in the twenty volumes of the National Edition of Galileo's works, edited by Antonio Favaro and first published in 1890–1909. The exception is the recently discovered "Anonymous Complaint About *The Assayer*," whose original was discovered and first published in 1983 by Pietro Redondi; this document is also contained in the critical edition of the Inquisition proceedings edited by Sergio M. Pagano and published in 1984 by the Pontifical Academy of Sciences. My selection was affected partly by the criterion of importance insofar as I chose documents that I felt to be (more or less) essential. Since I was also influenced by the double focus of this documentary history on events *and* issues, I therefore included two types of documents: the first consists of relatively short documents which are mostly either Inquisition proceedings (Chapters V and IX) or letters (Chapters I, VII, and VIII) and which primarily (though not exclusively) record various occurrences; the second type consists of longer essays by Galileo (Chapters II, III, IV, and VI) which discuss many of the central scientific and philosophical issues and have intrinsic importance independent of the affair. Finally, my goal of maximizing the autonomy of this volume suggested another reason for including some of these longer informative essays on the scientific issues (Chapters IV and VI).

The arrangement of the documents is primarily chronological, but a secondary criterion pertains to origin, authorship, or provenance. That is, while the sequence of documents within each chapter is in absolute chronological order (at least insofar as dates are ascertainable), the chronological order of the various chapters is compromised somewhat to ensure the grouping together of documents written by the same individual, or deriving from the same institution, or pertaining to the same subtopic. For example, the dates at the end of Chapter I overlap with those at the beginning of Chapter V, but the separation is advantageous because Chapter V contains the earlier Inquisition proceedings and other directly related texts, and the incisiveness of their sequence would be unnecessarily perturbed by the inclusion of indirectly related letters. There are similar overlappings between the end of Chapter VII and the beginning of Chapter VIII and between the end of Chapter VIII and the beginning of Chapter IX. This arrangement should not, however, cause the slightest confusion because each selection is dated, and thus readers can easily reconstruct the chronology from the documents themselves; moreover, in the Chronology of Events and the Concordance to the Documents the absolute chronological order is explicitly provided.

All documents (except for the Inquisition Minutes of 3 March 1616) have been translated in their entirety as found in the printed sources. In most cases the printed texts are complete, but some of the Inquisition Minutes (in Chapter V) and some of the letters in the "Diplomatic Correspondence (1632–1633)" (Chapter VIII) are published in abridged form in the National Edition of Galileo's works; thus, my translations are abridged to exactly the same extent. While attempting to use idiomatic English and avoid excessive awkwardness, I have tried to be as literal as possible in my translations; I have also tried to keep as much as possible of the original sentence and paragraph structure, use of capital letters, wording of person's titles, etc. In my translations I have also retained whatever repetitions are present within the original documents; for example, parts I and II of "Pasqualigo's Report on the *Dialogue*" (in Chapter IX) are almost (though not entirely) identical. I felt that in a documentary history there was no other choice and that to do otherwise would have violated the guiding principle of minimizing interpretation and evaluation.

The annotations found in the Notes have been compiled by the editor, and none of them are in the original documents. Occasionally, the originals contain references to various authors and books, an example being the "Letter to the Grand Duchess Christina" (Chapter III); in these cases I have inserted the original references in parentheses in the text. My notes are meant to be historical, explanatory, and informative, rather than critical, analytical, or evaluative. However, they contain extensive references to various interpretative and evaluative issues, as well as to the relevant secondary literature. References to the latter are in the abbreviated form, now standard, citing only the author's last name, the year of publication, and the pages. Full bibliographical information is given in the Selected Bibliography. References to the National Edition of Galileo's works have been given by citing the editor's name, Favaro, rather than Galileo.

There is another type of editorial annotation that deserves mention. This occurs within the text of the translations and consists of the corresponding page numbers in the printed sources from which the translations have been made. Since the Concordance to the Documents gives the exact references to the various volumes, mere page numbers were deemed sufficient in these parenthetical references. These references are denoted by numerals enclosed in pointed brackets (that is, the characters ⟨ and ⟩), and they are placed at the point where the page bearing that number begins in the original; whenever possible I have tried to be precise up to the word, but in some cases the different grammatical structure of the respective languages yields only an approximate place, though never more than a few words off. In other cases, these references were best placed at the end of a sentence or paragraph to indicate that the page so numbered begins with the following sentence or paragraph.

Besides the Notes, the Selected Bibliography, and the Concordance to the Documents, there are several other items in the critical apparatus. The Biographical Glossary gives brief sketches of the persons named or mentioned in the documents and in the Introduction, many of whom are too obscure to be described in standard scholarly sources; the sketches of well-known figures are given as a help for students. No such biographical information is given in the Notes, except to establish the identity of persons whose full names are not given when mentioned in the documents, or to give additional details relevant in the particular context. Therefore, the Biographical Glossary is the place to consult whenever readers want information about a name they do not recognize.

The Chronology of Events provides a list of dates followed by brief descriptions of relevant occurrences on those dates; these descriptions are followed by references to sources where documentation can be found, as well as explicit references to the documents translated in this volume. Thus the chronology can also serve as an introduction to the documents, insofar as it places them in a chronological-historical context.

The Introduction is partly a sketch of a philosophical approach to the study of the Galileo affair, partly an elementary account of some of its historical background, and partly a synthetic overview of its main events and issues. It is unavoidably interpretational, but the reader ought to remember that it was written after the documentary history had been compiled. It is *not* the interpretation in the light of which the documents were collected, but rather the one which I felt the duty of

writing at the end of my labors. Further, it contains all the simplifications necessary in an elementary presentation. While I do not think any
of these are *over*simplifications, I have often avoided many of the qualifications and documentations which are common in essays having a
different purpose.

It should be noted that Chapters I through IX consist exclusively of
translations without any introductory or intervening editorial commentary mixed with the text of the originals. The editorial hand is, of
course, present in such matters as the selection in the first place, the
translation, the arrangement, the title given to each document and to
the various chapters, the pagination from Favaro explained above, and
the Notes. However, the intent of this collection is to let the documents
themselves tell the story, as much as possible. Therefore, it may be useful to preview those chapters individually here.

Chapter I consists of the initial correspondence in which is recorded
the immediate background to the earlier Inquisition proceedings. These
letters set the stage for what is to follow by providing explicit information about many of the earlier developments, as well as some of the
central issues.

In Chapter II Galileo discusses at length some of the key epistemological issues involved in the controversy, including answers or sketches
of answers to many of the objections contained in Cardinal Bellarmine's
letter to Foscarini (included in Chapter I). The chapter consists of three
parts: two relatively polished essays and a detailed outline of a third.

In Chapter III we have Galileo's classic discussion of the relationship
between science and the Bible, in the form of a letter to Lady Christina
of Lorraine, the grand duchess dowager of Tuscany. The essay is in part
an expansion of Galileo's earlier letter to Castelli (in Chapter I) and in
part an articulated response to some of the objections found in Bellarmine's letter to Foscarini (in Chapter I). Logically and rhetorically
speaking, it takes the form of an explicit critique of the biblical objection to the earth's motion (according to which this Copernican hypothesis must be wrong because it conflicts with the Bible); implicitly the
critique discusses some of the key epistemological and methodological
issues in the affair, such as the role of the Bible in scientific inquiry, the
relationship between biblical interpretation and scientific inquiry, and
the relationship between theology and science.

In Chapter IV Galileo formulates for the first time an argument in
favor of the earth's motion based on an explanation of the tides in terms
of this hypothesis. This is an argument he had conceived about two

decades earlier and which he elaborated further in the *Dialogue* in 1632. It thus provides information about one of the crucial scientific issues in the controversy.

Chapter V contains the most important of the earlier Inquisition proceedings in 1615–1616, together with a number of other nonofficial writings that are directly related to them. The set provides an explicit account of the developments during the first phase of the legal case and an implicit definition of the legal issues that would become crucial in the second phase in 1632–1633.

In Chapter VI Galileo provides a critique of most of the prevailing scientific arguments against the earth's motion. It is written in the form of a letter to one Francesco Ingoli, who in 1616 had written an essay against the earth's motion; Galileo had been unable to answer it in view of the Church's decisions that year, but in 1624 he felt safe in doing so due to the election of Maffeo Barberini as Pope Urban VIII. The essay served as a first draft of many parts of the *Dialogue* of 1632, and it can serve readers now as a worthy substitute for that classic work, providing useful information about most of the scientific issues of the affair.

Chapter VII contains several miscellaneous documents that record the main relevant developments between the first and second phases of the affair. The key development was, of course, the writing, licensing, and publication of the *Dialogue*, and so, by focusing on that process, this set provides the immediate background to the trial of 1633. It should be noted that this chapter includes one document (the "Special Commission Report on the *Dialogue*") which is in a sense formally a part of the later Inquisition proceedings, but which (more precisely speaking) is not really an Inquisition document but one that served as the official basis for the actual Inquisition proceedings. Note also that this chapter contains the recently discovered "Anonymous Complaint About *The Assayer*," whose import has been exaggerated but which ought not to be totally ignored.

Chapter VIII consists of twenty-nine letters from the Tuscan ambassador to Rome, Francesco Niccolini, to the Tuscan secretary of state in Florence, Andrea Cioli. Their dates (15 August 1632 to 19 June 1633) overlap with both Chapter VII and Chapter IX, but this correspondence provides both a distinct point of view and an independent source of information about the ongoing events and the issues. In particular it provides a unique access to the state of mind of Pope Urban VIII.

In Chapter IX we find the climax of the affair: the more or less formal proceedings of the trial of 1633. Besides the documents originating from the special Galileo file of the Vatican Secret Archives, this chapter contains other crucial texts collected in the National Edition of Galileo's works—namely the letter by the commissary general of the Inquisition to Cardinal Barberini (about an out-of-court settlement), the sentence, and the abjuration. Moreover, it should be noted that this chapter includes the three Inquisition consultants' reports about the *Dialogue*, which not only constitute a step in the formal proceedings but also shed further light on the scientific and philosophical issues.

Finally, in creating this work, I have benefited from a number of persons, institutions, and books, and here I should like to express my acknowledgment to some of them.

First of all, this book would probably not have come into being without the initial encouragement and support of Stillman Drake (Toronto). I am also enormously indebted to John L. Heilbron (Berkeley) for his constant support. Paul K. Feyerabend (Berkeley and Zurich) has provided constant encouragement, support, and comments. I also received helpful comments, suggestions, and constructive criticism from several scholars who read all or part of the manuscript at various stages of its development: James MacLachlan (Ryerson Polytechnic), William A. Wallace (Catholic University), Richard S. Westfall (Indiana), and Robert S. Westman (UCLA). Some of the original suggestions by Winifred L. Wisan (Oneonta, N.Y.) were helpful and were put into practice. My student Kelly Witcraft provided invaluable assistance, both by being a meticulous proofreader of the penultimate manuscript and by playing the role of an intelligent nonexpert reader. My College Dean Tom Wright and my Department Chair Craig Walton provided valuable administrative and material support.

Acknowledgments are also due to the Vatican Secret Archives for granting me the privilege of consulting the special Galileo file kept there; to Mons. Charles Burns from that office for the courtesy shown me during the process; and to Father George V. Coyne, director of the Vatican Observatory, for being instrumental in obtaining that privilege. At my own institution, the University of Nevada–Las Vegas, my gratitude goes to the following: the Sabbatical Leave Committee, for a sabbatical leave in Fall 1986, during which most of this book was written; the University Research Council, which granted me a Barrick Distinguished Scholar award for the academic year 1986–1987; and the College of Arts and Letters and the Department of Philosophy for reductions in teaching load and assignment to the Center for Advanced Research during several subsequent semesters, which allowed me to bring this project to completion.

I am especially thankful to the Pontifical Academy of Sciences for permission to publish my English translation of the document entitled "Anonymous Complaint About *The Assayer* (1624 or 1625)," in regard to which it holds the copyright to the original Italian text.

Finally, I wish to acknowledge my debt to several books I have utilized in my work. In the compilation of my notes, I have borrowed freely from the notes in the National Edition of Galileo's works, in Stillman Drake's works, in Flora (1953), and in Montinari (1983). In the compilation of my Biographical Glossary, my dependence on the biographical index of the National Edition is obvious. And, in my translations, I have often consulted previous versions when available, not only those by Stillman Drake but also those found in the works of Gebler (1879), Koestler (1959), Langford (1966), and Santillana (1955).

Introduction

To facilitate the independent reading, interpretation, and critical evaluation of the documents collected in the body of this work, this Introduction provides an elementary overview of the events and issues of the Galileo affair, together with some of its historical background and a sketch of a philosophical approach to its study. By the "Galileo affair" [1] is meant the sequence of developments which began in 1613 and culminated with the trial and condemnation of Galileo Galilei by the Roman Catholic Inquisition in 1633. Galileo Galilei is the Italian scientist and philosopher whose contributions to astronomy, physics, and scientific instrumentation and methodology in general were so numerous and crucial that, of the several founders of modern science, he is usually singled out as the "Father of Modern Science." The approach to be sketched here is philosophical in a double sense. First (section 1), the study of the affair is motivated for the light it may shed on a number of general cultural and interdisciplinary problems that have their own intrinsic theoretical interest—for example, the problem of the nature of science and how it relates to other human activities such as religion and politics. Second (section 2), a framework of general and conceptual distinctions is provided to enable the student and the scholar to avoid a number of pitfalls, while at the same time giving them the widest possible freedom to arrive at their own interpretation and evaluation. [2] The historical background consists primarily of two bodies of information, one pertaining to nonintellectual factors (section 3), the other involving a set of arguments which for centuries had convinced scientists and phi-

1

losophers that the earth must be standing still at the center of the universe (section 4). Finally, the overview focuses on the earlier and the later phases of the affair (sections 6 and 7), but includes an account of how Galileo became involved in it (section 5) and a discussion of its aftermath (section 8).

1. THE SCIENCE-RELIGION CONTROVERSY AND OTHER PROBLEMS

The question of the relationship between science and religion is one of the most basic and persistent problems in interdisciplinary thinking and general culture. It may also be regarded as a key philosophical problem, if we take philosophy to be centrally concerned with the critical understanding of human experience; that is, if we take philosophy not as a technical discipline which studies its own special topics and problems, but rather as a way of thinking that has been applied throughout the ages to the study of the most diverse subject matter originating from other disciplines or from human life in general. The controversy ranks with such problems as those of the relationship between science and the humanities, between science and literature, between science and art, as well as the problems of the relationship between religion and morality, between morality and politics, and between politics and economics.

The problem could be formulated in terms of questions like the following. Are science and religion compatible, incompatible, mutually reinforcing, mutually indifferent, or incommensurable? Are they perhaps complex entities which exhibit different relationships at different times and in different contexts? Can a church impose any restrictions on the freedom of scientific investigation of its adherents? Is theology the queen of the sciences? Can theology be scientific, or is it intrinsically nonscientific? Can science provide any information about the nature of ultimate reality and the meaning of life, or are such questions necessarily the province of religion? What is the role of faith in science and in religion? Is it something essential to religion but foreign to science, or is it essential to both? What is the role of authority in science and in religion? Again, is authority the essence of religion but incompatible with science? Are there different kinds of faith and authority, each appropriate to one but not to the other? What is the role of values in science and in religion? Do they have a different role in each, or are they a common element of both? Is science itself a religion? That is, does science, or do certain attitudes toward science, have a religious component? Does science promote secular humanism, and is secular humanism a religion? What is the proper relationship between scientific and religious beliefs? Is it ever proper to accept or reject scientific theories on the basis of religious belief? Can one ever prove or disprove religious beliefs scientifically? What is the proper attitude toward scientific knowledge in the light of religious beliefs, and toward religious beliefs in the light of scientific knowledge?

One way to come to grips with this problem is to follow a historical approach by examining various episodes that involve the interaction between science and religion to see what lessons if any may be derived from them. One of the most instructive and fascinating of these historical episodes is, of course, the Galileo affair. It is certainly not the only episode that can shed light (and heat) on the science–religion controversy. That the problem is not unique to Catholicism may be seen from Protestant reactions to Charles Darwin's theory of evolution in nineteenth-century England and from the persistent Fundamentalist uneasiness about evolution in twentieth-century America. The problem is not even restricted to revealed, supernatural religions such as Christianity; episodes like the Lysenko affair in the Soviet Union suggest that secular religions are not contradictions in terms and that modern totalitarian societies are counterparts of the older, traditional religions.

Nevertheless, the Galileo affair remains the cause célèbre par excellence for reflecting on the science-religion relationship. This is so in part because the incident was the first to illustrate the problem as it exists today, in the sense that the facts of the case do indeed show all the external signs of a clash between a symbol of science and a symbol of religion. And in part its classic relevance to the same problem is due to the impression which the trial of Galileo made on subsequent history. In fact, the most common interpretation of the event continues to be cast in terms of what it shows about the relationship between science and religion;[3] and here we have such an overwhelming abundance of perceptions that they acquire the strength of a material force and must be dealt with, even if they should be ultimately incorrect.

The Galileo affair can also be studied in other ways, reflecting the viewpoint of problems and issues that are distinct from the science–religion question. For example, since Galileo's contributions were such as to earn him the label of Father of Modern Science, it is obvious that one may see the event as a microcosm of the Scientific Revolution itself and scientific investigation in general and therefore study it primarily for what it tells us about the nature of scientific knowledge per se and how it develops.[4] One may then try to derive lessons about such issues as the relative importance of facts, theories, and instruments; the difference between observation and experimentation; the role of quantitative analysis, causal inquiry, hypotheses, and authority in science; the rationality of replacing one scientific theory by another; and so on. The focus on this aspect of the Galileo affair may be called the *epistemological* approach, if epistemology is defined as the study of the nature of knowledge and how it develops.

Others like to focus on the fact that in the historical context in which Galileo found himself, the Catholic church was not merely a religious institution but performed an important political function and social role. They then stress that it was the sociopolitical side of the Church which interfered with Galileo, and thus they see the episode as illustrative of the interaction between science and society, the problem of the social control of science, the conflict between scientific truth and political expediency, and so on.[5]

The Galileo affair is also a classic example of controversy in a double sense: first, it was itself an episode of controversy involving scientific, epistemological, philosophical, theological, religious, social, and political issues; and second, it has always generated controversial historical interpretations and critical evaluations—so much so that relevant accounts can often give us insight into the historical conditions and philosophical presuppositions of the corresponding authors and times. Thus the Galileo affair can be studied for what it tells us about the nature of controversy as such and it can be compared with other controversies, whether they be the Pelagian controversy in fifth-century Christianity or the Spasmodic controversy in nineteenth-century English literature.

Finally, the Galileo affair is a human tragedy of universal significance, and thus some have attempted to portray it as human drama; that is, the episode illustrates the interaction among such elements as vanity, friendship, betrayal, compromise, loyalty, piety, courage, sacrifice, humiliation, and duty. The most famous of these portrayals is perhaps Bertolt Brecht's play *Galileo*.

These aspects of the affair, and the consequent orientations in studying it, are not mutually exclusive, but rather interrelated and somewhat overlapping. For example, a full understanding of the nature of scientific knowledge would require understanding its relationship to other important human activities and institutions such as religion, politics, society, and controversy; therefore, the epistemological approach could be taken as encompassing all the others. Nevertheless, the various orientations should not be *confused* with each other; nor can it be denied that the science–religion dimension is the most striking one at first.

All of this perspective is essential, but it is not yet sufficient. We need two other signposts in order not to lose our bearings. One of these involves a pair of critical interpretations that are opposite extremes and must be avoided; the other involves a set of distinctions which provide merely the framework for various critical interpretations, but which need to be applied for a proper account of the episode.

2. THE ANTI-CATHOLIC AND ANTI-GALILEAN EXTREMES AND OTHER DISTINCTIONS

Villa Medici in Rome is one of the most impressive palaces in the city. Its name derives from the fact that for a long time it was the embassy of the Grand Duchy of Tuscany, which was ruled by the House of Medici. After receiving in 1610 the title of "Philosopher and Chief Mathematician to the Most Serene Grand Duke of Tuscany," Galileo normally resided there during his visits to Rome. His last stay at the villa was, however, rather unfortunate: it was in fact the period during which his trial of 1633 was taking place. Galileo's visit on that particular occasion thus had the status of imprisonment, a privileged imprisonment to be sure, but a forced residence nonetheless, as I shall relate below.[6] Next to the building, at the edge of the street, stands a commemorative column, erected at the end of the nineteenth century, which says expressly that "it was here that Galileo was kept prisoner by the Holy Office, being guilty of having seen that the earth moves around the sun." The historical and cultural importance of this lesser tourist attraction is that it expresses one of the most common myths widely held about the Galileo affair, namely that he was condemned by the Catholic church for having discovered the truth. Now, since to condemn a person for such a reason can only be the result of ignorance and narrowmindedness, this is also the myth which is used to justify the incompatibility between science and religion.

The fact that I have described this interpretation of the affair as a myth reveals part of my attitude. In fact, I believe that such a thesis is erroneous, misleading, and simplistic. Nevertheless, this myth continues to circulate. For example, a formulation of the myth, not much more sophisticated than the one on the Villa Medici column, can be found in Albert Einstein's foreword to the standard English translation of Galileo's *Dialogue*.[7]

The reason for identifying this first anti-Catholic myth about the Galileo affair is that it may be usefully contrasted to a second myth at the opposite extreme. It seems that some found it appropriate to fight a disagreeable myth by constructing another one. The anti-Galilean myth maintains that Galileo deserved condemnation because his 1632 book violated not only various ecclesiastical norms but also various rules of scientific methodology and logical reasoning; he is thus portrayed as a master of cunning and knavery, and it is in fact difficult to find a misdeed of which the proponents of this myth have not accused him. The history of this myth too has its own fascination, and it too includes illustrious names, such as Arthur Koestler (author of *The Sleepwalkers*) and French physicist, philosopher, and historian Pierre Duhem.[8]

These two opposite myths are useful as reference points in order to orient oneself in the study of the controversy, since it is impossible to evaluate the affair adequately unless one admits that both of these interpretations are mythological and thus rejects both. Avoiding them is easier said than done, however; for example, one cannot simply follow a mechanical approach of mediating a compromise by dividing in half the difference that separates them. A helpful way of proceeding is to read the relevant texts with care and with an awareness of a number of important conceptual distinctions.

One of the most important points to keep in mind is that the Galileo affair involved both questions about the truth of nature and the nature of truth, to use Owen Gingerich's eloquent expression.[9] That is, the controversy was at least two-sided: it involved *scientific issues* about physical facts, natural phenomena, and astronomical and cosmological matters; and it also involved methodological and *epistemological questions* about what truth is and the proper way to search for it, and about what knowledge is and how to acquire it.

The overarching scientific issue was whether the earth stands still at the center of the universe, with all the heavenly bodies revolving around it, or whether the earth is itself a heavenly body that rotates on its axis every day and revolves around the sun once a year. There were several distinct but interrelated questions here. One was whether the whole universe revolves daily from east to west around a motionless earth, or whether the earth rotates daily on its axis in the opposite direction (west to east); this was the problem of whether the so-called *diurnal motion* belongs to the earth or to the rest of the universe. Another question was whether the sun revolves yearly from west to east around the earth, or whether the earth revolves in the same way around the sun; this was the issue of whether the so-called *annual motion* belongs to the sun or to the earth. Another aspect of the controversy was whether the center of the universe, or at least the center of the revolutions of the planets, is the earth or the sun. And there was also the problem of whether or not the universe is divided into two very different regions, containing bodies made of different elements, having different properties, and moving and behaving in different ways: the terrestrial or sublunary part where the earth, including water and air, are located; and the celestial, heavenly, or superlunary region, beginning at the moon and extending beyond to include the sun, planets, and fixed stars.

The traditional view may be labeled *geostatic*, insofar as it claims the earth to be motionless; or *geocentric*, insofar as it locates the earth at the center of the universe; or *Ptolemaic*, insofar as in the second century A.D. the Greek astronomer Ptolemy had elaborated it in sufficient detail to make it a workable theoretical system; or *Aristotelian*, insofar as it corresponded to the worldview advanced in the fourth century B.C. by the Greek philosopher Aristotle, whose ideas in a wide variety of fields had become predominant in the sixteenth century; or *Peripatetic*, insofar as this was a nickname given to followers of Aristotle. The other view may be called either *geokinetic*, insofar as it holds the earth to be in motion; or *heliocentric*, insofar as it places the sun at the center; or *Copernican*, named after the Polish astronomer Nicolaus Copernicus who in the first half of the sixteenth century elaborated its details into a workable theoretical system; or *Pythagorean*, named after the ancient Greek pre-Socratic Pythagoras, one of the earliest thinkers (sixth century B.C) to advance the idea in a general way. We may thus say that the scientific issue was essentially whether the geostatic or the geokinetic theory is true, or at least whether one or the other is more likely to be true.

The epistemological issues were several. There was the question of whether or not physical truth has to be directly observable, or whether any significant phenomenon (for example, the earth's motion) can be true even though our senses cannot detect it, but can detect only its effects; here, one should remember that even today the earth's motion cannot be seen directly by an observer on earth. Then there was the question of whether artificial instruments like the telescope have any legitimate role in the search for truth, or whether the proper way to proceed is to use only the natural senses; here, it should be mentioned that the telescope was the first artificial instrument ever used to learn new truths about the world. A third issue of this sort involved the question of the role of the Bible in scientific inquiry, whether its assertions about natural phenomena have any authority, or whether the search for truth about nature ought to be conducted completely independently of the claims mentioned in the Bible; in this regard it should be noted that it was widely believed that the new geokinetic theory contradicted the Bible. Fourth, there was the question of the epistemological status of the science of astronomy, its relationship to physics, and whether the hierarchy between these two sciences is such that cases of conflict between the two are to be resolved necessarily by letting physics prevail over astronomy; this problem arose because, as we shall see, the earth's motion contradicted the physics of the time. Let us call these four central issues, respectively, the problems of the observability of truth, the legitimacy of artificial instruments, the scientific authority of the Bible, and interdisciplinary hierarchy.

For the second conceptual clarification one needs to distinguish between *factual correctness* and *rational correctness*, that is, between being right about the truth of the matter and having the right reasons for believing the truth. Suppose we begin by asking who was right about the scientific issue. It is obvious that Galileo was right and his opponents wrong, since he preferred the geokinetic to the geostatic view, and today we know for a fact that the earth does move and is not standing still at the center of the universe. However, it is equally clear that his being right about this matter does not *necessarily* mean that his motivating reasons were correct, since it is conceivable that although he might have chanced to hit upon the truth, his supporting arguments may have been unsatisfactory. Hence, the evaluation of his arguments is a separate issue.

I am not saying that the various proponents of the anti-Galilean interpretation [10] are right when they try to show that his arguments left much to be desired, ranging from inconclusive to weak to fallacious to sophistical. In fact, this evaluation is in my opinion untenable, and I think I have adequately shown elsewhere that, when accurately reconstructed, Galileo's arguments are largely correct. [11] Rather, I am saying that Galileo's critics have raised a distinct and important issue *about* the Galileo affair—namely, whether or not, or to what extent, his *reasoning* was correct.

The next point that must be appreciated is also easy when stated in general terms but extremely difficult to apply in practice. It is that *essential correctness* must not be equated with either *total correctness* or *perfect conclusiveness*. Applied to our case, this means that even if Galileo's arguments were essentially correct, as I would hold, the possibility must be allowed that the reasoning of his opponents was neither worthless, nor irrelevant, nor completely unsound. This point is a consequence of the fact that we are dealing with nonapodictic arguments which are not completely conclusive but rather susceptible of degrees of rational correctness, and so it is entirely conceivable that there should sometimes be equally good arguments in support of opposite sides, as well as that the arguments for one side should be better than those for the opposite, without the latter being worthless. I believe this is the case for the Galileo affair, though the anti-Catholic critics do not seem to be able to understand this point. The proper antidote here is the study of the details of the relevant arguments.

To appreciate the next distinction, let us ask whether Galileo or the Church was right in regard to the epistemological aspect of the controversy. Since epistemological issues are normally more controversial than scientific ones, this is an area that some like to exploit by trying to argue that the Church's epistemological and philosophical insight was superior to Galileo's.[12] The argument is usually made in the context of a frank and explicit admission that Galileo was unquestionably right on the scientific issue. Thus, these anti-Galilean critics could be said to show evenhandedness and balanced judgment by contending that on the one hand Galileo was right from a scientific and factual point of view, but that on the other hand the Church was right from an epistemological or philosophical point of view. However, such interpretations could also be criticized for their exaggeration and one-sidedness in the analysis of the epistemological component of the affair. That is, I have already mentioned that there were at least four epistemological issues in the affair, and I am very doubtful that they can all be reduced to one. Moreover, it cannot be denied that Galileo turned out to be right on at least *some* of the epistemological issues—for example, those pertaining to the legitimacy of artificial instruments and to the Bible's lacking scientific authority. On this last point it should be mentioned that it is about one hundred years since the Catholic church officially adopted the Galilean principle that the Bible is an authority only in matters of faith and morals, and not in questions of natural science. Further, it seems to me that with the epistemological issues too one must introduce the question of the rationale underlying the two conflicting positions.

That is, we must examine their respective arguments and try to determine which were the better ones, although this is a more difficult matter than is the case with the scientific arguments. Therefore, the point of this last series of considerations is not to decide the initial question with which they began, but rather to underscore the *multiplicity* of the epistemological issues in the Galileo affair and to suggest avoiding any one-sided focus on a single issue.

Finally, one must bear in mind that this episode was not merely an *intellectual* affair. Besides the scientific, epistemological, methodological, theological, and philosophical issues, and besides the arguments pro and con, there were legal, political, social, economic, personal, and psychological factors involved. To be sure it would be a mistake to concentrate on these issues, or even to devote to them equal attention in comparison with the intellectual issues, for these latter constitute the heart of the event, and so they must have priority. Nevertheless, it would be equally a mistake to neglect these external factors altogether.

To summarize, a balanced approach to the study of the Galileo affair must avoid the two opposite extremes exemplified by the anti-Galilean and the anti-Catholic interpretations. There is no easy way of doing this, but it may help to distinguish scientific from epistemological issues, factual correctness from rational correctness, essential correctness from total correctness, the several epistemological issues from each other, intellectual from external factors, and the several external factors from each other (personal, psychological, social, economic, political). However, the same caution applies here as in the case of the various interdisciplinary and cultural problems mentioned earlier: these distinct entities are also interrelated, so the point is not to deny their interaction but to make sure they are not confused with one another.

3. NONINTELLECTUAL FACTORS

Beginning with personal or psychological factors, it is easy to see that Galileo had a penchant for controversy, was a master of wit and sarcasm, and wrote with unsurpassable eloquence. Interacting with each other and with his scientific and philosophical virtues, these qualities resulted in his making many enemies and getting involved in many other bitter disputes besides the main one that concerns us here. Typically these disputes involved questions of priority of invention or discovery and fundamental disagreements about the occurrence and interpretation of various natural phenomena. It may be of some interest to give a brief list of the other major controversies: a successful lawsuit against another inventor for plagiarism in regard to Galileo's invention of a calculating device and in regard to its accompanying instructions; a dispute with his philosophy colleagues at the University of Padua, where he taught mathematics, about the exact location of the novas that became visible in the heavens in October 1604; a dispute with other philosophers in Florence in 1612 about the reason why bodies float in water; a dispute with an astronomer named Christopher Scheiner about priority in the discovery of sunspots and about their proper interpretation; and a dispute with an astronomer named Orazio Grassi about the nature of comets, occasioned by the appearance of some of these phenomena in 1618. If we remember all this, and what it indicates about Galileo's personality, we may wonder how he managed to acquire and keep the many friends and admirers he did.

In regard to social and economic factors, it should be noted that Galileo was not wealthy. He had to earn his living, first as a university professor and then under the patronage of the grand duke of Tuscany.[13] During his university career in the first half of his life, his economic condition was always precarious. His university salary was very modest, and this was especially so given that he taught mathematics and thus received only a fraction of the remuneration given to a professor of philosophy.[14] This only compounded other unlucky family circumstances, such as having to provide dowries for his sisters. Galileo was forced to supplement his salary by giving private lessons, by taking on boarders at his house, and by working in and managing a profitable workshop which built various devices, some of his own invention. These financial difficulties eased in the second half of his life when he attained the position of "philosopher and chief mathematician" to the grand duke of Tuscany. In this position he was constantly facing a different problem, however, stemming from the nature of patronage and his relationship to his patron: since the fame and accomplishments of an artist or scientist were meant to reflect on the magnificence of his patron, Galileo was in constant need to prove himself scientifically and philosophically, either by surpassing the original accomplishments that had earned him the position or by giving new evidence for that original worth.[15]

Let us now go on to the politics of the Galileo affair. Here we have first the political background of the Catholic Counter-Reformation. Martin Luther had started the Protestant Reformation in 1517, and the Catholic church had convened the Council of Trent in 1545–1563. So

Galileo's troubles developed and climaxed during a time of violent struggle between Catholics and Protestants. Since he was a Catholic living in a Catholic country, it was also a period when the decisions of that council were being taken seriously and implemented and thus affected him directly. Aside from the question of papal authority, one main issue dividing the two camps was the interpretation of the Bible—both how specific points were to be interpreted and who was entitled to do the interpreting. The Protestants, of course, were inclined toward relatively novel and individualistic or pluralistic interpretations, whereas the Catholics were committed to relatively traditional interpretations by the appropriate authorities. In this regard, it is instructive to see exactly what the most relevant decrees of the Council of Trent stated. At its Fourth Session (8 April 1546), the council had issued two decrees about Holy Scripture, one of which contains the following paragraph:

> Furthermore, to check unbridled spirits, it decrees that no one relying on his own judgment shall, in matters of faith and morals pertaining to the edification of Christian doctrine, distorting the Holy Scriptures in accordance with his own conceptions, presume to interpret them contrary to that sense which holy mother Church, to whom it belongs to judge of their true sense and interpretation, has held and holds, or even contrary to the unanimous teaching of the Fathers, even though such interpretations should never at any time be published. Those who act contrary to this shall be made known to ordinaries and punished in accordance with the penalties prescribed by the law.[16]

And the Fifth Session (17 June 1546) issued a decree regulating the teaching of Holy Scripture, stating in part: "[so] that under the semblance of piety impiety may not be disseminated, the same holy council has decreed that no one be admitted to this office of instructor, whether such instruction be public or private, who has not been previously examined and approved by the bishop of the locality as to his life, morals and knowledge."[17]

A more specific element of religious politics concerns the fact that the final climax of the affair in 1632–1633 was taking place during the so-called Thirty Years War between Catholics and Protestants (1618–1648). At that particular juncture Pope Urban VIII, who had earlier been an admirer and supporter of Galileo, was in an especially vulnerable position; thus not only could he not continue to protect Galileo, but he had to use Galileo as a scapegoat to reassert, exhibit, and test his authority and power. The problem stemmed from the fact that in 1632 the Catholic side led by the king of Spain and by the Bohemian Holy Roman Emperor was disastrously losing the war to the Protestant side led by the king of Sweden Gustavus Adolphus. Religion was not the only issue in the war, however, which was being fought also over dynastic rights and territorial disputes. In fact, ever since his election in 1623, the pope's policy had been motivated primarily by political considerations, such as his wish to limit and balance the power of the Hapsburg dynasty which ruled Spain and the Holy Roman Empire. And it had also been motivated by personal interest—that is, by cooperation with the French, whose support had been instrumental in his election, and who for nationalistic reasons also opposed the Hapsburg hegemony. However, in the wake of Gustavus Adolphus's spectacular victories, the Spanish and imperial ambassadors were accusing Urban of having in effect favored and helped the Protestant cause. They mentioned such matters as his failure to send the kind of military and financial support that popes had usually provided on such occasions and his refusal to declare the war a holy war. There were even suspicions of a more direct understanding with the Protestants. Thus the pope's own religious credentials were being questioned, and there were rumors of convening a council to depose him.[18]

Then there was what may be called the Tuscan factor, which had at least two aspects. One was that the Grand Duchy of Tuscany whose ruler Galileo served was closely allied with Spain, and so the pope's intransigence with him was in part a way of getting back at Spain. The other was related to the fact that almost all the leading protagonists and many of the secondary figures in the Galileo affair were Tuscan. Pope Urban VIII himself was a Florentine of the House of Barberini; Tuscan also was Cardinal Robert Bellarmine, the key figure in the earlier phase of the affair; and so was the commissary general of the Inquisition during the later phase. Thus the entire episode has some of the flavor of a family squabble.

Finally, another political element involved the internal power struggle within the Church, on the part of various religious orders, such as the Jesuits and the Dominicans. Although such details are beyond the scope of this Introduction, it is interesting to note that in the earlier phase of the affair climaxing in 1616, Galileo seems to have been attacked by Dominicans and defended by Jesuits, whereas in the later phase, in 1632–1633, it seems that the two religious orders had exchanged roles.

Just as the political background of the affair involved primarily matters of religious politics, so the legal background involved essentially questions of ecclesiastical, or "canon," law. In Catholic countries, the activities of intellectuals like Galileo were subject to the jurisdiction of the Congregation of the Index and the Congregation of the Holy Office, or Inquisition. In the administration of the Catholic church a "congregation" is a committee of cardinals charged with some department of Church business. The Congregation of the Index was instituted by Pope Pius V in 1571 with the purpose of book censorship; one of its main responsibilities was the compilation of a list of forbidden books (called *Index librorum prohibitorum*); this congregation was abolished by Pope Benedict XV in 1917, and book censorship was then handled once again by the Congregation of the Holy Office, which had been in charge of the matter before 1571. The Congregation of the Holy Office, in turn, had been instituted in 1542 by Pope Paul III with the purpose of defending and upholding Catholic faith and morals; one of its specific duties was to take over the suppression of heresies and heretics which had been handled by the Medieval Inquisition; hence, from that time onward, the Holy Office and the Inquisition became practically synonymous. In 1965 at the Second Vatican Council, its name was officially changed to Congregation for the Doctrine of the Faith. The Holy Office or Inquisition was, therefore, more important and authoritative than the Index. By the time Galileo got into religious trouble, the notion of heresy had been given something of a legal definition, and inquisitorial procedures had been more or less codified. Let us examine some of the most relevant details.[19]

Although the Inquisition dealt with other offenses such as witchcraft, it was primarily interested in two main categories of crimes: formal heresy and suspicion of heresy. Here, the term *suspicion* did not have the modern legal connotation pertaining to allegation and contrasting it to proof. One difference between formal heresy and suspicion of heresy was the seriousness of the offense. For example, a standard Inquisition manual of the time stated that "heretics are those who say, teach, preach, or write things against Holy Scripture; against the articles of the Holy Faith; . . . against the decrees of the Sacred Councils and the determinations made by the Supreme Pontiffs; . . . those who reject the Holy Faith and become Moslems, Jews, or members of other sects, and who praise their practices and live in accordance with them. . . ."[20] The same manual stated that "suspects of heresy are those who occasionally utter propositions that offend the listeners . . . those who keep, write, read, or give others to read books forbidden in the Index and in other particular Decrees; ... those who receive the holy orders even though they have a wife, or who take another wife even though they are already married; ... those who listen, even once, to sermons by heretics. . . ."[21] Another difference between formal heresy and suspicion of heresy was whether or not the culprit, having confessed the incriminating facts, admitted having an evil intention.[22] Furthermore, within the major category of suspicion of heresy, two main subcategories were distinguished: vehement suspicion of heresy and slight suspicion of heresy;[23] their difference depended on the seriousness of the crime. Thus, in effect there were three main types of religious crimes, in descending order of seriousness: formal heresy, vehement suspicion of heresy, and slight suspicion of heresy.

In regard to procedure, there were two ways in which legal proceedings could begin: either by the initiative of an inquisitor, responding to publicly available knowledge or publicly expressed opinion; or in response to a complaint filed by some third party, who was required to make a declaration of the purity of his motivation and to give a deposition. Then there were specific rules about the interrogation of defendants and witnesses; how injunctions and decrees were to be worded; how, when, and why interrogation by torture was to be used; and the various judicial sentences and defendant's abjurations with which to conclude the proceedings.

However important all this psychological, social, economic, political, and legal background is, the intellectual background is even more important. To this we now turn.

4. COPERNICUS'S CHALLENGE TO TRADITIONAL IDEAS

In 1543 Copernicus published his epoch-making book *On the Revolutions of the Heavenly Spheres.* In it he updated an idea which had been advanced as early as Pythagoras in ancient Greece but had been almost universally rejected—that is, the idea that the earth moves by rotating on its own axis daily and by revolving around the sun once a year. This means that the earth is not standing still at the center of the universe, with all the heavenly bodies revolving around it. In its essentials this geokinetic idea turned out to be true, as we know today beyond any reasonable doubt, after five centuries of accumulating evidence. At the time, however, the situation was very different. In fact, Copernicus's accomplishment was really to give a *new* argument in support of the *old* idea that had been considered and rejected earlier: he was able to demonstrate that the *known* details about the motions of the heavenly bodies could be explained *more simply* and *more coherently* if the sun rather than the earth is assumed to be at the center and the earth is taken to be the third planet circling the sun.

In regard to simplicity, for example, there are thousands fewer moving parts in the geokinetic system since the apparent daily motion of all heavenly bodies around the earth is explained by the earth's axial rotation, and thus there is only one thing moving daily (the earth), rather than thousands of stars. Moreover, in the geostatic system there are *two* opposite directions of motion, whereas in the geokinetic system all bodies rotate or revolve in the same direction. That is, in the geostatic system, while all the heavenly bodies revolve around the earth with the diurnal motion from *east to west*, the seven planets (moon, Mercury, Venus, sun, Mars, Jupiter, and Saturn) also simultaneously revolve around it from *west to east*, each in a different period of time (ranging from one month for the moon, to one year for the sun, to many years for Saturn). On the other hand, in the geokinetic system there is only one direction of motion since, for example, if the apparent diurnal motion from *east to west* is explained by attributing to the earth an axial rotation, then the direction of the latter has to be reversed (*west to east*), whereas if the apparent annual motion of the sun from *west to east* is explained by attributing to the earth an orbital revolution around the sun then the same direction has to be retained.[24]

In regard to explanatory coherence, without giving examples, I mean that Copernicus was able to explain many phenomena in detail by means of his basic assumption of a moving earth, without having to add artificial and ad hoc assumptions; the phenomena in question were primarily the various known facts about the motions and orbits of the planets. On the other hand, in the previous geostatic system, the thesis of a motionless earth had to be combined with a whole series of unrelated assumptions in order to explain what is observed to happen.[25]

Despite these geokinetic advantages, however, as a proof of the earth's motion Copernicus's argument was far from conclusive. Notice first that his argument is hypothetical. That is, it is based on the claim that *if* the earth were in motion *then* the observed phenomena would result. But from this it does not follow with logical necessity that the earth is in motion; all we would be entitled to infer is that the earth's motion offers an explanation of observed facts. Now, given the greater simplicity and coherence just mentioned, we could add that the earth's motion offered a simpler and more coherent explanation of heavenly phenomena. This does indeed provide a reason for preferring the geokinetic idea, but it is not a *decisive* reason. It would be decisive only in the absence of reasons for rejecting the idea or for preferring the opposite. In short, one has to look at counterarguments, and there were plenty of them.

The arguments against the earth's motion can be classified into various groups, depending on the branch of learning or type of principle from which they stemmed. In fact, these objections reflected the various traditional beliefs which contradicted or seemed to contradict the Copernican system. Thus, there were epistemological, philosophical, theological, religious, physical, mechanical, astronomical, and empirical objections. It is instructive to begin with the last type to underscore the fact that the opposition to Copernicanism was neither all mindless nor simply religious; however, to set the stage for the empirical details, it is best to begin with an argument which is empirical in the sense of involving observation and sensory experience, but which does so in such a way that it amounts to an epistemological objection.

The argument was aptly called the *objection from the deception of the senses*. To understand the deception involved, note that Copernicus did not claim that he could feel, see, or otherwise perceive the earth's motion by means of the senses. Like everyone else, Copernicus's senses told him that the earth is at rest. Therefore, if his theory were true, then the human senses would not be reporting the truth, or would be lying to us. But it was regarded as absurd that the senses should deceive us about such a basic phenomenon as the state of rest or motion of the terrestrial globe on which we live. In other words, the geokinetic theory seemed to be in flat contradiction with direct sense experience and appeared to violate the fundamental epistemological principle that under normal conditions the senses are reliable and provide us with one of the best instruments to learn the truth about reality.[26]

One could begin trying to answer this difficulty by saying that deceptions of the senses are neither unknown nor uncommon, as shown, for example, by the straight stick half immersed in water which appears bent or by the shore appearing to move away from a ship to an observer standing on the ship and looking at the shore.[27] However, the difference was that these perceptual illusions involve relatively minor and secondary experiences, whereas to live all one's life on a moving globe without noticing it would be a gigantic and radical deception. Moreover, it was added, the former illusions are corrigible, since we have other ways of discovering what really happens, whereas there is no way of correcting the perception of the earth being at rest.

This general empirical difficulty is in a sense the reverse side of the coin of the fundamental advantage of the geostatic system, which was that it corresponds to direct observation and sensory experience. (The same applies, of course, to all the other anti-geokinetic objections, which may thus be easily turned into pro-geostatic arguments.) This difficulty may be labeled an epistemological objection because the real issues are whether the earth's motion ought to be observable and whether the human senses ought to be capable of revealing the fundamental features of physical reality. The other empirical difficulties are more specific. They are based primarily on effects in the heavens which ought to be observed in a Copernican universe, but which in fact were not; these *specific empirical difficulties* may therefore be called the *astronomical objections*.

The *objection from the earth–heaven dichotomy*[28] argued that if Copernicus were right then the earth would share many physical properties with the other heavenly bodies, especially the planets, since the earth would itself be a planet, the third one circling the sun. However, it was widely believed that whereas the heavenly bodies are made of the element aether which is weightless, luminous, and changeless, the earth is made of rocks, water, and air which have (positive or negative) weight, are dark, and subject to constant changes. Before the telescope this belief had considerable empirical support.

The *appearance of the planet Venus*[29] was the basis of another objection. For if the Copernican system were correct, then this planet should exhibit phases similar to those of the moon but with a different period; however, none were visible (before the telescope). The reason why Venus would have to show such phases stems from the fact that in the Copernican system it is the second planet circling the sun, the earth is the third, and these two planets have different periods of revolution. Therefore, the relative positions of the sun, Venus, and the earth would be changing periodically and so would the amount of Venus's surface visible from the earth: when Venus is on the far side of the sun from the earth, its entire hemisphere lit by the sun is visible from the earth, and the planet should appear as a disk full of light (like a full moon, though much smaller); when Venus comes between the sun and the earth, none of its hemisphere lit by the sun is visible from the earth, and the planet would be invisible (as in the case of a new moon); and at intermediate locations, when the three bodies are so positioned that the line connecting them forms a noticeable angle, then different amounts would be visible, giving Venus an appearance ranging from nearly fully lit, to half lit, to a crescent shape.

The *apparent brightness and size of the planet Mars* involved another problematic issue. In the Copernican system this planet is the next outer one after the earth, and, since they also revolve at different rates, they are relatively close to each other when their orbital revolutions align both on the same side of the sun and relatively far when they are on opposite sides of the sun. Because the variation in distance is considerable, this would cause a corresponding variation in apparent size and an ever greater change in brightness, since the intensity of light varies as the square of the distance. Now, the difficulty was that although Mars did indeed exhibit a noticeable change in brightness with periodic regularity, this change was not nearly as much as it should be; further, there was practically no variation in apparent size.[30]

The last of the empirical arguments to be discussed here was based on the fact that observation revealed no change in the *apparent position of the fixed stars*;[31] this is commonly known as the objection from stellar *parallax*, a term that denotes a change in the apparent position of an observed object due to a change in the observer's location. At its simplest level, the apparent position of a star may be thought of as its location on the (imaginary) celestial sphere, which in a sense is its position relative to all the other stars (also located on that sphere); or, from the viewpoint of the Copernican system, it may be conceived as measured by the angular position of the star above the plane of the earth's orbit. Now, if the earth were revolving around the sun, then in the course of a year its position in space would change by a considerable amount defined by the size of the earth's orbit. Therefore, a terrestrial observer looking at the same star at six-month intervals would be observing it from different positions, the difference being a distance equal to the diameter of the earth's orbit. Consequently, the same star should appear as having shifted its position either on the celestial sphere or in terms of its angular distance above the plane of the earth's orbit. It follows that if Copernicanism were correct, we should be able to see stellar parallaxes with a periodic regularity of one year; but none were observed.

It should be noted that the first three of these empirical-astronomical objections were not answered until Galileo's telescopic discoveries and that the stellar parallax was not detected until much later (1838). In fact, the magnitude of parallax varies inversely as the distance of the observed object; the stars are so far away that their parallax is exceedingly small; and for about two centuries telescopes were not sufficiently powerful to make the fine discriminations required. One may then begin to sympathize with Copernicus's contemporaries, who found his idea very hard to accept. However, as mentioned earlier, there were many other reasons for their opposition. The next group may be labeled *mechanical* or *physical*, in the sense that they are based directly or indirectly on a number of principles of the branch of physics which today we call mechanics, the study of how bodies move. We shall first mention four objections which hinge indirectly (though crucially) on the laws of motion and may therefore initially appear as empirical objections; later we shall present two others where the appeal to such physical principles is direct and explicit.

The *objection from vertical fall* began with the fact that bodies fall vertically. This is something that everyone can easily observe by looking at rainfall when there is no disturbing wind; or one can take a small rock into one's hand, throw it directly upward, and notice that it then falls back to the place from which it was thrown; or one can drop a rock from the top of a building or tower and notice that it moves perpendicularly downward, landing directly below. Then it was argued that this could not happen if the earth were rotating; for, while the body is falling through the air, the ground below would move a considerable distance to the east (due to the earth's axial rotation), and although the building and person would be carried along, the unattached falling body would be left behind; so that on a rotating earth the body would land to the west of where it was dropped and would appear to be falling along a slanted path. Since this is not seen, but rather bodies fall vertically, it was concluded that the earth does not rotate.[32]

An analogous argument was advanced by the *objection from the motion of projectiles*. The relevant observation here was that, when ejected with equal force, projectiles range an equal distance to the east and to the west. This can be most easily observed by throwing a rock with the same exertion in opposite directions in turn and then measuring the two distances; one could also use bow and arrow to have a slightly better measure of the propulsive force; or one could use a gun, and shoot it first to the east and then to the west with the same amount of charge. Now, the argument claimed that on a rotating earth such projectiles should instead range further toward the west than toward the east. The reason given was that in its westward flight the projectile would be moving against the earth's rotation, which would carry the place of ejection and the ejector some extra distance to the east; whereas, in its eastward flight, the projectile would be traveling in the same direction as the ejector, due to the latter being carried eastward by the earth's rotation. Therefore, on a rotating earth the westward projectiles would range further by a distance equal to the amount of the earth's motion, while the eastward ones would fall short by the same amount. Again, since observation reveals that this is not so, it supposedly follows that the earth does not rotate.[33]

Of course, today these arguments can be refuted. However, their refutation requires knowledge of at least two fundamental principles of mechanics. One is the law of conservation of momentum, or more simply the principle of conservation of motion, according to which the motion acquired by a body is conserved unless an external force interferes with it; the other is the principle of superposition, which specifies how motions in different directions are to be added to each other to yield a resultant motion. The details of these mechanical principles and their application are beyond the scope of this Introduction. The point that needs to be stressed here is that since the phenomena appealed to are indeed true, the two objections just presented raised issues about how bodies would move on a rotating earth, and the resolution of these issues depended on the possession of more accurate mechanical principles. The next objection raised these same issues, but also the question of the true facts of the case; however, to establish these facts was not so easy as it might seem.

The *ship analogy objection*[34] referred to an experiment to be made on a ship, and it then drew an analogy between the earth and the ship. The experiment was that of dropping a rock from the top of a ship's mast, both when the ship is motionless and when it is advancing, and then checking the place where the rock hits the deck. It was asserted that the experiment yielded different results in the two cases—that when the ship is standing still the rock falls at the foot of the mast, but when the ship is moving forward the rock hits the deck some distance to the back. Then the moving ship was compared to a portion of land on a rotating earth, and a tower on the earth was regarded as the analogue of the ship's mast. From this it was inferred that if the earth were rotating, then a rock dropped from a tower would land to the west of the foot of the tower, just as it happens on a ship moving forward; since the rock can be observed to land at the foot of the tower, however, they concluded that the earth must be standing still.

This objection partly involves the empirical issue of exactly what happens when the experiment is made on a moving ship. If the experiment is properly made, the rock still falls at the foot of the mast. However, it is easy to get the wrong result due to extraneous causes, such as wind and any rocking motion the boat may have in addition to its forward motion. Therefore, it is not surprising that there were common reports of the experiment having been made and yielding anti-Copernican results.[35] Nor is it surprising that when Galileo tried to refute the objection, although he disputed the results of the actual experiment, he emphasized a more theoretical answer in terms of the principles of conservation and superposition of motion.[36] The principles are needed to determine what will happen to the horizontal motion the rock had before it was dropped from the mast of the moving ship, and how it is to be combined with the new vertical motion of fall it acquires.

The last of the indirectly mechanical objections to be discussed here is the *argument from the extruding power of whirling*,[37] or, as we might say today, the *centrifugal force argument*. The basis of this objection was the fact that in a rotating system, or in motion along a curve, bodies have a tendency to move away from the center of rotation or the center of the curve. For example, if one is in a vehicle traveling at a high rate of speed, whenever the vehicle makes a turn one experiences a force pushing one away from the center of the curve defined by the turn: if the vehicle turns right, one experiences a push to the left and vice versa. Or one could make the simple experiment of tying a small pail of water at the end of a string and whirling the pail in a vertical circle by the motion of one's hand. Now suppose a small hole is made in the bottom of the pail; as the pail is whirled one will see water rushing out of the hole always in a direction away from one's hand. Then the argument called attention to the fact that, if the earth rotates, bodies on its surface are traveling in circles around its axis at different speeds depending on the latitude, the greatest being about 1000 miles per hour at the equator. This sounds like a very high rate of speed, which would generate such a strong extruding power that all bodies would fly off the earth's surface and the earth itself might disintegrate. Since this obviously does not happen, it was concluded that the earth must not be rotating.

This objection raised issues whose resolution involves the correct laws of centrifugal force. At the time, therefore, it was felt to be a very strong objection. Next, we come to the objections according to which the conflict with physical principles was so explicit that the earth's motion seemed a straightforward physical impossibility.

The *natural motion argument*[38] claimed that the earth's motion (whether of axial rotation or orbital revolution) is physically impossible because the natural motion of earthly bodies (rocks and water) is to move in a *straight* line toward the center of the *universe*. The context of this argument was a science of physics which contrasted natural motion to forced or violent motion, which postulated three basic types of natural motion, and which attributed each type to one or more of the basic elements: circular motion around the center of the universe was attributed to aether, the element of the heavenly bodies; straight motion away from the center of the universe was ascribed to the elements air and fire; and straight motion toward the center of the universe was given to the elements earth and water. Thus, unlike natural circular motion which can last forever, straight natural motion, especially straight downward, cannot: once the center (of the universe) is reached, the body will no longer have any natural tendency to move. Now, the terrestrial globe on which we live is essentially the collection of all things made of the elements earth and water, which have collected at the center (of the universe) or as close to it as possible. Therefore, this whole collection cannot move around the center (in an orbital revolution as Copernicus would have it), because such a motion would be unnatural, could not last forever, and would in any case be overcome by the tendency to move naturally in a straight line toward the center; further, for the same reasons, once at the center the whole collection could not even acquire any axial rotation.

The Copernican system was also deemed physically impossible because it was in direct violation of the principle according to which *every simple body can have one and only one natural motion*.[39] This principle was another aspect of the laws of motion of Aristotelian physics just sketched, whereas Copernicanism seemed to attribute to the earth at least three natural motions: the revolution of the whole earth in an orbit around the sun, the rotation of the earth around its own axis, and the downward motion of parts of the earth in free fall.

Just as the last two objections are essentially unanswerable as long as one accepts the two principles of traditional physics just mentioned, they are easily answerable by rejecting these two principles. However, rejecting them is easier said than done since, to be effective, the rejection should be accompanied by the formulation of some alternatives. In short, what was really required was the construction of a new science of motion, a new physics. In fact, the alternatives were such corner-stones of modern physics as the law of inertia, the law of gravitational force, and the law of conservation of (linear and angular) momentum.

For example, according to the law of inertia, the natural motion of *all* bodies is uniform and rectilinear; and according to the law of gravitation, all bodies attract each other with a force that makes them accelerate toward each other or diverge from their natural inertial motion in a measurable way. Thus, the earth's orbital motion becomes a forced motion under the influence of the sun's gravitational attraction; the axial rotation of the whole earth becomes a type of natural motion in accordance with conservation laws; and the downward fall of heavy bodies near the earth's surface becomes a forced motion under the influence of the earth's gravitational attraction.

Finally, there were theological and religious objections. One of these appealed to the *authority of the Bible*.[40] It claimed that the idea of the earth moving is heretical because it conflicts with many biblical passages which state or imply that the earth stands still. For example, Psalm 104:5 says that the Lord "laid the foundations of the earth, that it should not be removed for ever"; and this seems to say rather explicitly that the earth is motionless. Other passages were less explicit, but they seemed to attribute motion to the sun and thus to presuppose the geostatic system. For example, Ecclesiastes 1:5 states that "the sun also riseth, and the sun goeth down, and hasteth to the place where he ariseth"; and Joshua 10:12–13 asserts: "Then spake Joshua to the Lord in the day when the Lord delivered up the Amorites before the children of Israel, and he said in the sight of Israel, 'Sun, stand thou still upon Gibeon; and thou, Moon, in the valley of Ajalon.' And the sun stood still, and the moon stayed, until the people had avenged themselves upon their enemies."

The biblical objection had greater appeal to those (such as Protestants) who took a literal interpretation of the Bible more seriously. However, for those (such as Catholics) less inclined in this direction, the same conclusion could be reinforced by appeal to the *consensus of Church Fathers*;[41] these were the saints, theologians, and churchmen who had played an influential and formative role in the establishment and development of Christianity. The argument claimed that all Church Fathers were unanimous in interpreting relevant biblical passages (such as those just mentioned) in accordance with the geostatic view; therefore, the geostatic system is binding on all believers, and to claim otherwise (as Copernicans did) is heretical.

In summary, the idea updated by Copernicus was vulnerable to a host of counterarguments and counterevidence. The earth's motion seemed epistemologically absurd because it flatly contradicted direct sensory experience and thus undermined the normal procedure in the search for truth; it seemed empirically untrue because it had astronomical consequences that were not seen to happen; it seemed a physical impossibility because it seemed to have consequences that contradicted the most incontrovertible mechanical phenomena, and because it directly violated many of the most basic principles of the available physics; and it seemed a religious heresy because it conflicted with the words of the Bible and the biblical interpretations of the Church Fathers.

Copernicus was aware of these difficulties.[42] He realized that his novel argument did not conclusively prove the earth's motion and that there were many counterarguments of apparently greater strength. I believe that these were the main reasons why he delayed publication of his book until he was almost on his deathbed.

5. GALILEO'S REASSESSMENT OF COPERNICANISM

Galileo's attitude was at first similar to that of Copernicus. The main difference was that Galileo's central interest was physics, mechanics, and mathematics, rather than astronomy. He began university teaching in 1589, almost fifty years after Copernicus's death. In his official position as professor of mathematics, his duties included the teaching of astronomy and physics, as well as mathematics. Although acquainted with the Copernican theory, Galileo did not regard it as sufficiently well established to teach it in his courses; instead he covered traditional geostatic astronomy. Nor was he directly pursuing the geokinetic theory in his research. Rather his research consisted of investigations into the nature of motion and the laws in accordance with which bodies move. Here his work was original and revolutionary, for he was critical of the traditional physics of motion and was attempting to construct a new mechanics and physics. It did not take him long to realize that the physics he was building was very much in line with the geokinetic theory: what he was discovering about the motion of bodies in general made it possible for the earth to move and rendered unlikely its rest at the center of the universe. In short, Galileo soon realized that his physical research had important consequences in the astronomical field—namely, to strengthen the Copernican theory by removing the physical and mechanical objections against it and providing new arguments in its favor.

We have evidence that he was aware of this reinforcement but still was dissatisfied with the Copernican theory. After all, the other objections were still there, especially the all-important empirical-astronomical difficulties. It was only the invention of the telescope, and the astronomical discoveries it made possible, which removed most of them and paved the way for removing the others.

The invention of the telescope and Galileo's role in it is a fascinating story in its own right, but too long and complex to be told here.[43] Suffice it to say the following. This instrument was first invented by others in 1608, but in 1609 Galileo was able to improve its quality and magnification sufficiently to produce an astronomically useful instrument that could not be duplicated by others for quite some time. With this instrument, in the next few years he made a number of startling discoveries: that, like the earth, the moon has a rough surface covered with mountains and valleys and appears to be made of the same rocky, opaque, and nonluminous substance; that the planet Jupiter has at least four satellites and thus constitutes a miniature planetary system; that the planet Venus does show phases similar to those of the moon; that the planet Mars does show changes in brightness and apparent size of about sixtyfold; and that there are dark spots on the surface of the sun which appear and disappear at irregular intervals but move in such a way as to indicate that the sun rotates on its own axis once a month.

Some of these discoveries were published in 1610 in a book entitled *The Starry Messenger*; others were published in 1613 in a book entitled *Sunspot Letters*. His 1610 book allowed Galileo to leave the position of professor of mathematics at the University of Padua which he had held for some twenty years. He went back to his native Tuscany to be under the patronage of its ruling grand duke and to devote full time to research and writing. Galileo requested that his official title should include the word philosopher as well as mathematician, and so for the rest of his life he held the position of philosopher and chief mathematician to the grand duke of Tuscany.[44]

The new telescopic evidence led Galileo to reassess the status of Copernicanism, for it removed most of the empirical-astronomical objections against the earth's motion and added new arguments in its favor. Thus, Galileo now believed not only that the geokinetic theory was simpler and more coherent than the geostatic theory (as Copernicus had shown), not only that it was more physically and mechanically adequate (as he himself had been discovering in his twenty years of university research), but also that it was empirically superior in astronomy (as the telescope now revealed). However, he had not yet published anything of his new physics, and the epistemological and especially theological objections had not yet been dealt with; so the case in favor of the earth's motion was still not conclusive.

Thus, Galileo's new attitude toward the geokinetic theory could be described as one of direct pursuit and tentative acceptance, by contrast with the indirect pursuit and contextually divided loyalty of the pre-telescopic situation.[45] However, in Galileo's publications of this period we find yet no explicit acceptance of or committed belief in the earth's motion. We do find a more favorable attitude expressed in his private correspondence, as well as a stronger endorsement in the *Sunspot Letters* of 1613 than in *The Starry Messenger* of 1610. But that is as one would expect.

Besides realizing that the pro-Copernican arguments were still not absolutely conclusive, Galileo must have also perceived the potentially explosive character of the biblical and religious objections. In fact, for a number of years he did not get involved even though his book of 1610 had been attacked by several authors on biblical grounds, among others. Eventually, however, he was dragged into the theological discussion. What happened was the following.

6. THE EARLIER INCIDENTS

In the years following the publication of *The Starry Messenger* in 1610, as support for the geostatic theory continued to dwindle, conservative scientists and philosophers began relying more and more heavily on biblical, theological, and religious arguments. These discussions became so frequent and widespread that the ducal family must have started to wonder whether they had a heretic in their employment. Thus, in December 1613 the Grand Duchess Dowager Christina confronted one of Galileo's friends and followers named Benedetto Castelli, who had succeeded him in the chair of mathematics at the University of Pisa, and she presented him with the biblical objection to the motion of the earth. This was done in an informal, gracious, and friendly manner, and clearly as much out of genuine curiosity as out of worry. Castelli answered in a way that satisfied both the duchess and Galileo, when Castelli informed him of the incident. However, Galileo felt the need to write a very long letter to his former pupil containing a detailed refutation of the biblical objection. Recall that this objection argued that the geokinetic theory must be wrong because many biblical passages state or imply that the geostatic theory is right, especially the passage describing the Joshua miracle (Joshua 10:12–13).

In this letter, Galileo suggested that the objection has three fatal flaws. First, it attempts to prove a conclusion (the earth's rest) on the basis of a premise (the Bible's commitment to the geostatic system) which can only be ascertained with a knowledge of that conclusion in the first place. In fact, the interpretation of the Bible is a serious business, and normally the proper meaning of its statements about natural phenomena can be determined only after we know what is true in nature. Thus, the business of biblical interpretation is dependent on physical investigation, and to base a controversial physical conclusion on the Bible is to put the cart before the horse. Second, the biblical objection is a non sequitur, since the Bible is an authority only in matters of faith and morals, not in scientific questions, and thus its saying something about a natural phenomenon does not make it so, and therefore its statements do not constitute valid reasons for drawing corresponding scientific conclusions. Finally, it is questionable whether the earth's motion really contradicts the Bible, and an analysis of the Joshua passage shows that it cannot be easily interpreted in accordance with the geostatic theory; rather, it accords better with the geokinetic view, especially as improved by Galileo's own discoveries. The biblical objection is therefore groundless, aside from its other faults.

Although unpublished, Galileo's letter to Castelli began circulating widely and copies were made. Some of these came into the hands of traditionalists, who soon passed to the counterattack. In December 1614, at a church in Florence, a Dominican friar named Tommaso Caccini preached a Sunday sermon against mathematicians in general, and Galileo in particular, on the grounds that their beliefs and practices contradicted the Bible and were thus heretical. In February 1615 another Dominican, named Niccolò Lorini, filed a written complaint against Galileo with the Inquisition in Rome, enclosing his letter to Castelli as incriminating evidence. Then in March of the same year, Caccini, who had attacked Galileo from the pulpit, made a personal appearance before the Roman Inquisition; in his deposition he charged Galileo with suspicion of heresy, based not only on the content of the letter to Castelli but also on the book of *Sunspot Letters* and on hearsay evidence of a general sort and a more specific type involving two other individuals, named Ferdinando Ximenes and Giannozzo Attavanti. The Roman Inquisition responded by ordering an examination of these two individuals and the two mentioned writings.

In the meantime Galileo was writing for advice and support to many friends and patrons who were either clergymen or had clerical connections. He had no way of knowing about the details of the Inquisition proceedings, which were a well-kept secret, but Caccini's original sermon had been public and he was able to learn about Lorini's initial complaint. Galileo also wrote and started to circulate privately three long essays on the issues. One (now known as "Galileo's Letter to the Grand Duchess Christina") dealt with the religious objections and was an elaboration of the letter to Castelli, which was thus expanded from eight to forty pages. Another (now known as "Galileo's Considerations on the Copernican Opinion") began to sketch a way of answering the epistemological and philosophical objections, which Galileo had never really addressed. And the third one (the "Discourse on the Tides") was an elementary discussion of the scientific issues in the form of a new physical argument in support of the earth's motion based on its alleged ability to explain the existence of the tides and the trade winds. He also received the unexpected but welcome support of a Neapolitan friar named Paolo Antonio Foscarini, who published a book arguing in detail for the specific thesis that the theory of the earth's motion is compatible with the Bible. Finally, in December 1615, after a long delay due to illness, Galileo went to Rome of his own initiative, to try to clear his name and prevent the condemnation of Copernicanism. He did succeed in the former, but not in the latter, undertaking.

In fact, the results of the Inquisition investigations were as follows. The consultant who examined the letter to Castelli reported that in its essence it did not deviate from Catholic doctrine. The cross-examination of the two witnesses, Ximenes and Attavanti, exonerated Galileo since the hearsay evidence of his utterance of heresies was found to be baseless. And the examination of his *Sunspot Letters* failed to reveal any explicit assertion of the earth's motion or other presumably heretical assertion, if indeed the inquisitors examined this book.[46] However, the Inquisition felt it necessary to consult its experts for an opinion on the status of Copernicanism.

On 24 February 1616 a committee of eleven consultants reported unanimously that Copernicanism was philosophically and scientifically untenable and theologically heretical. In a way, much of the tragedy of the Galileo affair stems from this opinion, which even Catholic apologists seldom if ever defend. Although it is indefensible, if one wants to understand how this opinion came about, one must recall all the traditional scientific and epistemological arguments against the earth's motion. Moreover, one must view the judgment of heresy in light of the two objections based on the words of the Bible and on the consensus of the Church Fathers and in light of the Catholic Counter-Reformation's rejection of new and individualistic interpretations of the Bible. At any rate, the Inquisition must have had some misgivings about the opinion of the committee of eleven consultants, for it issued no formal condemnation. Instead two milder consequences followed.

One was to give a private warning to Galileo to stop defending and to abandon his geokinetic views. The warning was conveyed to Galileo within a few days by Cardinal Robert Bellarmine, the most influential and highly respected theologian and churchman of the time, with whom Galileo was on very good terms despite their philosophical and scientific differences. The exact content, form, and circumstances of this warning are not completely known, but they are extremely complex and a subject of great controversy. Moreover, as we shall soon see, the occurrence and propriety of the later Inquisition proceedings in 1633 hinge on the nature of this warning. For now let us simply note that Bellarmine reported back to the Inquisition that he had warned Galileo to abandon his defense of and belief in the geokinetic thesis, and Galileo had promised to obey.

The other development was a public decree issued by the Congregation of the Index, the department of the Church in charge of book censorship. In March 1616 this congregation published a decree containing four main points. First, it stated that the doctrine of the earth's motion is false, contrary to the Bible, and a threat to Catholicism. Second, it condemned and prohibited completely Foscarini's book; this was the work that had tried to show that the earth's motion is compatible with the Bible. Third, it suspended circulation of Copernicus's book, pending correction and revision; these corrections were eventually issued in 1620, their gist being to delete or modify about a dozen passages containing either religious references or else language indicating that Copernicus was taking the earth's motion literally as a description of physical reality and not merely as a convenient manner of speaking in order to make astronomical calculations and predictions. Fourth, the decree ordered analogous censures for analogous books. Galileo was not mentioned at all.

It should be stressed that this was a decree issued by the Congregation of the Index, not a pronouncement by the Congregation of the Holy Office (or Inquisition); hence, though Catholics were still obliged to obey it, it did not carry the weight and generality of pronouncements which define the Catholic faith, just as even an Inquisition decree did not carry the authority of an official papal decree or a decree issued by a sacred council, such as the Council of Trent. Moreover, the actual wording in the decree was vague and unclear, which is a sign of having been some kind of compromise, and the exact offensive features of the prohibited books were not spelled out; thus the declaration that analogous works were analogously prohibited was not very informative and was liable to great abuse. Finally, one was left in the dark concerning what type of *discussion* of Copernicanism was indeed allowed.

In view of the confusing message in this decree of the Index, and in view of the even more confusing circumstances of Bellarmine's private warning to Galileo (which I shall analyze shortly), it is not difficult to sympathize with Galileo's next two moves before he left Rome to return home to Florence. He obtained an audience with Pope Paul V; the precise content of their discussion is not known, but in a letter to the Tuscan secretary of state Galileo reported that he had been warmly received and reassured during three-quarters of an hour with the pontiff. Moreover, at this time Galileo began receiving letters from friends in Venice and Pisa saying that there were rumors in those cities to the effect that he had been personally put on trial, condemned, forced to recant, and given appropriate penalties by the Inquisition. Having shown these letters to Cardinal Bellarmine, Galileo was able to persuade him to write a brief and clear statement of what had happened and how Galileo was affected. Thus in a document half a page long, the most authoritative churchman of his time declared the following: Galileo had been neither tried nor otherwise condemned; rather, he had been personally notified of the decree of the Index and informed that in view of this decree the geokinetic thesis could be neither held nor defended.

With this certificate in his possession, Galileo left Rome soon thereafter. But before we too leave this first phase of the affair let us pause to see where we are. Despite the fact that Galileo was not personally condemned, despite the fact that the geokinetic theory was not formally condemned by the Inquisition, despite the vagueness of the decree of the Index, and despite the pope's and Bellarmine's personal assurances to Galileo, it seems obvious that he lost the battle. He had been personally forbidden to defend the geokinetic theory, and to any Catholics who wanted to avoid trouble the decree of the Index meant that they too should avoid defending it. The earth's motion had not been formally declared a heresy (as the Inquisition consultants had judged in their report), but the practical effect of what happened was about the same. Moreover, since our story continues, let us recall the three different documents that will play a crucial role later: the private, oral warning given to Galileo by Cardinal Bellarmine in the name of the Inquisition; the public decree of the Index; and Bellarmine's certificate to Galileo. The propriety or impropriety of his subsequent behavior will depend on which of these three items is stressed.

7. THE LATER INCIDENTS

For the next several years Galileo did refrain from defending or explicitly discussing the geokinetic theory, though he did discuss it implicitly and indirectly in the context of a controversy about the nature of comets. Even the publication of the corrections to Copernicus's book in 1620, which gave one a better idea of what was allowed and what was not, did not motivate him to resume the earlier struggle. The event that put an end to the interlude took place in 1623, when the old pope died and Cardinal Maffeo Barberini was elected Pope Urban VIII. Urban was a well-educated Florentine, and in 1616 he had been instrumental in preventing the direct condemnation of Galileo and the formal condemnation of Copernicanism as a heresy. He was also a great admirer of Galileo and employed as his personal secretary one of Galileo's closest acquaintances. Further, at about this time Galileo's book on the comets entitled *The Assayer* was being published in Rome, so it was decided to dedicate the book to the new pope. Urban appreciated the gesture and liked the book very much. Finally, as soon as circumstances allowed, in the spring of 1624, Galileo went to Rome to pay his respects to the pontiff; he stayed about six weeks and was warmly received by Church officials in general and the pope in particular, who granted him weekly audiences.

The details of the conversation during these six audiences are not known. There is evidence, however, that Urban VIII did not think Copernicanism to be a heresy or to have been declared a heresy by the Church in 1616. He interpreted the decree of the Index to mean that the earth's motion was a dangerous doctrine whose study and discussion required special care and vigilance. He thought the theory could never be proved to be necessarily true, and here it is interesting to mention his favorite argument for this skepticism, an argument based on the omnipotence of God:[47] Urban liked to argue that since God is all-powerful, he could have created any one of a number of worlds, for example one in which the earth is motionless; therefore, regardless of how much evidence there is supporting the earth's motion, we can never assert that this must be so, for that would be to want to limit God's power to do otherwise. This argument, together with his interpretation of the decree of 1616, must have reinforced his liberal inclination that, as long as one exercised the proper care, there was nothing wrong with the hypothetical discussion of Copernicanism—that is to say, with treating the earth's motion as a hypothesis, to study its consequences, and to determine its utility in making astronomical calculations and predictions.

At any rate, Galileo must have gotten some such impression during his six conversations with Urban, for upon his return to Florence he began working on a book. This was in part the work on the system of the world which he had conceived at the time of his first telescopic discoveries, but it now acquired a new form and new dimensions in view of all that he had learned and experienced since. His first step was to write and circulate privately a lengthy reply to an anti-Copernican essay written in 1616 by Francesco Ingoli. This Galilean "Reply to Ingoli," as well as his earlier "Discourse on the Tides," were incorporated into the new book. After a number of delays in its writing, licensing, and printing, the work was finally published in Florence in February 1632 with the title *Dialogue on the Two Chief World Systems, Ptolemaic and Copernican.*

The author had done a number of things to avoid trouble, to ensure compliance with the many restrictions under which he was operating, and to satisfy the various censors who issued him permissions to print. To emphasize the hypothetical character of the discussion, he had originally entitled it "Dialogue on the Tides" and structured it accordingly. That is, it was to begin with a statement of the problem of the cause of tides, and then it would introduce the earth's motion as a hypothetical cause of the phenomenon; this would lead to the problem of the earth's motion, and to a discussion of the arguments pro and con, as a way of assessing the merits of this hypothetical explanation of the tides.[48] However, the book censors, partly interpreting the pope's wishes and partly in accordance with another impression that Galileo himself wanted to convey, decided to try to make the book look like a vindication of the Index's decree of 1616. The book's preface, whose compilation must be regarded as a joint effort by the author and the censors, claimed that the work was being published to prove to non-Catholics that Catholics knew all the arguments and evidence about the scientific issue, and therefore their decision to believe in the geostatic theory was motivated by religious reasons and not by scientific ignorance. It went on to add that the scientific arguments seemed to favor the geokinetic theory, but they were inconclusive, and thus the earth's motion remained a hypothesis.

Galileo also complied with the explicit request to end the book with a statement of the pope's favorite argument. Moreover, to ensure he would not be seen as holding or defending the geokinetic thesis (which he had been forbidden to do) the author did two things. He wrote the book in the form of a dialogue among three speakers, one defending the geostatic side, another taking the Copernican view, and the third being an uncommitted observer who listens to both sides and accepts the arguments that seem to survive critical scrutiny. And in many places throughout the book, usually at the end of a particular topic, the Copernican spokesman utters the qualification that the purpose of the discussion is information and enlightenment—not to decide the issue, which is a task to be reserved for the proper authorities. Finally, it should be mentioned that Galileo obtained written permission to print the book—first from the proper Church officials in Rome (when the plan was to publish the book there) and then from the proper officials in Florence (when a number of external circumstances dictated that the book be printed in the Tuscan capital).

The book was well received in scientific circles; however, a number of rumors and complaints began emerging and circulating in Rome. The most serious complaint involved a document found in the file of the Inquisition proceedings of 1615–1616. It reads like a report of what took place when Cardinal Bellarmine, in the name of the Inquisition, gave Galileo the private warning to abandon his geokinetic views. The cardinal had died in 1621 and thus was no longer available to clarify the situation. The document states that in February 1616 Galileo had not only been ordered to stop holding or defending the geokinetic thesis but had also been prohibited from discussing it in any way whatsoever, either orally or in writing. In other words, this document states that Galileo had been given a special injunction, above and beyond what bound Catholics in general. The charge, then, was that his book of 1632 was a clear violation of this special injunction, since whatever else the book did, and however else it might be described, it undeniably contained a discussion of the earth's motion. To be sure, the document does not bear Galileo's signature and thus was of questionable legal validity, and under different circumstances such a judicial technicality could have been taken seriously. However, too many other difficulties were being raised about the book.

One of these, the second major complaint, was that the work only paid lip service to the stipulation about a hypothetical discussion, which represented Urban's compromise; in reality, the book allegedly treated the earth's motion not as a hypothesis but in a factual, nonconditional, and realistic manner. This was a more or less legitimate complaint on the part of the pope, but the truth of the matter is that the concept of hypothesis was ambiguous and had not been sufficiently clarified in that historical context. By hypothetical treatment Urban meant a discussion that would treat the earth's motion merely as an instrument of prediction and calculation, rather than as a potentially true description of reality. Galileo, however, took a hypothesis to be an assumption about physical reality which accounts for what is known and which may be true, though it has not yet been proved to be true.[49]

Third, there was the problem that Galileo's book was in actuality a defense of the geokinetic theory. Despite the dialogue form, despite the repeated disclaimers that no assertion of Copernicanism was being intended, despite the nonapodictic and nonconclusive character of the pro-Copernican arguments, and despite the presentation of the anti-Copernican and pro-geostatic arguments, it was readily apparent that the pro-geostatic arguments were being criticized and the pro-Copernican ones were being portrayed favorably; and to do this is one way of arguing in favor of a conclusion, which in turn is one way of defending a conclusion.

There were also complaints involving alleged irregularities in the various permissions to print that Galileo obtained; there were substantive criticisms of various specific points discussed in the book; there were hurt feelings about some of his rhetorical excesses and biting sarcasm; and there were malicious slanders suggesting that the book was in effect a personal caricature of the pope himself.

Although some of the complaints in the last miscellaneous group were easily cleared, the sheer number in the whole list and the seriousness of some charges were such that the pope might have been forced to take some action even under normal circumstances. But Urban was himself in political trouble due to his behavior in the Thirty Years War. Thus in the summer of 1632 sales of the book were stopped, unsold copies were confiscated, and a special commission was appointed to investigate the matter.[50] The pope did not immediately send the case to the Inquisition, but he took the unusual step of appointing a special commission first. This three-member panel issued its report in September 1632, and it listed as areas of concern about the book all of the above-mentioned problems, with the exception of the malicious slanders. In fact, it is from the report that we learn about these complaints that had been accumulating since the book's publication. In view of the report, the pope felt he had no choice but to forward the case to the Inquisition. So Galileo was summoned to Rome to stand trial.

The entire autumn was taken up by various attempts on the part of Galileo and the Tuscan government to prevent the inevitable. The Tuscan government got involved partly because of Galileo's position as philosopher and chief mathematician to the grand duke, partly because the book contained a dedication to the grand duke, and partly because the grand duke had been instrumental in getting the book finally printed in Florence. At first they tried to have the trial moved from Rome to Florence. Then they asked that Galileo be sent the charges in writing and be allowed to respond in writing. As a last resort, three physicians signed a medical certificate stating that Galileo was too ill to travel. This was true, and here it should be added that he was sixty-eight years old and there had been an outbreak of the plague for the past two years, which meant that travelers from Tuscany to the Papal States were subject to quarantine at the border. At the end of December the Inquisition sent Galileo an ultimatum: if he did not come to Rome of his own accord, they would send officers to arrest him and bring him to Rome in chains. On 20 January 1633, after making a last will and testament, Galileo began the journey. When he arrived in Rome three weeks later, he was not placed under arrest or imprisoned by the Inquisition but was allowed to lodge at the Tuscan embassy, though he was ordered not to socialize and to keep himself in seclusion until he was called for interrogation.

This was slow in coming, as if the Inquisition wanted to use the torment of the uncertainty, suspense, and anxiety as part of the punishment to be administered to the old man. This was very much in line with one reason mentioned earlier by officials why Galileo had to make the journey to Rome, despite his old age, ill health, and the epidemic of the plague. That is, he had to do it as an advance partial punishment or penance, and if he did this the Inquisitors might take it into consideration when the time of the actual proceedings came.

The first interrogation was held on 12 April. In answer to various questions about the *Dialogue* of 1632 and the events of 1616, the defendant claimed the following. He admitted having been given a warning by Cardinal Bellarmine in February 1616 and described this as an oral warning that the geokinetic theory could be neither held nor defended but only discussed hypothetically. He denied having received any

special injunction not to discuss the earth's motion in any way whatsoever, and he introduced Bellarmine's certificate as supporting evidence.
His third main claim was made in answer to the question why he had
not obtained any permission to write the book in the first place and
why he had not mentioned Bellarmine's warning when obtaining permission to print the book; these omissions had angered the pope and
left him feeling deceived. Galileo answered that he had not done so
because the book did not hold or defend the earth's motion; rather, it
showed that the arguments in its favor were not conclusive and thus it
did not violate Bellarmine's warning.

This was a very strong and practicable line of defense. In particular,
just as the special injunction was news to Galileo, so Bellarmine's certificate must have surprised and disoriented the inquisitors. Thus it took
another three weeks before they finally decided on the next step in the
proceedings. In the meantime Galileo was detained at the headquarters
of the Inquisition but allowed to lodge in the chief prosecutor's apartment. What the inquisitors finally decided was essentially what today
we would call an out-of-court settlement. That is, they would not press
the most serious charge (of having violated the special injunction), but
Galileo would have to plead guilty to the lesser charge of having inadvertently transgressed the order not to defend Copernicanism, in regard
to which his defense was the weakest.

The deal was worked out as follows. The Inquisition asked three
consultants to determine whether or not Galileo's *Dialogue* held or defended the geokinetic theory; in separate reports all three concluded
that the book clearly defended the doctrine and came close to holding
it. Then the executive secretary of the Inquisition (who, incidentally,
was another Tuscan) talked privately with Galileo to try to arrange the
deal, and after lengthy discussions he succeeded. Galileo requested and
obtained a few days to think of a dignified way of pleading guilty to the
lesser charge. Thus on 30 April, the defendant appeared before the Inquisition for the second time and signed a deposition stating the following. Ever since the first hearing he had reflected about whether, without
meaning to, he might have done anything wrong; it dawned on him to
reread his book, which he had not done for the past three years since
completing the manuscript. He was surprised by what he found, since
the book did give the reader the impression that the author was defending the geokinetic theory, even though this had not been his intention.
To explain how this could have happened, Galileo attributed it to
vanity, literary flamboyance, and an excessive desire to appear clever by

making the weaker side look stronger. He was deeply sorry for this transgression and was ready to make amends.

After this deposition, Galileo was allowed to return to the Tuscan embassy. On 10 May there was a third formal hearing at which Galileo presented his entire case, including the original copy of Bellarmine's certificate, repeating his recent admission of some wrongdoing together with a denial of any malicious intent, and adding a plea for clemency and pity. The trial might have ended here, but it was not concluded for another six weeks. The new development was one of those things that make the Galileo affair such an unending source of controversy and such rich material for tragedy.

Obviously the pope would have to approve the final disposition of the case. Thus a report to him was compiled by the Inquisition, summarizing the events from 1615 to Galileo's third deposition just completed. Reading it did not resolve Urban's doubts about Galileo's intention, so he directed that the defendant be interrogated under the verbal threat of torture in order to determine his intention, a standard Inquisition practice.[51] The pope added that even if his intention was found to have been pure, Galileo had to make a public abjuration and had to be held under formal arrest at the pleasure of the Inquisition, and his book had to be banned.

On 21 June Galileo was subjected to interrogation under the formal threat of torture. The result was favorable, in the sense that, even under such a formal threat, Galileo denied any malicious intention and showed his readiness to die rather than admit that. The following day, at a public ceremony in the convent of Santa Maria sopra Minerva in Rome, he was read the sentence and then recited the formal abjuration.

The sentence states that Galileo had been found "vehemently suspected of heresy." We have seen that this was a technical legal term which meant much more than it may sound to modern ears. In fact, it seems that "suspicion of heresy" was not merely suspicion of having committed a crime but was itself a specific category of crime; it seems that he was in effect being convicted of the second most serious offense handled by the Inquisition. Two distinct heretical views were mentioned: the cosmological thesis that the earth rotates daily on its axis and circles the sun once a year and the methodological principle that one may believe and defend as probable a thesis contrary to the Bible.[52] There is one other interesting detail about the sentence: only seven of the ten cardinal-inquisitors signed it; two of the three who did not were Cardinal Francesco Barberini, the pope's nephew, who was the most powerful man in Rome after the pope himself, and Cardinal Gaspare Borgia, the Spanish ambassador and leader of the Spanish party, who a year earlier had threatened the pope with impeachment on account of his behavior in the Thirty Years War.

It took another six months before Galileo was allowed to return home, to remain under arrest in his own house. He was first sent back to the Tuscan embassy, where he stayed for another ten days; then for about five months he was under house arrest at the residence of the archbishop of Siena, who proved to be a very congenial and sympathetic host.

8. AFTERMATH

Thus the original Galileo affair ended and the new one began—that is, the unresolved and perhaps unresolvable controversy about the events, the issues, the causes, and the lessons of the original episode. This is not the proper place to relate the story of this second affair, even though its fascination rivals that of the original one. Instead we shall end this Introduction with a few details about the historical and the historiographical aftermath of the original episode.

The old man died in 1642, but not before having won a curious and ironical kind of revenge. In fact, when his condition of house arrest forced him to stay away from all controversy, he went back to his relatively noncontroversial studies of mechanics and the laws of motion. He had carried out the relevant work earlier during his university career, but that research was interrupted by his telescopic discoveries. The suffering and humiliation he had experienced in his dealings with the Inquisition renewed the determination and strengthened the motivation of his indomitable spirit. Thus, in 1638 he published in Holland his most important scientific book; entitled *Two New Sciences*, it laid the foundations of modern physics and engineering. Even though the forbidden topic of the earth's motion was not so much as mentioned, the conceptual framework and laws of motion he elaborated therein could be used later to provide a more effective proof of that phenomenon than any he had been able to formulate before. Here we have one of those ironies of history in which the temporary victor of a battle, in this case the Church, creates conditions that pave the way for his eventually losing the war.

The condemned man was buried at the Church of Santa Croce in Florence, in a grave without decoration or inscription. About a century later, in 1734, the Roman Inquisition agreed to a Florentine request to erect a mausoleum for Galileo in that same church. In 1744 the Inquisition allowed the *Dialogue* to be published as part of an edition of his collected works, though accompanied by a number of qualifications and disclaimers. In 1822 the Inquisition decided to allow the publication of books treating of the earth's motion in accordance with modern science, and so the 1835 edition of the Index of Prohibited Books for the first time omitted the *Dialogue* from the list.

Coming to our own lifetime, in November 1979, the first compatriot of Copernicus ever to become pope, and the first non-Italian to do so since the Galileo affair, added the latest twist. In a speech to the Pontifical Academy of Sciences commemorating the centenary of Albert Einstein's birth, John Paul II on the one hand apologetically admitted that Galileo "had to suffer a great deal—we cannot conceal the fact—at the hands of men and organisms of the Church."[53] On the other hand, he boldly tried to reverse the traditional interpretation in terms of a conflict between science and religion by arguing that "in this affair the agreements between religion and science are more numerous and above all more important than the incomprehensions which led to the bitter and painful conflict that continued in the course of the following centuries."[54]

Besides these subsequent footnotes to the original Galileo affair, the last three and one-half centuries have witnessed a rich and complex series of developments in regard to its documentation, interpretation, and evaluation—the new Galileo affair, as it were. Since our focus is on the documents, here it will have to suffice to give a brief account of their origin.

The first selection and collection of documents relating to the affair was made by Church officials for internal administrative reasons. The process began even before the trial was concluded in 1633, for a report was being made for Pope Urban VIII in late May or early June of that year, before the final disposition of the case. Besides writing a summary of the events, the author of this report paginated in one uniform sequence a set of documents relating to the proceedings of 1615–1616 and a set relating to 1632–1633, and he included page references to these documents.[55] This report was soon added to the collection, along with certain other documents directly relating to the trial. After the trial, the collection was enlarged primarily by adding to it dozens of letters received in Rome from inquisitors and papal nuncios of cities throughout Italy and Europe, reporting on the actions they had taken or were taking in view of the condemnation of Galileo. Since the middle of the nineteenth century this set of manuscripts has been kept in the Vatican Secret Archives,[56] and so it may be called the Vatican File on Galileo.

For a long time the mere existence of such a file was not generally known, and at any rate it was not open to inspection. After a number of vicissitudes related to nineteenth-century political and military history,[57] and after the publication of a number of excerpts, selections, and partial editions,[58] the complete file was first published in the 1870s.[59] It is now generally available in volume 19 of Galileo's collected works edited by Favaro,[60] which was first published in 1902[61] and has been reprinted various times[62] and may be regarded as the definitive edition. However, a more accessible and more readable edition was recently published by the Pontifical Academy of Sciences, under the editorship of Sergio M. Pagano (1984).

More important than the history of the Vatican File is for us its content. It consists primarily of three sets of documents: a selection of Inquisition proceedings from 1615–1616, a selection of Inquisition proceedings including the actual trial from 1632–1633, and a voluminous posttrial correspondence originating from inquisitors and papal nuncios throughout Europe. The earlier and later proceedings each take up about one-quarter of the 228 folios in the file, while the posttrial correspondence takes up the remaining half. This collection of documents is invaluable and irreplaceable, and its importance as a source is beyond question. Nevertheless, if one is looking for a collection that provides information about the key events and issues in the affair, it is both too inclusive and too incomplete. It is too inclusive essentially because the posttrial correspondence, while interesting and relevant, has only secondary importance at best. And it is too incomplete because it lacks too many crucial documents; for example, it does not even include the sentence and the abjuration, which were part of the formal Inquisition proceedings, not to mention documents relating to other key characters, which are to be found in other sources.

One of these other sources is the archives of the Inquisition in Rome, which are distinct from the so-called Vatican Secret Archives. Indeed, before the existence of a special Vatican File on Galileo became generally known at the beginning of the nineteenth century, it was the Inquisition Archives that were the object of primary interest and curiosity. The most relevant documents were found among a series of Inquisition Minutes (called *Decreta*), which are summaries of the actions and de-

cisions taken at its regular meetings, usually held several times a week.
A collection of these minutes was first published by Gherardi (1870),
and it reflected the political and military history of nineteenth-century
Europe. This type of archival search culminated in the National Edition
of Galileo's collected works edited by Favaro; volume 19 contains a set
of thirty-five Inquisition Minutes primarily from the years 1615–1616
and 1632–1633, but beginning as early as 1611 and going up to
1734.[63] Favaro was given a privilege not usually granted to scholars, for
to this day the archives of the Roman Inquisition are not generally open
to researchers. This fact remains a source of constant complaint in some
quarters,[64] and it also accounts for the temptation to engage in speculative construction when some new relevant document is accidentally
discovered.[65] More recently Pagano (1984) has had free access to these
archives, and he reports that he can confirm the essential completeness
of Favaro's collection of Inquisition Minutes. However, he has found
about half a dozen new minor items, which he published in his collection of documents.[66] The most important point to remember here is that
the available collection of Inquisition Minutes confirms the story as related by the Vatican File; indeed the latter includes copies of Inquisition
Minutes which correspond to the originals verbatim.

Favaro's accomplishment in publishing material from the Vatican
File and the Inquisition Minutes was matched by his success in discovering and publishing other relevant documents. These fall primarily into
three categories: correspondence to and from Galileo, correspondence
between third parties, and Galilean essays discussing relevant issues.
The twenty volumes of the National Edition of Galileo's collected works
were published in the course of a twenty-year period from 1890 to
1909. The documentation they provide is so rich and exhaustive that
since that time the emphasis has been on interpretative and critical studies of the Galileo affair. Apparently it was felt that Favaro's edition lifted
the interpretation and evaluation of the episode onto a new plane,
and that nothing significant was any longer possible in the area of
documentation.

This view is largely correct. But one crucial task involving documentation had not yet been done.[67] That was to compile a documentary
history, by which I mean a collection of the essential documents and
writings, judiciously selected, properly arranged, of manageable length,
intended to provide a reader with the essential information about the
key events and the key issues and avoiding as much as possible both

controversial interpretation and evaluation. In fact, the Galileo affair is too controversial for it to be read about merely in interpretative or critical accounts; it is of the utmost importance to provide a book where any educated person can read firsthand what happened and what the dispute was all about. These have been the ideas guiding the present work.

In summary, although the Galileo affair climaxed with the trial of 1633, the trial ought not to be equated with the whole affair, any more than the condemnation in June ought to be equated with the whole trial; we are taking the affair to be a sequence of occurrences spanning the twenty-year period from 1613 to 1633. Relevant events prior to 1613 are regarded here as part of the causes of the affair, rather than part of the affair as such, and thus they fall outside the scope of this documentary history.

Further, the most striking component of the opposition to Galileo was obviously the religious and theological element since we are dealing here with proceedings against him by the Inquisition; however, it is undeniable that there were epistemological and scientific aspects to the anti-Galilean opposition, and these were of comparable crucial importance. Thus this documentary history aims to exhibit all three aspects of the controversy: scientific, epistemological, and theological.

I would not deny that the affair was in part a political episode, involving power politics and state interests; nor that it had a social component involving institutional change, class struggle, and social hierarchies; nor that it had a judicial side involving matters of legal procedure and canon law. Although not without interest, these aspects seem to me to be secondary; I believe they are dependent on the scientific, epistemological, and theological elements in a way these elements are not dependent on them. Thus my documentary history does not focus on these secondary matters.

In regard to the scientific aspects of the controversy, I believe the available documents make it overwhelmingly obvious that the paramount issue was the earth's motion. However, this is not to deny that during the same period Galileo was involved in other scientific disagreements (such as the interpretation of sunspots, the nature of comets, and atomism) and that these disputes may have had an indirect bearing on what I am calling the Galileo affair. It is merely to say that there is little documentary evidence indicating that they are an integral part of the sequence of events that led to the trial of 1633.

DOCUMENTS

CHAPTER I

Correspondence (1613–1615)

Castelli to Galileo (14 December 1613)

Very Illustrious and Most Excellent Sir:

Thursday morning I had breakfast with our Lordships, and, when asked about school by the Grand Duke,[1] I gave him a detailed account of everything, and he seemed to be very satisfied. He asked me whether I had a telescope, and I told him Yes and so began to relate the observation of the Medicean planets[2] which I had made just the previous night. Then Her Most Serene Ladyship[3] inquired about their position and began saying to herself that they had better be real and not deceptions of the instrument. So their Highnesses asked Mr. Boscaglia, who answered that truly their existence could not be denied. I used the occasion to add whatever I knew and could say about ⟨606⟩ your wonderful invention and your proof of the motions of these planets. At the table was also Don Antonio,[4] who smiled at me in such a dignified manner that it was a clear sign that he was pleased with what I was saying. Finally, after many other things, all handled properly, the meal ended and I left. As soon as I had come out of the palace, the porter of Her Most Serene Ladyship caught up with me and called me back. However, before I say what followed, you must know that at the table Boscaglia had been whispering for a long time to the ear of Her Ladyship; he admitted as true all the celestial novelties you have discovered, but he said that the earth's motion was incredible and could not happen, especially since the Holy Scripture was clearly contrary to this claim.

Now, to get back to the story, I entered the chambers of Her Highness, where I found the Grand Duke, Her Ladyship, the Archduchess,[5] Mr. Antonio, Don Paolo Giordano,[6] and Boscaglia. At this point, after some questions about my views, Her Ladyship began to argue against me by means of the Holy Scripture. I first expressed the appropriate disclaimers, but then I began to play the theologian with such finesse and authority that you would have been especially pleased to hear. Don Antonio helped and encouraged me so much that I behaved like a champion, despite the fact that the majesty of their Highnesses was enough to frighten me. The Grand Duke and the Archduchess were on my side, and Don Paolo Giordano came to my defense with a very appropriate passage from the Holy Scripture. Only Her Ladyship contradicted me, but in such a way that I thought she was doing it in order to hear me. Mr. Boscaglia remained silent.

All the details that occurred in the course of at least two hours will be related to you by Mr. Niccolò Arrighetti, for I feel obliged to tell you only this. As the occasion came for me to express praise on your behalf while in those chambers, Don Antonio also did this in a manner you can imagine. Moreover, as I was leaving he made me many offers truly with the magnanimity of a prince; indeed yesterday he requested that I inform you of all these occurrences and of what he had said. He explicitly uttered these words: "Write Mr. Galileo about my having become acquainted with you and about what I said in the chambers of Her Highness." I answered this by saying that I would report to you about this beautiful opportunity I have had to serve you. Similarly, Don Paolo has done me all sorts of favors; hence, my affairs are proceeding so well that I do not know what else to want (thank God, who is helping me). Because I have no more time, I kiss your hands and pray that you receive all good things from Heaven.

Pisa, 14 December 1613.

To you Very Illustrious and Most Excellent Sir.

Your Most Obliged Servant and Disciple,
Don Benedetto Castelli.

Galileo to Castelli (21 December 1613)

⟨281⟩ Very Reverend Father and My Most Respectable Sir:

Yesterday Mr. Niccolò Arrighetti[7] came to visit me and told me about you. Thus I took infinite pleasure in hearing about what I did not doubt at all, namely about the great satisfaction you have been giving to the whole University, to its administrators as well as to the professors themselves and to the students from all countries. This approval has not increased the number of your rivals, as it usually happens in similar cases, but rather they have been reduced to very few; and these few too will have to acquiesce unless they want this competition (which is sometimes called a virtue) to degenerate and to change into a blameworthy and harmful feeling, harmful ultimately more to those who practice it than to anyone else. However, the seal of my pleasure was to hear him relate the arguments which, through the great kindness of their Most Serene Highnesses, you had the occasion of advancing at their table and then of continuing in the chambers of the Most Serene Ladyship, in the presence also of the Grand Duke and the Most Serene Archduchess, the Most Illustrious and Excellent Don Antonio and Don Paolo Giordano, and some of the very excellent philosophers there. What greater fortune can you wish than to see their Highnesses themselves enjoying discussing with you, putting forth doubts, listening to your solutions, and finally remaining satisfied with your answers?

⟨282⟩ After Mr. Arrighetti related the details you had mentioned, they gave me the occasion to go back to examine some general questions about the use of the Holy Scripture in disputes involving physical conclusions and some particular other ones about Joshua's passage, which was presented in opposition to the earth's motion and sun's stability by the Grand Duchess Dowager with some support by the Most Serene Archduchess.

In regard to the first general point of the Most Serene Ladyship, it seems to me very prudent of her to propose and of you to concede and to agree that the Holy Scripture can never lie or err, and that its declarations are absolutely and inviolably true. I should have added only that, though the Scripture cannot err, nevertheless some of its interpreters and expositors can sometimes err in various ways. One of these would be very serious and very frequent, namely to want to limit oneself always to the literal meaning of the words; for there would thus emerge not only various contradictions but also serious heresies and blasphemies, and it would be necessary to attribute to God feet, hands, and eyes, as well as bodily and human feelings like anger, regret, hate, and sometimes even forgetfulness of things past and ignorance of future ones. Thus in the Scripture one finds many propositions which look different from the truth if one goes by the literal meaning of the words, but which are expressed in this manner to accommodate the incapacity of common people; likewise, for the few who deserve to be separated from the masses, it is necessary that wise interpreters produce their true meaning and indicate the particular reasons why they have been expressed by means of such words.

Thus, given that in many places the Scripture is not only capable but necessarily in need of interpretations different from the apparent meaning of the words, it seems to me that in disputes about natural phenomena it should be reserved to the last place. For the Holy Scripture and nature both equally derive from the divine Word, the former as the dictation of the Holy Spirit, the latter as the most obedient executrix of God's commands; moreover, in order to adapt itself to the understanding of all people, it was appropriate for the Scripture to say many things ⟨283⟩ which are different from absolute truth, in appearance and in regard to the meaning of the words; on the other hand, nature is inexorable and immutable, and she does not care at all whether or not her recondite reasons and modes of operations are revealed to human understanding, and so she never transgresses the terms of the laws imposed on her; therefore, whatever sensory experience places before our eyes or necessary demonstrations prove to us concerning natural effects should not in any way be called into question on account of scriptural passages whose words appear to have a different meaning, since not every statement of the Scripture is bound to obligations as severely as each effect of nature. Indeed, because of the aim of adapting itself to the capacity of unrefined and undisciplined peoples, the Scripture has not abstained from somewhat concealing its most basic dogmas, thus attributing to God himself properties contrary to and very far from his essence; so who will categorically maintain that, in speaking even incidentally of the earth or the sun or other creatures, it abandoned this aim and chose to restrict itself rigorously within the limited and narrow meanings of the words? This would have been especially problematic when saying about these creatures things which are very far from the primary function of the Holy Writ, indeed things which, if said and put forth in their naked and unadorned truth, would more likely harm its primary intention and make people more resistant to persuasion about the articles pertaining to salvation.

Given this, and moreover it being obvious that two truths can never contradict each other, the task of wise interpreters is to strive to find the true meanings of scriptural passages agreeing with those physical conclusions of which we are already certain and sure from clear sensory experience or from necessary demonstrations. Furthermore, as I already said, though the Scripture was inspired by the Holy Spirit, because of the mentioned reasons many passages admit of interpretations far removed from the literal meaning, and also we cannot assert with certainty that all interpreters speak by divine inspiration; hence, I should think it would be prudent not to allow anyone to oblige ⟨284⟩ scriptural passages to have to maintain the truth of any physical conclusions whose contrary could ever be proved to us by the senses and demonstrative and necessary reasons. Who wants to fix a limit for the human mind? Who wants to assert that everything which is knowable in the world is already known? Because of this, it would be most advisable not to add anything beyond necessity to the articles concerning salvation and the definition of the Faith, which are firm enough that there is no danger of any valid and effective doctrine ever rising against them. If this is so, what greater disorder would result from adding them upon request by persons of whom we do not know whether they speak with celestial inspiration, and of whom also we see clearly that they are completely lacking in the intelligence needed to understand, let alone to criticize, the demonstrations by means of which the most exact sciences proceed in the confirmation of some of their conclusions?

I should believe that the authority of the Holy Writ has merely the aim of persuading men of those articles and propositions which are necessary for their salvation and surpass all human reason, and so could not become credible through some other science or any other means except the mouth of the Holy Spirit itself. However, I do not think it necessary to believe that the same God who has furnished us with senses, language, and intellect would want to bypass their use and give us by other means the information we can obtain with them. This applies especially to those sciences about which one can read only very small phrases and scattered conclusions in the Scripture, as is particularly the case for astronomy, of which it contains such a small portion that one does not even find in it the names of all the planets; but if the first sacred writers had been thinking of persuading the people about the arrangement and the movements of the heavenly bodies, they would not have treated of them so sparsely, which is to say almost ⟨285⟩ nothing in comparison to the infinity of very lofty and admirable conclusions contained in such a science.

So you see, if I am not mistaken, how disorderly is the procedure of those who in disputes about natural phenomena that do not directly involve the Faith give first place to scriptural passages, which they quite often misunderstand anyway. However, if these people really believe they have grasped the true meaning of a particular scriptural passage, and if they consequently feel sure of possessing the absolute truth on the question they intend to dispute about, then let them sincerely tell me whether they think that someone in a scientific dispute who happens to be right has a great advantage over another who happens to be wrong. I know they will answer Yes, and that the one who supports the true side will be able to provide a thousand experiments and a thousand necessary demonstrations for his side, whereas the other person can have nothing but sophisms, paralogisms, and fallacies. But if they know they have such an advantage over their opponents as long as the discussion is limited to physical questions and only philosophical weapons are used, why is it that when they come to the meeting they immediately introduce an irresistible and terrible weapon, the mere sight of which terrifies even the most skillful and expert champion? If I must tell the truth, I believe it is they who are the most terrified, and that they are trying to find a way of not letting the opponent approach because they feel unable to resist his assaults. However, consider that, as I just said, whoever has truth on his side has a great, indeed the greatest, advantage over the opponent, and that it is impossible for two truths to contradict each other; it follows therefore that we must not fear any assaults launched against us by anyone, as long as we are allowed to speak and to be heard by competent persons who are not excessively upset by their own emotions and interests.

To confirm this I now come to examining the specific passage of Joshua,[8] concerning which you put forth three theses for their Most Serene Highnesses. I take the third one, which you advanced as mine (as indeed it is), but I add some other consideration that I do not believe I have ever told you.

Let us then assume and concede to the opponent that the words ⟨286⟩ of the sacred text should be taken precisely in their literal meaning, namely that in answer to Joshua's prayers God made the sun stop and lengthened the day, so that as a result he achieved victory; but I request that the same rule should apply to both, so that the opponent should not pretend to tie me and to leave himself free to change or modify the meanings of the words. Given this, I say that this passage shows clearly the falsity and impossibility of the Aristotelian and Ptolemaic world system, and on the other hand agrees very well with the Copernican one.

I first ask the opponent whether he knows with how many motions the sun moves. If he knows, he must answer that it moves with two motions, namely with the annual motion from west to east and with the diurnal motion in the opposite direction from east to west.[9]

Then, secondly, I ask him whether these two motions, so different and almost contrary to each other, belong to the sun and are its own to an equal extent. The answer must be No, but that only one is specifically its own, namely the annual motion, whereas the other is not but belongs to the highest heaven, I mean the Prime Mobile;[10] the latter carries along with it the sun as well as the other planets and the stellar sphere, forcing them to make a revolution around the earth in twenty-four hours, with a motion, as I said, almost contrary to their own natural motion.

Coming to the third question, I ask him with which of these two motions the sun produces night and day, that is, whether with its own motion or else with that of the Prime Mobile. The answer must be that night and day are effects of the motion of the Prime Mobile, and that what depends on the sun's own motion is not night and day but the various seasons and the year itself.

Now, if the day derives not from the sun's motion but from that of the Prime Mobile, who does not see that to lengthen the day one must stop the Prime Mobile and not the sun? Indeed, is there anyone who understands these first elements of astronomy and does not know that, if God had stopped the sun's motion, He would have cut and shortened the day instead of lengthening it? For, the sun's motion being ⟨287⟩ contrary to the diurnal turning, the more the sun moves toward the east the more its progression toward the west is slowed down, whereas by its motion being diminished or annihilated the sun would set that much sooner; this phenomenon is observed in the moon, whose diurnal revolutions are slower than those of the sun inasmuch as its own motion is faster than that of the sun. It follows that it is absolutely impossible to stop the sun and lengthen the day in the system of Ptolemy and Aristotle, and therefore either the motions must not be arranged as Ptolemy says or we must modify the meaning of the words of the Scripture; we would have to claim that, when it says that God stopped the sun, it meant to say that He stopped the Prime Mobile, and that it said the contrary of what it would have said if speaking to educated men in order to adapt itself to the capacity of those who are barely able to understand the rising and setting of the sun.

Add to this that it is not believable that God would stop only the sun, letting the other spheres proceed; for He would have unnecessarily altered and upset all the order, appearances, and arrangements of the other stars in relation to the sun, and would have greatly disturbed the whole system of nature. On the other hand, it is believable that He would stop the whole system of celestial spheres, which could then together return to their operations without any confusion or change after the period of intervening rest.

However, we have already agreed not to change the meaning of the words in the text; therefore it is necessary to resort to another arrangement of the parts of the world, and to see whether the literal meaning of the words flows directly and without obstacle from its point of view. This is in fact what we see happening.

For I have discovered and conclusively demonstrated that the ⟨288⟩ solar globe turns on itself,[11] completing an entire rotation in about one lunar month, in exactly the same direction as all the other heavenly revolutions; moreover, it is very probable and reasonable that, as the chief instrument and minister of nature and almost the heart of the world, the sun gives not only light (as it obviously does) but also motion to all the planets that revolve around it; hence, if in conformity with Copernicus's position the diurnal motion is attributed to the earth, anyone can see that it sufficed stopping the sun to stop the whole system, and thus to lengthen the period of the diurnal illumination without altering in any way the rest of the mutual relationships of the planets; and that is exactly how the words of the sacred text sound. Here then is the manner in which by stopping the sun one can lengthen the day on the earth, without introducing any confusion among the parts of the world and without altering the words of the Scripture.

I have written much more than is appropriate in view of my slight illness. So I end by reminding you that I am at your service, and I kiss your hands and pray the Lord to give you happy holidays and all you desire.

Florence, 21 December 1613.

To Your Very Reverend Paternity.

Your Most Affectionate Servant,
Galileo Galilei.

Galileo to Monsignor Dini (16 February 1615) [12]

⟨291⟩ My Most Illustrious, Reverend, and Honorable Lord:

I know that Your Most Illustrious and Reverend Lordship was immediately notified of the repeated invectives which were thrown from the pulpit a few weeks ago against Copernicus's doctrine and its followers, and also against mathematicians and mathematics itself; [13] thus I shall not repeat anything about these details which you heard from others. However, I want very much for you to know how the rage of those people has not quieted down, although neither I nor the others took the least step or showed the least resentment for the insults with which we were very uncharitably burdened; indeed, upon his return from Pisa, the same Father who a few years ago expressed complaints in private conversation has hit me again. [14] He has come across, I do not know how, a copy of a letter which about a year ago I wrote to the Father Mathematician of Pisa [15] in connection with the use of sacred authorities in scientific disputes and the interpretation of Joshua's passage, and so they are making an uproar about it; from what I hear, they find many heresies in it and, in short, they have opened a new front to tear me to pieces. However, because I have not received the least sign of scruples from anyone else who has seen this letter, I suspect that perhaps whoever transcribed it may have inadvertently changed some word; [16] such a change, together with a little inclination toward censure, may make things look very different ⟨292⟩ from my intention. I hear that some of these Fathers, especially the same one who had complained earlier, have come there to try something else with his copy of this letter of mine, and so I did not consider it inappropriate to send you a copy of it in the correct version as I wrote it. I ask you to do me the favor of reading it along with the Jesuit Father Grienberger, a distinguished mathematician and a very good friend and patron of mine, and, if you deem it appropriate, of having it somehow come into the hands of the Most Illustrious Cardinal Bellarmine. The latter is the one whom these Dominican Fathers seem to want to rally around, with the hope of bringing about at least the condemnation of Copernicus's book, opinion, and doctrine.

I wrote the letter with a quick pen. However, this recent excitement, and the reasons these Fathers adduce to show the faults of this doctrine for which it deserves to be abolished, have made me look at other writings on the topic. So, indeed, not only do I find that all I wrote has been said by them, but also much more: how carefully one must proceed in regard to physical conclusions which are not matters of faith and to which one may arrive through experience and necessary demonstrations, and how harmful it would be to take as a doctrine decided by Holy Scripture any proposition for which there could ever be a disproof. I have written a very long essay[17] on these topics, but I have not yet polished it so as to be able to send you a copy, though I shall do that soon. Whatever the efficacy of the arguments and reasoning contained in it, of this I am very sure, namely that there is in it much more zeal toward the Holy Church and the dignity of Holy Scripture than in these persecutors of mine. For they try to prohibit a book which for many years has been permitted by the Holy Church, without their having ever seen it, let alone read or understood it; whereas I am only claiming that its doctrine be examined and its reasons assessed by the most Catholic and competent persons, that its views be compared with sensory experience, and that in short it should not ⟨293⟩ be condemned before being found false, if the truth is that a proposition cannot be simultaneously true and erroneous. There is no lack in Christendom of very competent professional men, whose opinion on the truth or falsity of the doctrine should not be discounted in favor of the whim of someone who is not in the least informed and who is very clearly known to be biased and emotionally upset; this is in fact very well known to many here, who see all the moves and are informed, at least in part, of the various machinations and deals.

Nicolaus Copernicus was not only a Catholic man, but a clergyman and a canon. He was called to Rome under Leo X, when the Lateran Council was discussing the emendation of the ecclesiastical calendar,[18] for everyone looked up to him as a very great astronomer. Nevertheless, this reform was not decided upon for the sole reason that the relationships between the year and month and the motions of sun and moon were not established with sufficient precision. Thus, at the request of the Bishop of Fossombrone,[19] who was in charge of this project, he set out to investigate these periods by means of new observations and very accurate studies. Eventually he acquired so much knowledge that not only did he bring order to all the motions of the heavenly bodies, but he earned the title of greatest astronomer, his doctrine was later followed by everyone, and the calendar was recently rectified in accordance with it.[20] He condensed his labors on the motions and arrangement of the heavenly bodies into six parts, which he published at the request of Nicolaus Schoenberg, Cardinal of Capua,[21] and dedicated to Pope Paul

III; and from that time on the work has been available to the public without any scruples. Now these good friars, merely on account of a sinister feeling against me and knowing that I have high regard for this author, are proud to reward his labors by having him declared a heretic.

However, what is more worthy of consideration is that their first move against this opinion was to let themselves be deceived by some enemies of mine, who portrayed it to them as my own work, without telling them that it had already been published seventy years ago; and they practice the same style with other persons, whom they try to convince of sinister things in my regard. They have become so successful at it that, when a few ⟨294⟩ days ago the Bishop of Fiesole Monsignor Gherardini came here, at his first public appearances where some friends of mine happened to be present, he burst out with the greatest vehemence against me, appearing deeply agitated, and saying he was going to mention the matter at great length to their Most Serene Highnesses, since my extravagant and erroneous opinion was causing much talk in Rome. Perhaps he has already kept his promise, if he was not already restrained by being promptly informed that the author of this doctrine is not a living Florentine but a dead German,[22] who printed it seventy years ago, dedicating the book to the Supreme Pontiff.

I am writing without noticing that I am speaking to a person who is very well informed about these occurrences, perhaps much more than I am, insofar as you are at the place where most of the noise is being generated. Excuse my prolixity, and, if you detect any justice in my cause, lend me your support, for I shall be eternally obliged. With this I humbly kiss your hands, I remain your most devout servant, and I pray the Lord God to grant you the greatest happiness.

Florence, 16 February 1615.

To Your Very Illustrious and Most Reverend Lordship.

> Your Most Obliged Servant,
> Galileo Galilei.

Postscript. Although I find it difficult to believe that anyone would rush to make the decision to annul this author, I know from other evidence how great is the power of my bad luck when combined with the wickedness and ignorance of my opponents; thus I think I have reason to be skeptical about the high prudence and holiness of those on whom the final decision depends, for they may be in part deceived by this fraud which is going around under the cloak of zeal and ⟨295⟩ charity. However, my upbringing and inclination are such that, rather than contradicting my superiors, "I should pluck my eye out so that it would not cause me to sin,"[23] if I could not do anything else, and if what I now believe and seem to grasp were prejudicial to my soul; thus, as much as it is within my power, I do not want to fail in my duty and in my true and purest zeal (zeal which you will shortly see from my essay),[24] and I desire that at least it could first be read and then they could make whatever decision will please God. I think the most immediate remedy would be to approach the Jesuit Fathers, as those whose knowledge is much above the common education of friars. Perhaps you can give them the copy of the letter, and read them this one I am writing you, if you wish; then, with your usual courtesy, you could let me know what you may have learned. I do not know whether it would be proper to approach Mr. Luca Valerio and give him a copy of the said letter; this man is very close to Cardinal Aldobrandini and might be able to be of some service. On this and on everything else I defer to your goodness and prudence; begging you to keep my reputation in mind, I again kiss your hands.

Monsignor Dini to Galileo (7 March 1615)

⟨151⟩ My Very Illustrious and Most Respectable Sir:
The many celebrations and festivities of this Mardis Gras period have made it impossible for me to find the persons I was supposed to; but instead I had many copies made of your letter to the Father Mathematician,[25] and I then gave it to Father Grienberger, to whom I also read the one you wrote to me. Then I did the same with many other people, including the Most Illustrious Bellarmine, with whom I spoke at length about the things you mention. He assured me he has not heard anyone talk about them at all, ever since he discussed them orally with you.[26] In regard to Copernicus, His Most Illustrious Lordship says he cannot believe that it is about to be prohibited; rather, in his view, the worst that could happen to the book is to have a note added to the effect that its doctrine is put forth in order to save the appearances,[27] in the manner of those who have put forth epicycles but do not really believe in them, or something similar; and so you could in any case speak of these things with such a qualification, for if they are interpreted in accordance with the new arrangement, it does not seem for now that they have any greater enemy in the Scripture than the verse "hath rejoiced as a giant to run the way"[28] and the following one,[29] which all interpreters so far have understood as attributing motion to the sun. I replied that even

this could be explained in terms of our usual way of understanding these things; but he answered that this is not something to jump into, just as one ought not to jump hurriedly into condemning any one of these opinions. If in that essay of yours[30] you have collected those interpretations which address the issue, His Most Illustrious Lordship will gladly look at them; since I know you will remember to defer to the decisions of the Holy Church, as you have indicated to me and to others, that cannot but benefit you a great deal. Furthermore, because the said Lord Cardinal had told me that he would meet with Father Grienberger to discuss these matters, this morning I went to this Father to hear whether there was any news. I found nothing important besides what I have already mentioned, except that he would have liked you to first carry out your demonstrations, and then get involved in discussing the Scripture; I replied that if you had proceeded in this manner, you could be said to have acted badly, for minding your own business first and thinking of the Holy Scripture later. In regard to the arguments advanced in favor of your position, ⟨152⟩ the said Father thinks they are more plausible than true, since he is worried about other passages of the Holy Writ.

This morning I sent one of the above-mentioned copies to Mr. Luca Valerio, with whom I have not yet spoken. I went to see the Lord Cardinal Del Monte in order to inform him, but I found there some persons I did not like, and so I discussed other things with him; nevertheless, I will go back since he likes you very much. I will also go to see the Lord Cardinal Barberini[31] in order to give him one of those copies, which he is expecting since I had the chance to quickly tell him part of the story. Perhaps by now he has been fully informed by Mr. Ciampoli, to whom I had spoken for this purpose. Thus I will continue taking similar steps whenever I see I can do some good; as you can see, things are rather unclear because everyone is on guard about a business of such importance, though mathematicians do not seem to be in as much doubt as the practitioners of other disciplines. This is all I can tell you for now; so, without going further, I kiss your hands and pray the Lord God to grant all your wishes.

Rome, 7 March 1615.

To You Very Illustrious Sir.

Your Most Affectionate Servant,
P. Dini.

Galileo to Monsignor Dini (23 March 1615)

⟨297⟩ My Most Illustrious, Reverend, and Honorable Lord:

I shall reply very briefly to the very courteous letter I received from Your Very Illustrious and Most Reverend Lordship, as the poor state of my health does not permit me to do otherwise.

Let me begin with the first detail you mention, namely that the most that might be decided about Copernicus's book would be to add some comments, to the effect that his doctrine was advanced to save the appearances, in the same way that others have advanced eccentrics and epicycles, without then believing that they really exist in nature. Let me add that I say what follows with the intention of always deferring to those who are wiser than I am, and out of zeal that what is about to be done should be done with the greatest possible caution. In regard to saving the appearances, Copernicus himself at first did the work and satisfied the interests of astronomers, in accordance with the usual and received technique of Ptolemy; but then, putting on philosophical garments, he considered whether such an arrangement of the parts of the universe could truly exist in physical reality, and he concluded No. Believing that the problem of its true structure deserved to be researched, he undertook the investigation of such a structure, with the knowledge that if a fictitious and untrue arrangement of parts could satisfy the ⟨298⟩ appearances, this could be done much better with the true and real one; at the same time philosophy would have gained sublime knowledge, for such is to know the true arrangement of the parts of the world. From many years of observation and study, he was abundantly in possession of all the details observed in the stars, for it is impossible to come to know this structure of the universe without having learned them all very diligently and having them very readily available in mind; and so, by repeated studies and very long labors, he accomplished what later earned him the admiration of all those who study him diligently enough to understand his discussions. Thus, to claim that Copernicus did not consider the earth's motion to be true could be accepted perhaps only by those who have not read him, in my opinion; for all six parts of his book are full of the doctrine of the earth's motion, and of explanations and confirmations of it. Furthermore, if in his dedicatory preface he understands very well and admits that the view of the earth's motion was going to make him appear insane to common people, whose judgment he says he disregards, he would have been even more insane to want to be regarded as such for an opinion introduced but not fully and truly believed by him.

Then, in regard to saying that the principal authors who have introduced eccentrics and epicycles did not regard them as true, I shall never believe this, especially since in our age they must be accepted with absolute necessity, as the senses themselves show us. For, an epicycle being nothing but a circle traced by the motion of a star and not enclosing the terrestrial globe, do we not see four such circles traced by four stars around Jupiter? And is it not clearer than sunlight that Venus traces its circle around the sun without enclosing the earth, and thus generates an epicycle? The same happens to Mercury. Furthermore, an eccentric being a circle which indeed surrounds the earth, but has it on one side rather than at the center, it cannot be doubted whether the path of Mars is eccentric to the earth, since it is seen now nearer and now farther, inasmuch as ⟨299⟩ we see it sometimes very small and other times with a surface sixty times larger; thus, whatever the details of its revolution, it surrounds the earth and in certain places is eight times closer to the latter than at other places. Recent discoveries have provided us with sensory experiences about all these things and other similar ones; so to want to grant the earth's motion only with the proviso and probability which are applied to eccentrics and epicycles is to grant it as very sure, very true, and incontrovertible.

It is indeed true that among those who have denied eccentrics and epicycles I find two groups. One consists of those who are completely ignorant of the observed stellar motions and the appearances that have to be saved, who deny without any foundation all that they do not understand; but these people deserve to be disregarded. Others, much more reasonably, do not deny circular motions traced by stellar bodies around centers other than the earth, which is such an obvious phenomenon that on the contrary it is clear that none of the planets perform their revolutions concentrically around the earth; they only deny that there exists in the heavens a structure of solid orbs, distinct and separated from each other, whose rotation and mutual friction carry the bodies of the planets, etc. I believe that these people reason very correctly. However, this is not to take away motions made by the stars in circles eccentric to the earth or in epicycles, which are the genuine and simple assumptions of Ptolemy and the other great astronomers; it is rather to repudiate the solid, material, and distinct orbs, introduced by the builders of models to facilitate understanding by beginners and computations by calculators; this is the only fictitious and unreal part, as God does not lack the means to make the stars move in the immense celestial spaces, within well-defined and definite paths, but without having them chained and forced.

As regards Copernicus, he is not, in my opinion, susceptible of compromise since the most essential point and general foundation of his whole doctrine is the earth's motion and the sun's stability; thus, either he must be wholly condemned, or he must be left alone, if I may speak as always in accordance with my own capacity. Whether in reaching such a decision it is advisable to consider, ponder, ⟨300⟩ and examine what he writes is something that I have done my best to show in an essay of mine.[32] I hope the blessed God has granted me this, for I have no other aim but the honor of the Holy Church and do not direct my small labors to any other goal. This very pure and zealous attitude will, I am very sure, be clearly noticed in that essay, even if it were otherwise full of errors or insignificant things. I would have already sent it to you, if to my many serious illnesses had not been added lately an attack of colic pains, which has bothered me a great deal; but I will send it soon. Indeed, out of the same zeal, I am in the process of collecting all of Copernicus's reasons and making them clearly intelligible to many people, for in his works they are very difficult; and I am adding to them many more considerations, always based on celestial observations, on sensory experiences, and on the support of physical effects.[33] Then I would offer them at the feet of the Supreme Pastor for the infallible decision by the Holy Church, to be used as their supreme prudence sees fit.

As for the opinion of the Very Reverend Father Grienberger, I truly praise him, and I gladly leave the labor of interpretation to those who understand infinitely more than I do; but the short composition I sent you is, as you see, a private letter, written more than a year ago to my friend, to be read only by him. Since he let it be copied without my knowledge, since I heard it had gotten into the hands of the same man who had attacked me so sourly even from the pulpit, and since I learned the latter had brought it to your city, I deemed it advisable to make another copy, so that they could be compared closely, especially in view of the fact that he and some of his theologian friends had spread the rumor here that this letter of mine is full of heresies. Thus, it is not my intention to undertake a task so superior to my forces, though one must not doubt that Divine Love may sometimes deign to inspire humble minds with a ray of His infinite wisdom, especially when they are full of sincere and holy zeal; furthermore, when sacred texts have to be reconciled with new and uncommon physical doctrines it is necessary to be completely informed about such doctrines, for one cannot tune two strings together by listening to just one. Now, if I knew I could ⟨301⟩ expect anything from the weakness of my intellect, I should dare say I find that some passages of the Holy Writ fit very well with this world system, whereas they do not seem to sound equally well in the popular philosophy. Your Most Reverend Lordship mentioned to me how a passage in Psalm 18 is among those regarded as most repugnant to this opinion, and so I was led to some new reflections on it; I send them to you all the less reluctantly, inasmuch as you tell me that the Most Illustrious and Most Reverend Cardinal Bellarmine will gladly see what I have to say. However, I have complied with a mere sign on your part, and so when you see this speculation of mine do with it whatever your great prudence deems appropriate, for my intention is only to revere and admire such sublime knowledge, to obey the hints of my superiors, and to submit all my work to their will.

Let it be understood that, whatever be the truth of the physical assumption, I do not presume that others could not more adequately interpret the words of the prophet; indeed I regard myself as the lowest of all, and I submit myself to all the wise men. So I should say it seems to me that there is in nature a very spirited, tenuous, and fast substance which spreads throughout the universe, penetrates everything without difficulty, and warms up, gives life, and renders fertile all living creatures. It also seems to me that the senses themselves show us the body of the sun to be by far the principal receptacle for this spirit, and that from there an immense amount of light spreads throughout the universe and, together with such a calorific and penetrating spirit, gives life and fertility to all vegetable bodies. One may reasonably believe that this is something different from light since it penetrates and spreads through all corporeal substances, even if very dense, many of which are not penetrated in the same way by light; thus, since we see and feel light and heat emanating from common fire, and the latter goes through all bodies even if opaque and very solid, and the former finds opposition in solidity and opacity, similarly the emanation of the sun is luminous and calorific, and the calorific part is the more penetrating one. ⟨302⟩ It seems to me that from the Holy Writ we can acquire evident certainty that the solar body is, as I have said, a receptacle and, so to speak, a reservoir of this spirit and this light, which it receives from elsewhere, rather than being their primary font and source from which they originally derive. For we read there, before the creation of the sun, about the spirit with its calorific and fertile power "skimming the waters or lying over the waters"[34] for future generations; similarly, we have the creation of light on the first day, whereas the solar body is created on the fourth day. Thus, we may affirm with great verisimilitude that this fertilizing spirit and this light diffused throughout the world come together to unite and be strengthened in the body of the sun, which because of this is located at the center of the universe; then, having been made more intense and vigorous, they are again diffused.

In Psalm 73:16[35] the prophet gives us testimony about the primordial light which was not very intense before its union and convergence into the solar body: "Thine is the day, and thine is the night: thou hast made the morning light and the sun."[36] This passage is interpreted as meaning that, before creating the sun, God made a light similar to that of dawn; furthermore, instead of "morning light" (or "dawn") the Hebrew text reads "light," to suggest to us the light which was created much before the sun, and is much weaker than the one received, fortified, and again diffused by the solar body. This is the view to which some ancient philosophers appear to be referring: they believed that the brilliance of the sun derives from the convergence at the world's center of the light from the stars arranged spherically around it; that their rays converge and intersect at this point, here are magnified, and have their light increased a thousand times; and that this intensified light is then spread in a form which is much more vigorous and, so to speak, full of virile and vivid warmth, and goes on to give life to all bodies that revolve around the center. Thus, there is a certain similarity: just as in an animal's heart there is a continuous regeneration of vital spirits which support and give life to all the limbs (while, however, from elsewhere the same heart receives the food and nourishment without which it would perish), so while the sun's food converges into it from the outside, we have the conservation of the principle whereby ⟨303⟩ the prolific light and heat, which give life to all the limbs that lie around it, are continuously produced and diffused. Although I could produce many quotations from philosophers and serious writers in favor of the wonderful force and energy of this solar spirit and light, I want to limit myself to a single passage from the book *The Divine Names* by the Blessed Dionysius the Areopagite.[37] It is this: "Light also gathers and attracts to itself all things that are seen, that move, that are illuminated, that are heated, and in a word that are surrounded by its splendor. Thus the sun is called Helios because it collects and gathers all things that are dispersed."[38] And a little below that he writes about it as follows:

"If in fact this sun, which we see and which (despite the multitude and dissimilarity of the essences and qualities of observed things) is nevertheless one, spreads its light equally and renews, nourishes, preserves, perfects, divides, joins, warms up, fertilizes, increases, changes, strengthens, produces, moves, and vitalizes all things; and if everything in this universe in accordance with its own power partakes of one and the same sun, and contains within itself an equal anticipation of the causes of the many things which are shared; then certainly all the more reason, etc."

This philosophical view is perhaps one of the principal doors by which one may enter the contemplation of nature. Now, I speak always with the humility and reverence due to the Holy Church and all its very learned Fathers, whom I respect and honor and to whose judgment I submit myself and every one of my thoughts. So I say I am inclined to believe that the passage in the Psalm could have this meaning, namely that "God hath set his tabernacle in the sun,"[39] as the most noble seat in the whole sensible world. Then where it says that "he, as a bridegroom coming out of his bride chamber, hath rejoiced as a giant to run the way,"[40] I should understand this as referring to the radiating sun, namely to the light and the already mentioned calorific spirit, which fertilizes all corporeal substances and which, starting from the solar body, spreads very fast throughout the world; all the words correspond exactly to this meaning. To begin with, in the word "bridegroom" we have the fertilizing and prolific power; the "rejoicing" indicates the emanation of these solar rays taking place somehow by jumps, as the senses clearly show us; "as a giant" or "as a strong man" denotes the very effective action and power of penetrating all bodies, as well as the very high speed with which it moves through immense distances, ⟨304⟩ the emanation of light being as if instantaneous. The words "coming out of his bride chamber" confirm that this emanation and motion must refer to the solar light and not to the solar body itself, for the body and globe of the sun are a receptacle and "like a bride chamber" for that light, and it does not sound right to say that "a bride chamber would come out of a bride chamber." In what follows,[41] the sentence "his going out is from the end of heaven" refers to the original departure of this spirit and light from the highest regions of the sky, namely from the stars of the firmament or from even more remote regions; the clause "and his circuit even to the end thereof" indicates the reflection and, so to speak, the re-emanation of the same light at the summit of the world; and the words after that, "and there is no one that can hide himself from his heat," point to the life-giving and fertilizing heat, which is distinct from the light and penetrates all corporeal substances (even the most dense) much more than the latter, for many things can defend and cover us from the penetration of light, but as regards this other power "there is no one that can hide himself from his heat." Nor should I be silent about another consideration of mine, which is not irrelevant to the present purpose. I have already discovered the constant generation on the solar body of some dark substances, which appear to the eye as very black spots and then are consumed and dissolved; and I have discussed how they could perhaps be regarded as part of the nourishment (or perhaps its excrements) which some ancient philosophers thought the sun needed for its sustenance. By constantly observing these dark substances, I have demonstrated how the solar body necessarily turns on itself, and I have also speculated how reasonable it is to believe that the motions of the planets around the sun depend on such turning. Furthermore, we know that the intention of this Psalm is to praise the divine law, as the prophet compares it to that heavenly body which is the most beautiful, useful, and powerful of all corporeal things. Therefore, after singing praises to the sun (and without being ignorant that it makes all the movable bodies in the world turn around itself), he goes on to the greater prerogatives of the divine law, and he wants to give it priority over the sun by adding, "the law of the Lord is unspotted, converting souls, etc.";[42] this is as if he wants to say that that law is as much greater than the sun itself, as ⟨305⟩ being spotless and having the power of making souls turn around itself is a more lofty condition than being covered with spots (as the sun is) and making corporeal and worldly globes turn around itself.

I know and confess my excessive audacity in wanting to open my mouth to explain meanings worthy of such elevated contemplation, given that I am not an expert in the Holy Writ. However, just as I may be excused because I submit totally to the judgment of my superiors, so the verse following the one already explained ("the testimony of the Lord is faithful, giving wisdom to little ones")[43] has given me hope that God's infinite love may direct toward my pure mind a very small ray of his grace, through which I may grasp some of the hidden meanings of his words. What I have written, my Lordship, is a small offspring in need of being put in better form by licking and polishing it with love and patience, as it is only a sketch and has at the moment disordered and rough limbs, however capable they may be of taking on a very well proportioned shape. If I have the opportunity, I shall give it a better symmetry; in the meantime, I beg you not to let it come into the hands of any person who would use the hard and sharp tooth of a beast rather than the delicate tongue of a mother, and so would completely mangle and tear it to pieces instead of polishing it. With this I humbly kiss your hands, together with Messrs. Buonarotti, Guiducci, Soldani, and Giraldi, present here at the sealing of this letter.

Florence, 23 March 1615.

To Your Very Illustrious and Most Reverend Lordship.

Your Most Obliged Servant,
Galileo Galilei.

Cardinal Bellarmine to Foscarini (12 April 1615)

To the Very Reverend Father Paolo Antonio Foscarini, Provincial of the Carmelites in the Province of Calabria:[44]

My Very Reverend Father,

I have read with interest the letter in Italian and the essay in Latin which Your Paternity sent me; I thank you for the one and for the other and confess that they are all full of intelligence and erudition. You ask for my opinion, and so I shall give it to you, but very briefly, since now you have little time for reading and I for writing.

First, I say that it seems to me that Your Paternity and Mr. Galileo are proceeding prudently by limiting yourselves to speaking suppositionally[45] and not absolutely, as I have always believed that Copernicus spoke. For there is no danger in saying that, by assuming the earth moves and the sun stands still, one saves all the appearances better than by postulating eccentrics and epicycles; and that is sufficient for the mathematician. However, it is different to want to affirm that in reality the sun is at the center of the world and only turns on itself without moving from east to west, and the earth is in the third heaven[46] and revolves with great speed around the sun; this is a very dangerous thing, likely not only to irritate all scholastic philosophers and theologians, but also to harm the Holy Faith by rendering Holy Scripture false. For Your Paternity has well shown many ways of interpreting Holy Scripture, but has not applied them to particular cases; without a doubt you would have encountered very great difficulties if you had wanted to interpret all those passages you yourself cited.

⟨172⟩ Second, I say that, as you know, the Council[47] prohibits interpreting Scripture against the common consensus of the Holy Fathers; and if Your Paternity wants to read not only the Holy Fathers, but also the modern commentaries on Genesis, the Psalms, Ecclesiastes, and Joshua, you will find[48] all agreeing in the literal interpretation that the sun is in heaven and turns around the earth with great speed, and that the earth is very far from heaven and sits motionless at the center of the world. Consider now, with your sense of prudence, whether the Church can tolerate giving Scripture a meaning contrary to the Holy Fathers and to all the Greek and Latin commentators. Nor can one answer that this is not a matter of faith, since if it is not a matter of faith "as regards the topic," it is a matter of faith "as regards the speaker"; and so it would be heretical to say that Abraham did not have two children and Jacob twelve, as well as to say that Christ was not born of a virgin, because both are said by the Holy Spirit through the mouth of the prophets and the apostles.

Third, I say that if there were a true demonstration that the sun is at the center of the world and the earth in the third heaven, and that the sun does not circle the earth but the earth circles the sun, then one would have to proceed with great care in explaining the Scriptures that appear contrary, and say rather that we do not understand them than that what is demonstrated is false. But I will not believe that there is such a demonstration, until it is shown me. Nor is it the same to demonstrate that by supposing the sun to be at the center and the earth in heaven one can save the appearances, and to demonstrate that in truth the sun is at the center and the earth in heaven; for I believe the first demonstration may be available, but I have very great doubts about the second, and in case of doubt one must not abandon the Holy Scripture as interpreted by the Holy Fathers. I add that the one who wrote, "The sun also ariseth, and the sun goeth down, and hasteth to his place where he arose,"[49] was Solomon, who not only spoke inspired by God, but was a man above all others wise and learned in the human sciences and in the knowledge of created things; he received all this wisdom from God; therefore it is not likely that he was affirming something that was contrary to truth already demonstrated or capable of being demonstrated. Now, suppose you say that Solomon speaks in accordance with appearances, since it seems to us that the sun moves (while the earth does so), just as to someone who moves away from the seashore on a ship it looks like the shore is moving. I shall answer that when someone moves away from the shore, although it appears to him that the shore is moving away from him, nevertheless he knows that this is an error and corrects it, seeing clearly that the ship moves and not the shore; but in regard to the sun and the earth, no scientist has any need to correct the error, since he clearly experiences that the earth stands still and that the eye is not in error when it judges that the sun moves, as it also is not in error when it judges that the moon and the stars move. And this is enough for now.

With this I greet dearly Your Paternity, and I pray to God to grant you all your wishes.

At home, 12 April 1615.

To Your Very Reverend Paternity.

As a Brother,
Cardinal Bellarmine.

CHAPTER II

Galileo's Considerations on the Copernican Opinion (1615)[1]

⟨351⟩ In order to remove (as much as the blessed God allows me) the occasion to deviate from the most correct judgment about the resolution of the pending controversy, I shall try to do away with two ideas. These are notions which I believe some are attempting to impress on the minds of those persons who are charged with the deliberations, and, if I am not mistaken, they are concepts far from the truth.

The first is that no one has any reason to fear that the outcome might be scandalous; for the earth's stability and sun's motion are so well demonstrated in philosophy that we can be sure and indubitably certain about them; on the other hand, the contrary position is such an immense paradox and obvious foolishness[3] that no one can doubt in any way that it cannot be demonstrated now or ever, or indeed that it can never find a place in the mind of sensible persons. The other idea which they try to spread is the following: although that contrary assumption has been used by Copernicus and other astronomers, they did this in a suppositional manner[4] and insofar as it can account more conveniently for the appearances of celestial motions and facilitate astronomical calculations and computations, and it is not the case that the same persons who assumed it believed it to be true de facto and in nature; so the conclusion is that one can safely proceed to condemn it. However, if I am not mistaken, these ideas are fallacious and far from the truth, as I can show with the following considerations. These will only be general and suitable to be understood without much effort and labor even by someone who is not well versed in the natural and astronomical sciences. For, if there were the opportunity to treat these ⟨352⟩ points with those who are very experienced in these studies, or at least who have the time to do the work required by the difficulty of the subject, then I should propose nothing but the reading of Copernicus's own book; from it and from the strength of his demonstrations one could clearly see how true or false are the two ideas we are discussing.

That it is not to be disparaged as ridiculous is, therefore, clearly shown by the quality of the men, both ancient and modern, who have held and do hold it. No one can regard it as ridiculous unless he considers ridiculous and foolish Pythagoras with all his school, Philolaus (teacher of Plato), Plato himself (as Aristotle testifies in his book *On the Heavens*), Heraclides of Pontus, Ecphantus, Aristarchus of Samos, Hicetas,[5] and Seleucus the mathematician. Seneca himself not only does not ridicule it, but he makes fun of those who do, writing in his book *On Comets:* "It is also important to study these questions in order to learn whether the universe goes around the motionless earth, or the earth rotates but the universe does not. For some have said that we are naturally unaware of motion, that sunrise and sunset are not due to the motion of the heavens, but that it is we ourselves who rise and set. The matter deserves consideration, so that we may know the conditions of our existence, whether we stand still or move very fast, whether God drives everything around us or drives us."[6] Regarding the moderns, Nicolaus Copernicus first accepted it and amply confirmed it in his whole book.[7] Then there were others: William Gilbert, a distinguished physician and philosopher, who treats it at length and confirms it in his book *On the Loadstone;*[8] Johannes Kepler, a living illustrious philosopher and mathematician in the service of the former and the current Emperor, follows the same opinion; Origanus (David Tost) at the beginning of his *Ephemerides*[9] supports the earth's motion with a very long discussion; and there is no lack of other authors who have published their reasons on the matter. Furthermore, though they have not published anything, I could name very many followers of this doctrine living in Rome, Florence, Venice, Padua, Naples, Pisa, Parma, and other places. This doctrine is not, therefore, ridiculous, having been accepted by great men; and, though their number is small compared to the followers of the common position, this is an indication of its being difficult to understand, rather than of its absurdity.

Moreover, that it is grounded on very powerful and effective ⟨353⟩

reasons may be shown from the fact that all its followers were previously of the contrary opinion, and indeed that for a long time they laughed at it and considered it foolish. Copernicus and I, and all others who are alive, are witnesses to this. Now, who will not believe that an opinion which is considered silly and indeed foolish, which has hardly one out of a thousand philosophers following it, and which is disapproved by the Prince[10] of the prevailing philosophy, can become acceptable through anything but very firm demonstrations, very clear experiences, and very subtle observations? Certainly no one will be dissuaded of an opinion imbibed with mother's milk from his earliest training, accepted by almost the whole world and supported by the authority of very serious writers, unless the contrary reasons are more than effective. If we reflect carefully, we find that there is more value in the authority of a single person who follows the Copernican opinion than in that of one hundred others who hold the contrary, since those who are persuaded of the truth of the Copernican system were in the beginning all very opposed. So I argue as follows.

Either those who are to be persuaded are capable of understanding the reasons of Copernicus and others who follow him, or they are not; moreover, either these reasons are true and demonstrative, or they are fallacious. If those who are to be persuaded are incapable, then they will never be persuaded by the true or by the false reasons; those who are capable of understanding the strength of the demonstrations will likewise never be persuaded if these demonstrations are fallacious; so neither those who do nor those who do not understand will be persuaded by fallacious reasons. Therefore, given that absolutely no one can be dissuaded from the first idea by fallacious reasons, it follows as a necessary consequence that, if anyone is persuaded of the contrary of what he previously believed, the reasons are persuasive and true. But as a matter of fact there are ⟨354⟩ many who are already persuaded by Copernican reasons. Therefore, it is true both that these reasons are effective and that the opinion does not deserve the label of ridiculous but the label of worthy of being very carefully considered and pondered.

Furthermore, how futile it is to argue for the plausibility of this or that opinion simply from the large number of followers may be easily inferred from this: no one follows this opinion who did not previously believe the contrary; but instead you will not find even a single person who, after holding this opinion, will pass to the other one, regardless of any discussion he hears; consequently, one may judge, even if he does not understand the reasons for one side or for the other, that probably the demonstrations for the earth's motion are stronger than those for the other side. But I shall say more, namely that if the probability of the two positions were something to be won by ballot, I would be willing to concede defeat when the opposite side had one more vote than I out of one hundred; not only that, but I would be willing to agree that every individual vote of the opponents was worth ten of mine, as long as the decision was made by persons who had perfectly heard, intimately penetrated, and subtly examined all the reasons and evidence of the two sides; indeed it is reasonable to expect that such would be those who cast the votes. Hence this opinion is not ridiculous and contemptible, but somewhat shaky is the position of whoever wanted to capitalize on the common opinion of the many who have not accurately studied these authors. What then should we say of the noises and the idle chatter of someone who has not understood even the first and simplest principles of these doctrines, and who is not qualified to understand them ever? What importance should we give him?

Consider now those who persist in wanting to say that as an astronomer Copernicus considered the earth's motion and the sun's stability only a hypothesis[11] which is more adequate to save celestial appearances and to calculate the motions of planets, but that he did not believe it to be true in reality and in nature. With all due respect, these people show that they have been too prone to believe the word of someone who speaks more out of whim than out of experience with Copernicus's book or understanding the nature of this business. For this reason they talk about it in a way that is not altogether right.

⟨355⟩ First, limiting ourselves to general considerations, let us see his preface to Pope Paul III, to whom he dedicates the work. We shall find, to begin with, as if to comply with what they call the astronomer's task, that he had done and completed the work in accordance with the hypothesis of the prevailing philosophy and of Ptolemy himself, so that there was in it nothing lacking. But then, taking off the clothes of a pure astronomer and putting on those of a contemplator of nature, he undertook to examine whether this astronomical assumption already introduced, which was completely satisfactory regarding the calculations and the appearances of the motions of all planets, could also truly happen in the world and in nature. He found that in no way could such an arrangement of parts exist: although each by itself was well proportioned, when they were put together the result was a very monstrous chimera.

And so he began to investigate what the system of the world could really be in nature, no longer for the sole convenience of the pure astronomer, whose calculations he had complied with, but in order to come to an understanding of such a noble physical problem; he was confident that, if one had been able to account for mere appearances by means of hypotheses which are not true, this could be done much better by means of the true and physical constitution of the world. Having at his disposal a very large number of physically true and real observations of the motions of the stars (and without this knowledge it is wholly impossible to solve the problem), he worked tirelessly in search of such a constitution. Encouraged by the authority of so many great men, he examined the motion of the earth and the stability of the sun. Without their encouragement and authority, by himself either he would not have conceived the idea, or he would have considered it a very great absurdity and paradox, as he confesses to have considered it at first. But then, through long sense observations, favorable results, and very firm demonstrations, he found it so consonant with the harmony of the world that he became completely certain of its truth. Hence this position is not introduced to satisfy the pure astronomer, but to satisfy the necessity of nature.

Furthermore, Copernicus knew and wrote in the same place that publishing this opinion would have made him look insane to the numberless followers of current philosophy, and especially to each and every ⟨356⟩ layman. Nevertheless, urged by the requests of the Cardinal of Capua [12] and the Bishop of Kulm, [13] he published it. Now, would he not have been truly mad if, considering this opinion physically false, he had published that he believed it to be true, with the certain consequence that he would be regarded as a fool by the whole world? And why would he not have declared that he was using it only as an astronomer, but that he denied it as a philosopher, thus escaping the universal label of foolishness, to the advantage of his common sense?

Moreover, Copernicus states in great detail the grounds and reasons why the ancients believed the earth to be motionless, and then, examining the value of each in turn, he shows them to be ineffective. Now, who ever saw a sensible author engaged in confuting the demonstrations that confirm a proposition he considers true and real? And what kind of judgment would it be to criticize and to condemn a conclusion while in reality he wanted the reader to believe that he accepted it? This sort of incoherence cannot be attributed to such a man.

Furthermore, note carefully that, since we are dealing with the motion or stability of the earth or of the sun, we are in a dilemma of contradictory propositions (one of which has to be true), and we cannot in any way resort to saying that perhaps it is neither this way nor that way. Now, if the earth's stability and sun's motion are de facto physically true and the contrary position is absurd, how can one reasonably say that the false view agrees better than the true one with the phenomena clearly visible and sensed in the movements and arrangement of the stars? Who does not know that there is a most agreeable harmony among all truths of nature, and a most sharp dissonance between false positions and true effects? Will it happen, then, that the earth's motion and sun's stability agree in every way with the arrangement of all other bodies in the universe and with all the phenomena, a thousand of them, which we and our predecessors have observed in great detail, and that this position is false? And can the earth's stability and sun's motion be considered true and not agree in any way with the other truths? If one could say that neither this nor that position is true, it might happen that one would be more convenient than the other in accounting for the appearances. But, given two ⟨357⟩ positions, one of which must be true and the other false, to say that the false one agrees better with the effects of nature is really something that surprises my imagination. I add: if Copernicus confesses to having fully satisfied astronomers by means of the hypothesis commonly accepted as true, how can one say that by means of the false and foolish one he could or would want to satisfy again the same astronomers?

However, I now go on to consider the nature of the business from an internal viewpoint, and to show with how much care one must discuss it.

Astronomers have so far made two sorts of suppositions: some are primary and pertain to the absolute truth of nature; others are secondary and are imagined in order to account for the appearances of stellar motions, which appearances seem not to agree with the primary and true assumptions. For example, before trying to account for the appearances, acting not as a pure astronomer but as a pure philosopher, Ptolemy supposes, indeed he takes from philosophers, that celestial movements are all circular and regular, namely uniform; that heaven has a spherical shape; that the earth is at the center of the celestial sphere, i spherical, motionless, etc. Turning then to the inequalities we see planetary movements and distances, which seem to clash with the mary physical suppositions already established, he goes on to an sort of supposition; these aim to identify the reasons why, without changing the primary ones, there is such a clear and sensible inequality in the movements of planets and in their approaching and their moving away from the earth. To do this he introduces some motions that are still circular, but around centers other than the earth's, tracing eccentric and epicyclic circles. This secondary supposition is the one of which it could be said that the astronomer supposes it to facilitate his computations, without committing himself to maintaining that it is true in reality and in nature.

Let us now see in what kind of hypothesis Copernicus places the earth's motion and sun's stability. There is no doubt whatever, if we reflect carefully, that he places them among the primary and necessary suppositions about nature. For, as I have already stated, it seems that he had already given satisfaction to astronomers by the other road, and that he takes this one only to try to solve the greatest problem ⟨358⟩ of nature. In fact, to say that he makes this supposition to facilitate astronomical calculations is so false that instead we can see him, when he comes to these calculations, leaving this supposition and returning to the old one, the latter being more readily and easily understood and still very quick even in computations. This may be seen as follows. Intrinsically, particular calculations can be made by taking one position as well as the other, that is, by making the earth or the heavens rotate; nevertheless, many geometers and astronomers in many books have already demonstrated the properties of orthogonal and oblique displacements of parts of the zodiac in relation to the equator, the declinations of the parts of the ecliptic, the variety of angles between it and both meridians and oblique horizons, and a thousand other specific details necessary to complete astronomical science. This ensures that, when he comes to examining these details of the primary motions, Copernicus himself examines them in the old manner, namely as occurring along circles traced in the heavens and around the motionless earth, even though stillness and stability should belong to the highest heaven, called the Prime Mobile, and motion to the earth. Thus in the introduction to Book Two he concludes: "People should not be surprised if we still use the ordinary terms for the rising and setting of the Sun and stars and similar occurrences, but should recognise that we are speaking in customary language, which is acceptable to everyone, yet always bearing in mind that 'For us who ride the Earth, the Sun and Moon are passing; patterns of stars return, and then again recede'." [14]

We should therefore understand clearly that Copernicus takes the earth's motion and sun's stability for no other reason and in no other way than to establish it, in the manner of the natural philosopher, as a hypothesis of the primary sort; on the contrary, when he comes to astronomical computations, he goes back to the old hypothesis, which takes the circles of the basic motions with their details to be located in the highest heaven around the motionless earth, being easier for everyone to understand on account of ingrained habit. But what am I saying? Such is the strength of truth and the weakness of falsehood, that those who speak this way reveal themselves not completely capable of understanding these subjects and not well versed in them; this happens when they let themselves be persuaded that the secondary kind of hypothesis is considered chimerical and fictional by Ptolemy and by other serious astronomers, ⟨359⟩ and that they really regard them as physically false and introduced only for the sake of astronomical computations. The only support they give for this very fanciful opinion is a passage in Ptolemy where, unable to observe more than one simple anomaly in the sun, he wrote that to account for it one could take the hypothesis of a simple eccentric as well as that of an epicycle on a concentric, and he added he preferred the first for being simpler than the second; from these words some very superficially argue that Ptolemy did not consider necessary, but rather wholly fictional, both this and that supposition, since he said they are both equally convenient, while one and only one can be attributed to the sun's behavior. But what kind of superficiality is this? Who can do both of the following? First, to suppose as true the primary suppositions that planetary motions are circular and regular, and to admit (as the senses themselves necessarily force us) that in running through the zodiac all planets are now slow and now fast, indeed that most of them can be not only slow but also stationary and retrograde, and that we see them now very large and very near the earth and now very small and very far; and then, having understood these former points, to deny that eccentrics and epicycles can really exist in nature? This is wholly excusable for men who are not specialists in these sciences, but for others who would claim to be experts in them it would be an indication that they do not even understand the meaning of the terms *eccentric* and *epicycle*. One might just as well first admit that there are three letters, the first of which is *G*, the second *O*, and the third *D*, and then at the end deny that their combination yields *GOD* and claim that the result is *SHADOW*. But if rational arguments were not sufficient to make one understand the necessity of having to place eccentrics and epicycles really in nature, at least the senses themselves would have to persuade him: for we see the four Medicean planets[15] trace four small circles around Jupiter which are very far from enclosing the earth, in short, four epicycles; Venus, which is seen now full of light and now very thinly crescent, provides conclusive evidence that its revolution is around the sun and not around the earth, and consequently that its orbit is an epicycle; and the same may be argued for the case of Mercury. Moreover, the three outer planets are ⟨360⟩ very near the earth when they are in opposition to the sun, and very far when in conjunction; for example, Mars at its closest appears to the senses more than fifty times larger than at its farthest, so that some have occasionally feared that it had gotten lost or had vanished, being really invisible because of its great distance; now, what else can one conclude but that their revolution is made in eccentric circles, or in epicycles, or in a combination of the two, if we take the second anomaly into consideration? So, to deny eccentrics and epicycles in the motions of planets is like denying the light of the sun, or else it is to contradict oneself. Let us apply what I am saying more directly to our purpose: some say that modern astronomers introduce the earth's motion and sun's stability suppositionally in order to account for the phenomena and to facilitate calculations, just as epicycles and eccentrics are assumed in the same manner, though the same astronomers consider them physically chimerical and repugnant; I answer that I shall gladly agree with all this talk, as long as they limit themselves to staying within their own conceptions, namely that the earth's motion and sun's stability are as false or true in nature as epicycles and eccentrics. Let them, then, make every effort to do away with the true and real existence of these circles, for if they succeed in demonstrating their nonexistence in nature, I shall immediately surrender and admit the earth's motion to be a great absurdity. But if, on the contrary, they are forced to accept them, let them also accept the earth's motion, and let them admit to have been convinced by their own contradictions.

I could present many other things for this same purpose. However, since I think that whoever is not persuaded by what I have said would not be persuaded by many more reasons either, I want these to suffice. I shall only add something about what could have been the motive why some have concluded with any plausibility that Copernicus himself did not really believe his own hypothesis.

There is on the reverse side of the title page of Copernicus's book a certain preface to the reader, which is not by the author since it mentions him in the third person and is without signature.[16] It clearly states that no one should believe in the least that Copernicus regarded his position as true, but only that he feigned ⟨361⟩ and introduced it for the calculation of celestial motions; it ends its discussion by concluding that to hold it as true and real would be foolish. This conclusion is so explicit that whoever reads no further, and believes it to have been placed at least with the author's consent, deserves to be somewhat excused for his error. But what weight to give to the opinion of those who would judge a book without reading anything but a brief preface by the printer or publisher, I let each one decide for himself. I say that this preface can only have originated from the publisher to facilitate the sale of a book which common people would have regarded as a fanciful chimera if a similar preface had not been added; for most of the time buyers are in the habit of reading such prefaces before buying the work. Not only was this preface not written by the author, but it was included without his consent, and also without his knowledge; this is shown by the errors it contains, which the author would never have committed.

This preface says no one can consider it verisimilar, unless he is completely ignorant of geometry and optics, that Venus has such a large epicycle enabling it now to precede and now to follow the sun by 40 degrees or more; for it would have to happen that when it is highest its diameter should appear only one-fourth of what it appears when it is lowest, and that in the latter location its body should be seen as sixteen times bigger than in the former; but these things, he says, are repugnant to the observations made throughout the centuries. In these assertions we see, first, that the writer does not know that Venus departs on one side and on the other of the sun by about 48 degrees, and not 40 as he says. Moreover, he asserts that its diameter should appear four times, and its body sixteen times, larger in one position than in the other. Here, first, due to a geometrical oversight he does not understand that when one globe has a diameter four times larger than another, its body is sixty-four times bigger, and not sixteen, as he stated. Hence, if he considered such an epicycle absurd and wanted to declare it to be physically impossible, if he had understood this subject, he could have made the absurdity much greater; for, according to the position he wants to refute (well known to astronomers), Venus digresses from the sun almost 48 degrees, and when farthest from the earth its distance ⟨362⟩ must be six times greater than when closest, and consequently its apparent diameter in the latter position is more than six times larger than in the former (not four times), and its body more than two hundred and sixteen times greater (and not just sixteen). These errors are so gross that it is impossible to believe they were committed by Copernicus, or by anyone else but the most unqualified persons. Moreover, why label such a large epicycle most absurd, so that because of such an absurdity we would conclude that Copernicus did not regard his assumptions as true, and that neither should others so regard them? He should have remembered that in chapter 10 of the first book Copernicus is speaking *ad hominem*[17] and is attacking other astronomers who allege that it is a great absurdity to give Venus such an epicycle, which is so large as to exceed the whole lunar orbit by more than two hundred times, and which does not contain anything inside; he then removes the absurdity when he shows that inside Venus's orbit is contained the orbit of Mercury and, placed at the center, the body of the sun itself. What frivolity is this, then, to want to show a position mistaken and false on account of a difficulty which that position not only does not introduce in nature, but completely removes? Similarly it removes the immense epicycles which out of necessity other astronomers assumed in the other system. This only touches the writer of Copernicus's preface; so we may argue that, if he had included something else professionally relevant, he would have committed other errors.

But finally, to remove any shadow of a doubt, if the failure to observe such great variations in the apparent sizes of the body of Venus should cast doubt on its circular revolution around the sun (from the viewpoint of the Copernican system), then let us make careful observations with a suitable instrument, namely with a good telescope, and we shall find all effects and experiences exactly agreeing; that is, we shall see Venus crescent when it is nearest to the earth, and with a diameter six times larger than when it is at its maximum distance, namely above the sun, where it is seen round and very small. I have discussed elsewhere the reasons for not detecting these variations with our simple eyesight, but just as from this failure we could reasonably deny that supposition, so now, from seeing the very exact correspondence in this and every other detail, we should abandon any doubt and consider the supposition true and real. As for the rest of this admirable ⟨363⟩ system, whoever desires to ascertain the opinion of Copernicus himself should not read the fanciful preface of the printer, but the whole work of the author himself; without a doubt he will grasp firsthand that Copernicus held as very true the stability of the sun and the motion of the earth.

[II]

⟨364⟩ The motion of the earth and stability of the sun could never be against Faith or Holy Scripture, if this proposition were correctly proved to be physically true by philosophers, astronomers, and mathematicians, with the help of sense experiences, accurate observations, and necessary demonstrations. However, in this case, if some passages of Scripture were to sound contrary, we would have to say that this is due to the weakness of our mind, which is unable to grasp the true meaning of Scripture in this particular case. This is the common doctrine, and it is entirely right, since one truth cannot contradict another truth. On the other hand, whoever wants to condemn it judicially must first demonstrate it to be physically false by collecting the reasons against it.

Now, one wants to know where to begin in order to ascertain its falsity, that is, whether from the authority of Scripture or from the refutation of the demonstrations and observations of philosophers and astronomers. I answer that one must start from the place which is safest and least likely to bring about a scandal; this means beginning with physical and mathematical arguments. For if the reasons proving the earth's motion are found fallacious, and the contrary ones conclusive, then we have already become certain of the falsity of this proposition and the truth of the opposite, which we now say corresponds to the meaning of Scripture; so one would be free to condemn the false proposition and there would be no danger. But if those reasons are found true and necessary, this will not bring any harm to the authority of Scripture; instead we shall have been cautioned that due to our ignorance we had not grasped the true sense of Scripture, and that we can learn this meaning with the help of the newly acquired physical truth. Therefore, beginning with the arguments is safe in any case. On the other hand, if we were to fix only on what seemed to us the true and certain meaning of Scripture, and we were to go on to condemn such a proposition without examining the strength of the arguments, what a scandal would follow if sense experiences and reasons were to show the opposite? And who would have brought confusion to ⟨365⟩ the Holy Church? Those who had suggested the greatest consideration of the arguments, or those who had disparaged them? One can see, then, which road is safer.

Moreover, we admit that a physical proposition which has been proved true by physical and mathematical demonstrations can never contradict Scripture, but that in such a case it is the weakness of our mind which prevents us from grasping its true meaning. On the other hand, whoever wants to use the authority of the same passages of Scripture to confute and prove false the same proposition would commit the error called "begging the question."[18] For, the true meaning of Scripture being in doubt in the light of the arguments, one cannot take it as clear and certain in order to refute the same proposition; instead one must cripple the arguments and find the fallacies with the help of other reasons and experiences and more certain observations. When the factual and physical truth has been found in this manner, then, and not before, can one be assured of the true meaning of Scripture and safely use it. Thus the safe road is to begin with the arguments, confirming the true and refuting the fallacious ones.

If the earth de facto moves, we cannot change nature and arrange for it not to move. But we can rather easily remove the opposition of Scripture with the mere admission that we do not grasp its true meaning. Therefore, the way to be sure not to err is to begin with astronomical and physical investigations, and not with scriptural ones.

I am always told that, in interpreting the passages of Scripture relevant to this point, all Fathers agree to the meaning which is simplest and corresponds to the literal meaning; hence, presumably, it is improper to give them another meaning or to change the common interpretation, because this would amount to accusing the Fathers of carelessness or negligence. I answer by admitting that the Fathers indeed deserve reasonable and proper respect, but I add that we have an excuse for them very readily: it is that on this subject they never interpreted Scripture differently from the literal meaning, because at their time the opinion of the earth's motion was totally buried and no one even talked about it, let alone wrote about it or maintained it. But there is no trace of negligence by the Fathers for not thinking about what was completely hidden. That they did not think about it is ⟨366⟩ clear from the fact that in their writings one cannot find even a word about this opinion. And if anyone were to say that they considered it, this would make its condemnation more dangerous; for after considering it, not only did they not condemn it, but they did not express any doubt about it.

Thus the defense of the Fathers is readily available and very easy. On the contrary, it would be very difficult or impossible to excuse or exonerate from a similar charge of carelessness the Popes, Councils, and Congregations of the Index of the last eighty years, if this doctrine were erroneous and deserving of condemnation; for they have let this opinion circulate[19] in a book which was first written on orders from a Pope, and then printed on orders from a cardinal and a bishop, dedicated to another Pope, and, most important, received by the Holy Church, so that one cannot say that it had remained unknown. If, then, the inappropriateness of charging our highest authorities with negligence is to be taken into account, as it should, let us make sure that in trying to escape one absurdity we do not fall into a greater one.

But assume now that someone regards it as inappropriate to abandon the unanimous interpretation of the Fathers, even in the case of physical propositions not discussed by them and whose opposite they did not even consider; I then ask what one should do if necessary demonstrations showed the facts of nature to be the opposite. Which of the two decrees should be changed? The one which stipulates that no proposition can be both true and erroneous, or the other one which obliges us to regard as articles of faith physical propositions supported by the unanimous interpretation of the Fathers? It seems to me, if I am not mistaken, that it would be safer to modify this second decree than to be forced to hold as an article of faith a physical proposition which had been demonstrated with conclusive reasons to be factually false in nature. It also seems to me that one could say that the unanimous interpretation of the Fathers should have absolute authority in the case of propositions which they aired, and for which no contrary demonstrations exist and it is certain that none could ever exist. I do not bring in the fact that it is very clear that the Council[20] requires only that one agree with the unanimous interpretation of the Fathers "in matters of faith and morals, etc." ⟨367⟩

[III]

1. Copernicus uses eccentrics and epicycles, but these were not the reason for rejecting the Ptolemaic system, since they undoubtedly exist in the heavens; it was other difficulties.

2. In regard to philosophers, if they were true philosophers, namely lovers of truth, they should not get irritated,[21] but, learning that they were wrong, they should thank whoever shows them the truth; and if their opinion were to stand up, they would have reason to take pride in it, rather than being irritated. Theologians should not get irritated because, if this opinion were found false then they could freely prohibit it, and if it were discovered true then they should rejoice that others have found the way to understand the true meaning of Scripture and have restrained them from perpetrating a serious scandal by condemning a true proposition.

In regard to falsifying Scripture, this is not and will never be the intention of Catholic astronomers such as ourselves; rather our view is that Scripture corresponds very well to truths demonstrated about nature. Moreover, certain theologians who are not astronomers should be careful about falsifying Scripture by wanting to interpret it as opposed to propositions which may be true and demonstrable.

3. It might happen that we could have difficulties in interpreting Scripture, but this would occur because of our ignorance and not be-

cause there really are or can be insuperable difficulties in reconciling Scripture with demonstrated truths.

4. The Council speaks "about matters of faith and morals, etc." So there is an answer to the claim that such a proposition is "an article of faith by reason of the speaker," though not "by reason of the topic,"[22] and that therefore it is among those covered by the Council. The answer is that everything in Scripture is "an article of faith by reason of the speaker," so that in this regard it should be included in the rule of the Council; but this clearly has not been done because in that case it would have said that "the interpretation of the Fathers is to be followed for every word of Scripture, etc.," and not "for matters of faith and morals"; having thus said "for matters of faith," we see that its intention was to mean "for matters of faith by reason of the topic."

Then consider that ⟨368⟩ it is much more a matter of faith to hold that Abraham had some children[23] and that Tobias had a dog, because Scripture says it, than it would be to hold that the earth moves, even if this were found in the same Scripture, and further that to deny the former is a heresy, but not to deny the latter. It seems to me that this depends on the following reason. There have always been in the world men who had two, four, six children, etc., or none, and similarly people who have dogs and who do not, so that it is equally credible that some have children or dogs and others do not; hence there appears to be no reason why in such propositions the Holy Spirit should speak differently from the truth, the negative and the affirmative sides being equally credible to all men. But it is not so with the motion of the earth and the stability of the sun; these propositions are very far removed from the understanding of the masses, for on these matters not relevant to their eternal life the Holy Spirit chose to conform its pronouncements with their abilities, even when facts are otherwise from the point of view of the thing in itself.

5. In regard to placing the sun in heaven and the earth outside it, as Scripture seems to affirm, etc., this truly seems to me to be a simple perception of ours and a manner of speaking only for our convenience. For, in reality all that is surrounded by heaven is in heaven, just as all that is surrounded by the city walls is in the city; indeed, if one were to express a preference, what is in the middle is more in heaven and in the city, being, as it were, at the heart of the city and of heaven. That difference exists because one takes the elemental region surrounding the earth as being very different from the celestial region. But such a difference will always exist regardless of where these elements are placed; and

it will always be true that from the viewpoint of our convenience the earth is below us and heaven above, since all the inhabitants of the earth have heaven above our heads, which is our upwards, and the center of the earth under our feet, which is our downwards; so, in relation to us the center of the earth and the surface of heaven are the farthest places, that is, the endpoints of our up and down, which are diametrically opposite points.

6. Not to believe that there is a demonstration of the earth's mobility until it is shown[24] is very prudent, nor do we ask that anyone believe such a thing without a demonstration. On the contrary, we only seek that, for the advantage of the Holy Church, one examine with ⟨369⟩ the utmost severity what the followers of this doctrine know and can advance, and that nothing be granted them unless the strength of their arguments greatly exceeds that of the reasons for the opposite side. Now, if they are not more than ninety percent right, they may be dismissed; but if all that is produced by philosophers and astronomers on the opposite side is shown to be mostly false and wholly inconsequential, then the other side should not be disparaged, nor deemed paradoxical, so as to think that it could never be clearly proved. It is proper to make such a generous offer since it is clear that those who hold the false side cannot have in their favor any valid reason or experiment, whereas it is necessary that all things agree and correspond with the true side.

7. It is true that it is not the same to show that one can save the appearances[25] with the earth's motion and the sun's stability, and to demonstrate that these hypotheses are really true in nature.[26] But it is equally true, or even more so, that one cannot account for such appearances with the other commonly accepted system. The latter is undoubtedly false, while it is clear that the former, which can account for them, may be true. Nor can one or should one seek any greater truth in a position than that it corresponds with all particular appearances.

8. One is not asking that in case of doubt the interpretation of the Fathers should be abandoned, but only that an attempt be made to gain certainty regarding what is in doubt, and that therefore no one disparage what attracts and has attracted very great philosophers and astronomers. Then, after all necessary care has been taken, the decision may be made.

9. We believe that Solomon,[27] Moses, and all other sacred writers knew perfectly the constitution of the world, as they also knew that God has no hands, no feet, and no experience of anger, forgetfulness, or

regret; nor will we ever doubt this. But we say what the Holy Fathers and in particular St. Augustine say about these matters, namely that the Holy Spirit inspired them to write what they wrote for various reasons, etc.

10. The error of the apparent movement of the shore and stability of the ship[28] is known by us after having many times observed the motion of boats from the shore, and many other times observed the shore from a boat; and so, if we could now stay on earth and now go to the sun ⟨370⟩ or other star, perhaps we would acquire sensible and certain knowledge of which one of them moves. To be sure, if we looked only at these two bodies, it would always seem to us that the one we were on was standing still, just as looking only at the water and the boat always gives the appearance that the water is flowing and the boat is standing still. Moreover, the two situations are very different: there is great disparity between a small boat, separable from its environment, and the immense shore, known by us through thousands of experiences to be motionless, that is, motionless in relation to the water and the boat; but the other comparison is between two bodies both of which are substantial and equally inclined toward motion and toward rest. Thus it would be more relevant to compare between themselves two boats, in which case it is absolutely certain that the one we were on would always appear to us as motionless, as long as we could not consider any other relationship but that which holds between these two ships.

There is, therefore, a very great need to correct the error about observing whether the earth or else the sun moves, for it is clear that to someone on the moon or any other planet it would always appear that it was standing still and the other stars were moving. But these and many other more plausible reasons of the followers of the common opinion are the ones that must be untied very openly, before one can pretend even to be heard, let alone approved; unfortunately we have not done a very detailed examination of what is produced against us. Moreover, neither Copernicus nor his followers will ever use this phenomenon of the shore and the boat to prove that the earth is in motion and the sun at rest. They only adduce it as an example that serves to show, not the truth of their position, but the absence of contradiction between the appearance of a stable earth and moving sun to our simple sense experience, and the reality of the contrary. For if this were one of Copernicus's demonstrations, or if his others did not argue more effectively, I really think that no one would agree with him.

CHAPTER III

Galileo's Letter to the Grand Duchess Christina (1615)

⟨309⟩ To the Most Serene Ladyship the Grand Duchess Dowager: [1]

[1] [2] As Your Most Serene Highness knows very well, a few years ago I discovered in the heavens many particulars which had been invisible until our time. [3] Because of their novelty, and because of some consequences deriving from them which contradict certain physical propositions [4] commonly accepted in philosophical schools, they roused against me no small number of such professors, as if I had placed these things in heaven with my hands in order to confound nature and the sciences. These people seemed to forget that a multitude of truths contribute to inquiry and to the growth and strength of disciplines rather than to their diminution or destruction, and at the same time they showed greater affection for their own opinions than for the true ones; thus they proceeded to deny and to try to nullify those novelties, about which the senses themselves could have rendered them certain, if they had wanted to look at those novelties carefully. To this end they produced various matters, and they published some writings full of useless discussions and sprinkled with quotations from the Holy Scripture, taken from passages which they do not properly understand and which they inappropriately adduce. [6] This was a very serious error, and they might not have fallen into it had they paid attention to St. Augustine's very useful advice ⟨310⟩ concerning how to proceed with care in reaching definite decisions about matters which are obscure and difficult to understand by means of reason alone. For, speaking also about a particular physical

conclusion pertaining to heavenly bodies, he writes this (*On the Literal Interpretation of Genesis*, book 2, at the end):[7] "Now then, always practicing a pious and serious moderation, we ought not to believe anything lightly about an obscure subject, lest we reject (out of love for our error) something which later may be truly shown not to be in any way contrary to the holy books of either the Old or New Testament."[8]

Then it developed that the passage of time disclosed to everyone the truths I had first pointed out, and, along with the truth of the matter, the difference in attitude between those who sincerely and without envy did not accept these discoveries as true and those who added emotional agitation to disbelief. Thus, just as those who were most competent in astronomical and physical science were convinced by my first announcement,[9] so gradually there has been a calming down of all the others whose denials and doubts were not sustained by anything other than the unexpected novelty and the lack of opportunity to see them and to experience them with the senses. However, there are those who are rendered ill-disposed, not so much toward the things as much as toward the author, by the love of their first error and by some interest which they imagine having but which escapes me. Unable to deny them any longer, these people became silent about them; but, embittered more than before by what has mellowed and quieted the others, they divert their thinking to other fictions and try to harm me in other ways. These would not really worry me any more than I was disturbed by the other oppositions, which I always laughed off, certain of the result that the business would have; I should not worry if I did not see that the new calumnies and persecutions are not limited to matters of greater or less theoretical understanding, which are relatively unimportant, but that they go further and try to damage me with stains which I do abhor and must abhor more than death. Nor can I be satisfied that these charges be known as false only by those who know me and them; their falsity must be known to every other person. These people are aware that in my ⟨311⟩ astronomical and philosophical studies, on the question of the constitution of the world's parts, I hold that the sun is located at the center of the revolutions of the heavenly orbs and does not change place, and that the earth rotates on itself and moves around it. Moreover, they hear how I confirm this view not only by refuting Ptolemy's and Aristotle's arguments, but also by producing many for the other side, especially some pertaining to physical effects whose causes perhaps cannot be determined in any other way, and other astronomical ones dependent on many features of the new celestial discoveries; these discoveries clearly confute the Ptolemaic system, and they agree admirably with this other position and confirm it. Now, these people are perhaps confounded by the known truth of the other propositions different from the ordinary which I hold, and so they may lack confidence to defend themselves as long as they remain in the philosophical field. Therefore, since they persist in their original self-appointed task of beating down me and my findings by every imaginable means, they have decided to try to shield the fallacies of their arguments with the cloak of simulated religiousness and with the authority of Holy Scripture, unintelligently using the latter for the confutation of arguments they neither understand nor have heard.

At first, they tried on their own to spread among common people the idea that such propositions are against Holy Scripture, and consequently damnable and heretical. Then they realized how by and large human nature is more inclined to join those ventures which result in the oppression of other people (even if unjustly) than those which result in their just improvement, and so it was not difficult for them to find someone who with unusual confidence did preach even from the pulpit that it is damnable and heretical;[10] and this was done with little compassion and with little consideration of the injury not only to this doctrine and its followers, but also to mathematics and all mathematicians. Thus, having acquired more confidence, and with the vain hope that the seed which first took root in their insincere mind would grow into a tree and rise toward the sky, they are spreading among the people the rumor that it will shortly be declared heretical by the supreme authority. They also know that such a declaration not only would uproot these two conclusions, but also would render damnable all the other astronomical and physical observations and propositions ⟨312⟩ which correspond and are necessarily connected with them; hence, they alleviate their task as much as they can by making it look, at least among common people, as if this opinion were new and especially mine, pretending not to know that Nicolaus Copernicus was its author or rather its reformer and confirmer.[11] Now, Copernicus was not only a Catholic but also a clergyman and a canon, and he was so highly regarded that he was called to Rome[12] from the remotest parts of Germany[13] when under Leo X the Lateran Council was discussing the reform of the ecclesiastical calendar; at that time this reform remained unfinished only because there was still no exact knowledge of the precise length of the year and the lunar month. Thus he was charged by the Bishop of Fossombrone,[14] who was then supervising this undertaking, to try by repeated studies and efforts to acquire more understanding and certainty about those celestial motions; and so he undertook this study, and, by truly Herculean labor and by his admirable mind, he made so much progress in this science and acquired such an exact knowledge of the periods of celestial motions that he earned the title of supreme astronomer; then in accordance with his doctrine not only was the calendar regularized,[15] but tables of all planetary motions were constructed. Having expounded this doctrine in six parts, he published it at the request of the Cardinal of Capua[16] and the Bishop of Kulm;[17] and since he had undertaken this task and these labors on orders from the Supreme Pontiff, he dedicated his book *On Heavenly Revolutions* to the successor of the latter, Paul III. Once printed this book was accepted by the Holy Church, and it was read and studied all over the world without anyone ever having had the least scruple about its doctrine.[18] Finally, now that one is discovering how well founded upon clear observations and necessary demonstrations[19] this doctrine is, some persons come along who, without having even seen the book, give its author the reward of so much work by trying to have him declared a heretic; this they do only in order to satisfy their special animosity, groundlessly conceived ⟨313⟩ against someone else who has no greater connection with Copernicus than the endorsement of his doctrine.

Now, in matters of religion and reputation I have the greatest regard for how common people judge and view me; so, because of the false aspersions my enemies so unjustly try to cast upon me, I have thought it necessary to justify myself by discussing the details of what they produce to detest and abolish this opinion, in short, to declare it not just false but heretical. They always shield themselves with a simulated religious zeal, and they also try to involve Holy Scripture and to make it somehow subservient to their insincere objectives; against the intention of Scripture and the Holy Fathers (if I am not mistaken), they want to extend, not to say abuse, its authority, so that even for purely physical conclusions which are not matters of faith one must totally abandon the senses and demonstrative arguments in favor of any scriptural passage whose apparent words may contain a different indication. Here I hope to demonstrate that I proceed with much more pious and religious zeal than they when I propose not that this book should not be condemned, but that it should not be condemned without understanding, examining, or even seeing it, as they would like. This is especially true since the author never treats of matters pertaining to religion and faith, nor uses arguments dependent in any way on the authority of Holy Scripture, in which case he might have interpreted it incorrectly; instead, he always limits himself to physical conclusions pertaining to celestial motions, and he treats of them with astronomical and geometrical demonstrations based above all on sensory experience and very accurate observations. He proceeded in this manner not because he did not pay any attention to the passages of Holy Scripture, but because he understood very well that ⟨314⟩ if his doctrine was demonstrated it could not contradict the properly interpreted Scripture. Hence, at the end of the dedication, speaking to the Supreme Pontiff, he says: "There may be triflers who though wholly ignorant of mathematics nevertheless abrogate the right to make judgements about it because of some passage in Scripture wrongly twisted to their purpose, and will dare to criticise and censure this undertaking of mine. I waste no time on them, and indeed I despise their judgement as thoughtless. For it is known that Lactantius, a distinguished writer in other ways but no mathematician, speaks very childishly about the shape of the Earth when he makes fun of those who reported that it has the shape of a globe. Mathematics is written for mathematicians, to whom this work of mine, if my judgement does not deceive me, will seem to be of value to the ecclesiastical Commonwealth over which your Holiness now holds dominion."[20]

Of this sort are also those who try to argue that this author should be condemned, without examining him; and to show that this is not only legitimate but a good thing, they use the authority of Scripture, experts in sacred theology, and Sacred Councils. I feel reverence for these authorities and hold them supreme, so that I should consider it most reckless to want to contradict them when they are used in accordance with the purpose of the Holy Church; similarly, I do not think it is wrong to speak out when it seems that someone, out of personal interest, wants to use them in a way different from the holiest intention of the Holy Church. Thus, while also believing that my sincerity will become self-evident, I declare not only that I intend to submit freely to the correction of any errors in matters pertaining to religion which I may have committed in this essay due to my ignorance, but I also declare that on these subjects I do not wish to quarrel with anyone, even if the points are debatable. For my purpose is nothing but the following: if these reflections, which are far from my own profession, should contain (besides errors) anything that may lead someone to advance a useful caution for the Holy Church in her deliberations about the ⟨315⟩ Copernican system, then let it be accepted with whatever profit superiors will deem appropriate; if not, let my essay be torn up and burned, for I do not intend or pretend to gain from it any advantage that is not pious or Catholic. Moreover, although I have heard with my own ears many of the things which I mention, I freely grant to whoever said them that they did not say them, if they so wish, and I admit that I may have misunderstood them; thus what I answer should not apply to them, but to whoever holds that opinion.

So the reason they advance to condemn the opinion of the earth's mobility and sun's stability is this: since in many places in Holy Scripture[21] one reads that the sun moves and the earth stands still, and since Scripture can never lie or err, it follows as a necessary consequence that the opinion of those who want to assert the sun to be motionless and the earth moving is erroneous and damnable.

[2] The first thing to note about this argument is the following. It is most pious to say and most prudent to take for granted that Holy Scripture can never lie, as long as its true meaning has been grasped; but I do not think one can deny that this is frequently recondite and very different from what appears to be the literal meaning of the words. From this it follows that, if in interpreting it someone were to limit himself always to the pure literal meaning, and if the latter were wrong, then he could make Scripture appear to be full not only of contradictions and false propositions but also of serious heresies and blasphemies; for one would have to attribute to God feet, hands, eyes, and bodily sensations, as well as human feelings like anger, contrition, and hatred, and such conditions as the forgetfulness of things past and the ignorance of future ones. Since these propositions dictated by the Holy Spirit were expressed by the sacred writers in such a way as to accommodate the capacities of the very unrefined and undisciplined masses, for those who deserve to rise above the common people it is therefore necessary that wise interpreters ⟨316⟩ formulate the true meaning and indicate the specific reasons why it is expressed by such words. This doctrine is so commonplace and so definite among all theologians that it would be superfluous to present any testimony for it.

From this I think one can very reasonably deduce that, whenever the same Holy Scripture has seen fit to assert any physical conclusion (especially on matters that are abstruse and difficult to understand), it has followed the same rule, in order not to sow confusion into the minds of the common people and make them more obstinate against dogmas involving higher mysteries. In fact, as I have said and as one can clearly see, for the sole purpose of accommodating popular understanding the Scripture has not abstained from concealing the most important truths, attributing even to God characteristics that are contrary to or very far from His essence; given this, who will categorically maintain that in speaking incidentally of the earth, water, sun, or other created thing the Scripture has set aside such regard and has chosen to limit itself rigorously to the literal and narrow meanings of the words? This would be especially implausible when mentioning features of these created things which are very remote from popular understanding and not at all pertinent to the primary purpose of the Holy Writ, that is, to the worship of God and the salvation of souls.

Therefore, I think that in disputes about natural phenomena one must begin not with the authority of scriptural passages but with sensory experience and necessary demonstrations. For the Holy Scripture and nature derive equally from the Godhead, the former as the dictation of the Holy Spirit and the latter as the most obedient executrix of God's orders; moreover, to accommodate the understanding of the common people it is appropriate for Scripture to say many things that are different (in appearance and in regard to the literal meaning of the words) from the absolute truth; on the other hand, nature is inexorable and immutable, never violates the terms of the laws imposed upon her, and does not care whether or not her recondite reasons and ways of operating are disclosed to human understanding; ⟨317⟩ but not every scriptural assertion is bound to obligations as severe as every natural phenomenon; finally, God reveals Himself to us no less excellently in the effects of nature than in the sacred words of Scripture, as Tertullian perhaps meant when he said, "We postulate that God ought first to be known by nature, and afterward further known by doctrine—by nature through His works, by doctrine through official teaching" (*Against Marcion*, I.18);[22] and so it seems that a natural phenomenon which is placed before our eyes by sensory experience or proved by necessary demonstrations should not be called into question, let alone condemned, on account of scriptural passages whose words appear to have a different meaning.

However, by this I do not wish to imply that one should not have the highest regard for passages of Holy Scripture; indeed, after becoming certain of some physical conclusions, we should use these as very appropriate aids to the correct interpretation of Scripture and to the investigation of the truths they must contain, for they are most true and agree with demonstrated truths. That is, I would say that the authority of Holy Scripture aims chiefly at persuading men about those articles and propositions which, surpassing all human reason, could not be discovered by scientific research or by any other means than through the mouth of the Holy Spirit himself. Moreover, even in regard to those propositions which are not articles of faith, the authority of the same Holy Writ should have priority over the authority of any human writings containing pure narration or even probable reasons, but no demonstrative proofs; this principle should be considered appropriate and necessary inasmuch as divine wisdom surpasses all human judgment and speculation. However, I do not think one has to believe that the same God who has given us senses, language, and intellect would want to set aside the use of these and give us by other means the information we can acquire with them, so that we would deny our senses and reason even in the case of those physical conclusions which are placed before our eyes and intellect by our sensory experiences or by necessary demonstrations. This is especially implausible for those sciences discussed in Scripture to a very minor extent and ⟨318⟩ with disconnected statements; such is precisely the case of astronomy, so little of which is contained therein that one does not find there even the names of the planets, except for the sun,[23] the moon, and only once or twice Venus, under the name of Morning Star. Thus if the sacred authors had had in mind to teach people about the arrangement and motions of the heavenly bodies, and consequently to have us acquire this information from Holy Scripture, then, in my opinion, they would not have discussed so little of the topic—that is to say, almost nothing in comparison with the innumerable admirable conclusions which are contained and demonstrated in this science. Indeed, it is the opinion of the holiest and most learned Fathers that the writers of Holy Scripture not only did not pretend to teach us about the structure and motions of the heavens and the stars, and their shape, size, and distance, but that they deliberately refrained from doing so, even though they knew all these things very well. For example, one reads the following words in St. Augustine (*On the Literal Interpretation of Genesis*, book 2, chapter 9): "It is also customary to ask what one should believe about the shape and arrangement of heaven according to our Scripture. In fact, many people argue a great deal about these things, which with greater prudence our authors omitted, which are of no use for eternal life to those who study them, and (what is worse) which take up a great deal of time that ought to be spent on matters pertaining to salvation. For what does it matter to me whether heaven, like a sphere, completely surrounds the earth, which is balanced at the center of the universe, or whether like a discus it covers the earth on one side from above? However, since the issue here is the authority of Scripture, let me repeat a point I have made more than once; that is, there is a danger that someone who does not understand the divine words may find in our books or infer from them something about these topics which seems to contradict received opinions, and *aug says* then he might not believe at all the other useful things contained in its precepts, stories, and assertions; therefore, briefly, it should be said that our authors did know the truth about the shape of heaven, but the Spirit of God, which was speaking through them, did not want to teach men these things which are of no use to salvation."[24] (The same opinion is found in Peter Lombard's *Book of Sentences*.) The same contempt which the sacred writers had for the investigation of such properties of heavenly bodies is repeated by St. Augustine in the following chapter 10, in regard to the question whether heaven should be thought to be in motion or standing still. He writes: "Some brethren have also advanced a question about the motion of heaven, ⟨319⟩ namely whether heaven moves or stands still. For if it moves, they say, how is it a firmament? But if it stands still, how do the stars which are thought to be fixed in it revolve from east to west, the northern ones completing shorter circuits near the pole, so that heaven seems to rotate like a sphere (if there is at the other end another pole invisible to us) or like a discus (if instead there is no other pole)? To them I answer that these things should be examined with very subtle and demanding arguments to determine truly whether or not it is so; but I do not have the time to undertake and pursue these investigations, nor should such time be available to those whom we desire to instruct for their salvation and for the needs and benefit of the Holy Church."[25]

Let us now come down from these matters to our particular point. We have seen that the Holy Spirit did not want to teach us whether heaven moves or stands still, nor whether its shape is spherical or like a discus or extended along a plane, nor whether the earth is located at its center or on one side. So it follows as a necessary consequence that the Holy Spirit also did not intend to teach us about other questions of the same kind and connected to those just mentioned in such a way that without knowing the truth about the former one cannot decide the latter, such as the question of the motion or rest of the earth or sun. But if the Holy Spirit deliberately avoided teaching us such propositions, inasmuch as they are of no relevance to His intention (that is, to our salvation), how can one now say that to hold this rather than that proposition on this topic is so important that one is a principle of faith and the other erroneous? Thus, can an opinion be both heretical and irrelevant to the salvation of souls? Or can one say that the Holy Spirit chose not to teach us something relevant to our salvation? Here I would say what I heard from an ecclesiastical person in a very eminent position (Cardinal Baronio), namely that the intention of the Holy Spirit is to teach us how one goes to heaven and not how heaven goes.

But let us go back and examine the importance of necessary demonstrations and sensory experiences in conclusions about natural phenomena, and how much weight has been assigned to them by learned and holy theologians. Among hundreds of instances of such testimony we have the following. Near the beginning of his work *On Genesis* Pererius asserts: ⟨320⟩ "In treating of Moses' doctrine, one must take diligent care to completely avoid holding and saying positively and categorically anything which contradicts the decisive observations and reasons of philosophy or other disciplines; in fact, since all truths always agree with one another, the truth of Holy Scripture cannot be contrary to the true reasons and observations of human doctrines." And in St. Augustine (Letter to Marcellinus, section 7), one reads: "If, against the most manifest and reliable testimony of reason, anything be set up claiming to have the authority of the Holy Scriptures, he who does this does it through a misapprehension of what he has read and is setting up against the truth not the real meaning of Scripture, which he has failed to discover, but an opinion of his own; he alleges not what he has found in the Scriptures, but what he has found in himself as their interpreter."[26]

Because of this, and because (as we said above) two truths cannot contradict one another, the task of a wise interpreter is to strive to fathom the true meaning of the sacred texts; this will undoubtedly agree with those physical conclusions of which we are already certain and sure through clear observations or necessary demonstrations. Indeed, besides saying (as we have) that in many places Scripture is open to interpretations far removed from the literal meaning of the words, we should add that we cannot assert with certainty that all interpreters speak with divine inspiration since if this were so then there would be no disagreement among them about the meaning of the same passages; therefore, I should think it would be very prudent not to allow anyone to commit and in a way oblige scriptural passages to have to maintain the truth of any physical conclusions whose contrary could ever be proved to us by the senses and demonstrative and necessary reasons. Indeed, who wants the human mind put to death? Who is going to claim that everything in the world which is observable and knowable has already been seen and discovered? Perhaps those who on other occasions admit, quite correctly, that the things we know are a very small part of the things we do not know? Indeed, we also have it from the mouth of the Holy Spirit that "God hath delivered the world to their consideration, so that man cannot find out the work which God hath made from the beginning to the end" (Ecclesiastes, chapter 3);[27] so one must not, in my opinion, contradict this statement and block the way of freedom of philosophizing about things ⟨321⟩ of the world and of nature, as if they had all already been discovered and disclosed with certainty. Nor should it be considered rash to be dissatisfied with opinions which are almost universally accepted; nor should people become indignant if in a dispute about natural phenomena someone disagrees with the opinion they favor, especially in regard to problems which have been controversial for thousands of years among very great philosophers, such as the sun's rest and earth's motion. This opinion has been held by Pythagoras and his whole school, by Heraclides of Pontus, by Philolaus (teacher of Plato), and by Plato himself (as Aristotle and Plutarch mention); the latter writes in the "Life of Numa" that when Plato was old he said it was very absurd to believe otherwise. The same opinion was accepted by Aristarchus of Samos (as Archimedes tells us), by the mathematician Seleucus, by the philosopher Hicetas[28] (according to Cicero),[29] and by many others. Finally, it was amplified and confirmed with many observations and demonstrations by Nicolaus Copernicus. Furthermore, in the book *On Comets*, the very distinguished philosopher Seneca tells us that one should attempt to ascertain with the greatest diligence whether the daily rotation belongs to the heavens or to the earth.

Therefore, it would perhaps be wise and useful advice not to add without necessity to the articles pertaining to salvation and to the definition of the faith, against the firmness of which there is no danger that any valid and effective doctrine could ever emerge. If this is so, it would really cause confusion to add them upon request from persons about whom not only do we not know whether they speak with heavenly inspiration, but we clearly see that they are deficient in the intelligence necessary first to understand and then to criticize the demonstrations by which the most acute sciences proceed in confirming similar conclusions. However, if I may be allowed to state my opinion, I should say further that it would be more appropriate to the dignity and majesty of Holy Writ to take steps to ensure that not every superficial and vulgar writer can lend credibility to his writings ⟨322⟩ (very often based on worthless fabrications) by sprinkling them with scriptural passages;

these are often interpreted, or rather distorted, in ways which are as remote from the true intention of Scripture as they are ridiculously close to the aims of those who ostentatiously adorn their writings with them. Many examples of such an abuse could be adduced, but I shall limit myself to two which are not far from these astronomical subjects. One of them consists of the writings that were published against the Medicean planets, which I recently discovered, and against the existence of which many passages of Holy Scripture were advanced;[30] now that these planets can be seen by the whole world, I should very much like to hear in what new ways those same opponents interpret Scripture and excuse their blunder. The other example involves someone who has recently argued in print against astronomers and philosophers, to the effect that the moon does not receive its light from the sun but is itself luminous; ultimately he confirms, or rather convinces himself to be confirming, this fancy with various scriptural passages which he thinks could not be accounted for if his opinion were not true and necessary.[31] Nevertheless, it is as clear as sunlight that the moon itself is dark.

It is thus obvious that, because these authors had not grasped the true meaning of Scripture, if they had commanded much authority they would have obliged it to compel others to hold as true conclusions repugnant to manifest reason and to the senses. This is an abuse which I hope God will prevent from taking root or gaining influence because it would in a short time require the prohibition of all ratiocinative sciences. In fact, the number of men ill-suited to understand adequately the Holy Scripture and the sciences is by nature much greater than the number of intelligent ones; thus the former, by superficially glancing through Scripture, would arrogate to themselves the authority of decreeing over all questions about nature in virtue of some word ill-understood by them and written by the sacred authors for some other purpose; nor could the small ⟨323⟩ number of the intelligent ones restrain the furious torrent of the others, who would find all the more followers, inasmuch as it is sweeter to be considered wise without study and labor than to wear oneself out unrelentingly in the pursuit of very arduous disciplines. However, we can render infinite thanks to the blessed God, whose benevolence frees us from this fear while it strips such persons of any authority. The deliberating, deciding, and decreeing about such important issues can be left to the excellent wisdom and goodness of very prudent Fathers and to the supreme authority of those who, guided by the Holy Spirit, can only behave in a holy manner and will not permit the irresponsibility of those others to gain influence. These sorts of men are, in my opinion, those toward whom serious and saintly writers become angry, not without reason. For instance, referring to the Holy Scripture, St. Jerome writes: "The chatty old woman, the doting old man, and the wordy sophist, one and all take in hand the Scriptures, rend them in pieces and teach them before they have learned them. Some with brows knit and bombastic words, balanced one against the other, philosophize concerning the sacred writings among weak women. Others—I blush to say it—learn of women what they are to teach men; and as if even this were not enough, they boldly explain to others what they themselves by no means understand. I say nothing of persons who, like myself, have been familiar with secular literature before they have come to the study of the Holy Scriptures. Such men when they charm the popular ear by the finish of their style suppose every word they say to be a law of God. They do not deign to notice what Prophets and apostles have intended but they adapt conflicting passages to suit their own meaning, as if it were a grand way of teaching—and not rather the faultiest of all—to misinterpret a writer's views and to force the Scriptures reluctantly to do their will" (Letter 53, to Paulinus).[32]

[3] Among such lay writers should not be numbered some theologians whom I regard as men of profound learning and of the holiest lifestyle, and whom I therefore hold in high esteem and reverence. However, I cannot deny having some qualms, which I consequently wish could be removed; for in disputes about natural phenomena they seem to claim the right to force others by means of the authority of Scripture to follow the opinion they think is most in accordance with its statements, and at the same time they believe they are not obliged to ⟨324⟩ answer observations and reasons to the contrary. As an explanation and a justification of this opinion of theirs, they say that theology is the queen of all the sciences and hence must not in any way lower herself to accommodate the principles of other less dignified disciplines subordinate to her; rather, these others must submit to her as to a supreme empress and change and revise their conclusions in accordance with theological rules and decrees; moreover, they add that whenever in the subordinate science there is a conclusion which is certain on the strength of demonstrations and observations, and which is repugnant to some other conclusion found in Scripture, the practitioners of that science must themselves undo their own demonstrations and disclose the fallacies of their own observations without help from theologians and scriptural experts; for, as stated, it is not proper to the dignity of theology to stoop to the investigation of the fallacies in the subordinate

sciences, but it is sufficient for it to determine the truth of a conclusion with absolute authority and with the certainty that it cannot err. Then they say that the physical conclusions in regard to which we must rely on Scripture, without glossing or interpreting it in nonliteral ways, are those of which Scripture always speaks in the same way and which all the Holy Fathers accept and interpret with the same meaning. Now, I happen to have some specific ideas on these claims, and I shall propose them in order to receive the proper advice from whoever is more competent than I in these subjects; I always defer to their judgment.

To begin with, I think one may fall into something of an equivocation if one does not distinguish the senses in which sacred theology is pre-eminent and worthy of the title of queen. For it could be such insofar as whatever is taught in all the other sciences is found explained and demonstrated in it by means of more excellent methods and more sublime principles, in the way that, for example, the rules for measuring fields and for accounting are better contained in Euclid's geometry and arithmetic[33] than they are ⟨325⟩ in the practices of surveyors and accountants; or else insofar as the topic on which theology focuses surpasses in dignity all the other topics which are the subject of the other sciences and also insofar as its teaching proceeds in more sublime ways. I do not believe that theologians who are acquainted with the other sciences can assert that theology deserves the royal title and authority in the first sense; I think no one will say that geometry, astronomy, music, and medicine are treated more excellently and exactly in the sacred books than in Archimedes, Ptolemy, Boethius, and Galen. So it seems that the royal preeminence belongs to it in the second sense, namely because of the eminence of the topic and because of the admirable teaching of divine revelation in conclusions which could not be learned by men in any other way and which concern chiefly the gaining of eternal bliss. Thus theology does deal with the loftiest divine contemplations and for this it does occupy the royal throne and command the highest authority; and it does not come down to the lower and humbler speculations of the inferior sciences, but rather (as stated above) it does not bother with them, inasmuch as they are irrelevant to salvation. If all this is so, then officials and experts of theology should not arrogate to themselves the authority to issue decrees in the professions they neither exercise nor study; for this would be the same as if an absolute prince, knowing he had unlimited power to issue orders and compel obedience, but being neither a physician nor an architect, wanted to direct medical treatment and the construction of buildings, resulting in serious danger to the life of the unfortunate sick and in the obvious collapse of structures.

Furthermore, to require astronomers to endeavor to protect themselves against their own observations and demonstrations, namely to show that these are nothing but fallacies and sophisms, is to demand they do the impossible; for ⟨326⟩ that would be to require that they not only should not see what they see and not understand what they understand, but also that in their research they should find the contrary of what they find. That is, before they can do this, they should be shown how to manage having the lower faculties of the soul direct the higher ones, so that the imagination and the will could and would believe the contrary of what the intellect thinks. (I am still speaking of purely physical propositions which are not matters of faith, rather than of supernatural propositions which are articles of faith.) I should like to ask these very prudent Fathers to agree to examine very diligently the difference between debatable and demonstrative doctrines. Keeping firmly in mind the compelling power of necessary deductions, they should come to see more clearly that it is not within the power of the practitioners of demonstrative sciences to change opinion at will, choosing now this and now that one; that there is a great difference between giving orders to a mathematician or a philosopher and giving them to a merchant or a lawyer; and that demonstrated conclusions about natural and celestial phenomena cannot be changed with the same ease as opinions about what is or is not legitimate in a contract, in a rental, or in commerce. This difference has been completely recognized by the holy and very learned Fathers, as shown by their having made ⟨327⟩ a great effort to confute many philosophical arguments or, to be more exact, fallacies, and it may be explicitly read in some of them. In particular, we read the following words in St. Augustine (*On the Literal Interpretation of Genesis*, book 1, chapter 21): "There should be no doubt about the following: whenever the experts of this world can truly demonstrate something about natural phenomena, we should show it not to be contrary to our Scripture; but whenever in their books they teach something contrary to the Holy Writ, we should without any doubt hold it to be most false and also show this by any means we can; and in this way we should keep the faith of our Lord, in whom are hidden all the treasures of knowledge, in order not to be seduced by the verbosity of false philosophy or frightened by the superstition of fake religion."[34]

These words imply, I think, the following doctrine: in the learned books of worldly authors are contained some propositions about nature which are truly demonstrated and others which are simply taught; in regard to the former, the task of wise theologians is to show that they are not contrary to Holy Scripture; as for the latter (which are taught but not demonstrated with necessity), if they contain anything contrary to the Holy Writ, then they must be considered indubitably false and must be demonstrated such by every possible means. So physical conclusions which have been truly demonstrated should not be given a lower place than scriptural passages, but rather one should clarify how such passages do not contradict those conclusions; therefore, before condemning a physical proposition, one must show that it is not conclusively demonstrated. Furthermore, it is much more reasonable and natural that this be done not by those who hold it to be true, but by those who regard it as false; for the fallacies of an argument can be found much more easily by those who regard it as false than by those who think it is true and conclusive, and indeed here it will happen that the more the followers of a given opinion thumb through books, examine the arguments, repeat the observations, and check the experiments, the more they will be testing ⟨328⟩ their belief. In fact, Your Highness knows what happened to the late mathematician of the University of Pisa:[35] in his old age he undertook an examination of Copernicus's doctrine with the hope of being able to refute it solidly since he considered it false, even though he had never examined it; but it so happened that as soon as he understood its foundations, procedures, and demonstrations he became convinced of it, and he turned from opponent to very strong supporter. I could also name other mathematicians (e.g., Clavius) who, influenced by my recent discoveries, have admitted the necessity of changing the previous conception of the constitution of the world, since it can no longer stand up in any way.

It would be very easy to remove from the world the new opinion and doctrine if it were sufficient to shut the mouth of only one person; this is perhaps the belief of those who measure the judgments of others in terms of their own, and who thus think it is impossible that such an opinion can stand up and find followers. However, this business proceeds otherwise. For in order to accomplish that objective, it would be necessary not only to prohibit Copernicus's book and the writings of the other authors who follow the same doctrine, but also to ban all astronomical science completely; moreover, one would have to forbid men to look toward the heavens, so that they would not see that Mars and Venus are sometimes very close to and sometimes very far from the earth (the difference being that the latter sometimes appears forty times greater than at other times and the former sixty times greater); nor should they be allowed to see the same Venus appear sometimes round and sometimes armed with very sharp horns[36] and many other observable phenomena which can in no way be adapted to the Ptolemaic system but provide very strong arguments for Copernicanism. At the moment, because of many new ⟨329⟩ observations and because of many scholars' contributions to its study, one is discovering daily that Copernicus's position is truer and truer and his doctrine firmer and firmer; so to prohibit Copernicus now, after being permitted for so many years when he was less widely followed and less well confirmed, would seem to me an encroachment on the truth and an attempt to step up its concealment and suppression in proportion to how much more it appears obvious and clear. Not to ban the whole book in its entirety, but to condemn as erroneous only this particular proposition, would cause greater harm to souls, if I am not mistaken; for it would expose them to the possibility of seeing the proof of a proposition which it would then be sinful to believe. To prohibit the entire science would be no different than to reject hundreds of statements from the Holy Writ, which teach us how the glory and the greatness of the supreme God are marvelously seen in all His works and by divine grace are read in the open book of the heavens. Nor should anyone think that the reading of the very lofty words written on those pages is completed by merely seeing the sun and the stars give off light, rise, and set, which is as far as the eyes of animals and common people reach; on the contrary, those pages contain such profound mysteries and such sublime concepts that the vigils, labors, and studies of hundreds of the sharpest minds in uninterrupted investigation for thousands of years have not yet completely fathomed them. Even idiots realize that what their eyes see when they look at the external appearance of a human body is very insignificant in comparison to the admirable contrivances found in it by a competent and diligent philosopher-anatomist when he investigates how so many muscles, tendons, nerves, and bones are used; when he examines the function of the heart and the other principal organs; when he searches for the seat of the vital faculties; when he observes the wonderful structures of the senses; and, with no end to his astonishment and curiosity, when he studies the location of the imagination, memory, ⟨330⟩ and reason. Likewise, what the unaided sense of sight shows is almost nothing in comparison to the sublime marvels which the mind of intelligent investigators reveals in the heavens through long and accurate observation. This is all I can think of in regard to this particular point.

[4] Let us now examine their other argument: that physical propositions concerning which the Scripture always says the same thing, and which all the Fathers unanimously accept in the same sense, should be understood in accordance with the literal meaning of the words, without glosses or interpretations, and should be accepted and held as most true; and that, since the sun's motion and earth's rest constitute a proposition of this sort, consequently it is an article of faith to hold it as true and the contrary opinion as erroneous. Here it should be noticed, first, that some physical propositions are of a type such that by any human speculation and reasoning one can only attain a probable opinion and a verisimilar conjecture about them, rather than a certain and demonstrated science; an example is whether the stars are animate. Others are of a type such that either one has, or one may firmly believe that it is possible to have, complete certainty on the basis of experiments, long observations, and necessary demonstrations; examples are whether or not the earth and the sun move and whether or not the earth is spherical. As for the first type, I have no doubt at all that, where human reason cannot reach, and where consequently one cannot have a science, but only opinion and faith, it is appropriate piously to conform absolutely to the literal meaning of Scripture. In regard to the others, however, I should think, as stated above, that it would be proper to ascertain the facts first, so that they could guide us in finding the true meaning of Scripture; this would be found to agree absolutely with demonstrated facts, even though prima facie the words would sound otherwise, since two truths can never contradict each other. This doctrine seems to me very ⟨331⟩ correct and certain, inasmuch as I find it exactly written in St. Augustine. At one point he discusses the shape of heaven and what one should believe it to be, given that what astronomers affirm seems to be contrary to Scripture, since the former consider it round while the latter calls it stretched out like hide.[37] He decides one should not have the slightest worry that Scripture may contradict astronomers: one should accept its authority if what they say is false and based only on conjecture typical of human weakness; however, if what they say is proved with indubitable reasons, this Holy Father does not say that astronomers themselves be ordered to refute their demonstrations and declare their conclusion false, but he says one must show that what Scripture asserts about the hide is not contrary to those true demonstra-

tions. Here are his words (*On the Literal Interpretation of Genesis*, book 2, chapter 9): "However, someone asks how what is written in our books, 'Who stretchest out the heavens like a hide,'[38] does not contradict those who attribute to heaven the shape of a sphere. Now, if what they say is false, let it contradict them by all means, for the truth lies in what is said by divine authority rather than what is conjectured by human weakness. But if, by chance, they can support it with such evidence that one cannot doubt it, then we have to demonstrate that what our books say about the hide is not contrary to those true reasons."[39] Then he goes on to warn us that we must not be less careful in reconciling a scriptural passage with a demonstrated physical proposition than with another scriptural passage that may appear contrary. Indeed I think the caution of this saint deserves to be admired and emulated; for even in the case of obscure conclusions concerning which one cannot be sure whether they can be the subject of a science based on human demonstrations, he is very careful in declaring what one should believe. This can be seen from what he writes at the end of the second book of *On the Literal Interpretation of Genesis*, when discussing whether stars should be considered animate: "Although at present this cannot be easily known, nevertheless I think that in the course of examining Scripture one may find more appropriate passages whereby we would be entitled, if not to prove something for certain, at least to believe something on this topic based on the words of the sacred authority. Now then, always practicing a pious and serious moderation, we ought not to believe anything lightly about an obscure subject, lest ⟨332⟩ we reject (out of love for our error) something which later may be truly shown not to be in any way contrary to the holy books of either the Old or New Testament."[40]

From this and other places it seems to me, if I am not mistaken, the intention of the Holy Fathers is that in questions about natural phenomena which do not involve articles of faith one must first consider whether they are demonstrated with certainty or known by sensory experience, or whether it is possible to have such knowledge and demonstration. When one is in possession of this, since it too is a gift from God, one must apply it to the investigation of the true meanings of the Holy Writ at those places which apparently seem to read differently. These meanings will undoubtedly be grasped by wise theologians, along with the reasons why the Holy Spirit has sometimes wanted to hide them under words with a different literal meaning, whether in order to test us or for some other reason unknown to me.

Returning to the preceding argument, if we keep in mind the primary aim of the Holy Writ, I do not think that its always saying the same thing should make us disregard this rule; for if to accommodate popular understanding the Scripture finds it necessary once to express a proposition with words whose meaning differs from the essence of the proposition, why should it not follow the same practice for the same reason every time it has to say the same thing? On the contrary, I think that to do otherwise would increase popular confusion and diminish the propensity to believe on the part of the people. Furthermore, in regard to the rest or motion of the sun and the earth, experience clearly shows that to accommodate popular understanding it is indeed necessary to assert what the words of Scripture say; for even in our age when people are more refined, they are kept in the same opinion by reasons which, when carefully examined and pondered, will be found to be most frivolous and by observations which are either completely false or totally irrelevant; nor can one try to move them since they are not capable of understanding the contrary reasons, which are dependent on extremely delicate observations and on subtle demonstrations ⟨333⟩ supported by abstractions whose understanding requires a very vivid imagination. Therefore, even if the sun's rest and the earth's motion were more than certain and demonstrated among the experts, it would still be necessary to utter the contrary in order to maintain credibility with large numbers of people; for among a thousand laymen who might be asked about these details, perhaps not even one will be found who would not answer that he firmly believes that the sun moves and the earth stands still. However, no one should take this very common popular consensus as an argument for the truth of what is being asserted; for if we ask the same men about the reasons and motives why they believe that way, and if on the other hand we listen to the observations and demonstrations which induce those other few to believe the opposite, we shall find that the latter are convinced by very solid reasons and the former by the simplest appearances and by empty and ridiculous considerations.

It is therefore clear that it was necessary to attribute motion to the sun and rest to the earth in order not to confuse the meager understanding of the people and not to make them obstinately reluctant to give assent to the principal dogmas which are absolutely articles of faith; but if it was necessary to do this, it is no wonder that this was most prudently done in divine Scripture. Indeed I shall say further that it was not only respect for popular inability, but also the current opinion of those times, that made the sacred writers accommodate themselves to received usage rather than to the essence of the matter in regard to subjects which are not necessary for eternal bliss. In fact, speaking of this St. Jerome writes: "As if in the Holy Scripture many things were not said in accordance with the opinion of the time when the facts are being reported, and not in accordance with the truth of the matter" (commentary on chapter 28 of Jeremiah). Elsewhere the same saint says: "In Scripture it is customary for the historian to report many opinions as they were accepted by everyone at that time" (commentary on chapter 13 of Matthew). Finally, on the words in chapter 27 of Job, "He stretched out the north ⟨334⟩ over the empty space, and hangeth the earth upon nothing,"[41] St. Thomas[42] notes that Scripture calls empty and nothing the space which embraces and surrounds the earth and which we know is not empty but full of air; nevertheless, he says that Scripture calls it empty and nothing in order to accommodate the belief of the people, who think there is nothing in this space. Here are St. Thomas's words: "The upper hemisphere of the heavens seems to us nothing but a space full of air, though common people consider it empty; thus it speaks in accordance with the judgment of common people, as is the custom in Holy Scripture." Now from this I think one can obviously argue that analogously the Holy Scripture had a much greater reason to call the sun moving and the earth motionless. For if we test the understanding of common people, we shall find them much more incapable of becoming convinced of the sun's rest and earth's motion than of the fact that the space surrounding us is full of air; therefore, if the sacred authors refrained from attempting to persuade the people about this point, which was not that difficult for their understanding, it seems very reasonable to think that they followed the same style in regard to other propositions which are much more recondite.

Indeed, Copernicus himself knew how much our imagination is dominated by an old habit and by a way of conceiving things which is already familiar to us since infancy, and so he did not want to increase the confusion and difficulty of his abstraction. Thus, after first demonstrating that the motions which appear to us as belonging to the sun or the firmament ⟨335⟩ really belong to the earth, then, in the process of compiling their tables and applying them in practice, he speaks of them as belonging to the sun and to the part of heaven above the planets; for example, he speaks of the rising and setting of the sun and the stars, of changes in the obliquity of the zodiac and in the equinoctial points, of the mean motion and the anomaly and the prosthaphaeresis[43] of the sun, and other similar things, which really belong to the earth. We call facts these things which appear to us as facts because, being attached to the earth, we are part of all its motions, and consequently we cannot directly detect these things in it but find it useful to consider it in relation to the heavenly bodies in which they appear to us. Therefore, note how appropriate it is to accommodate our usual manner of thinking.

Next consider the principle that the collective consensus of the Fathers, when they all accept in the same sense a physical proposition from Scripture, should authenticate it in such a way that it becomes an article of faith to hold it. I should think that at most this ought to apply only to those conclusions which the Fathers discussed and inspected with great diligence and debated on both sides of the issue and for which they then all agreed to reject one side and hold the other. However, the earth's motion and sun's rest are not of this sort, given that in those times this opinion was totally forgotten and far from academic dispute and was not examined, let alone followed, by anyone; thus one may believe that the Fathers did not even think of discussing it since the scriptural passages, their own opinion, and popular consensus were all in agreement, and no ⟨336⟩ contradiction by anyone was heard. Therefore, it is not enough to say that all the Fathers accept the earth's rest, etc., and so it is an article of faith to hold it; rather one would have to prove that they condemned the contrary opinion. For I can always say that their failure to reflect upon it and discuss it made them leave it stand as the current opinion, but not as something resolved and established. I think I can say this with very good reason: for either the Fathers reflected upon this conclusion as if it were controversial or they did not; if not, then they could not have decided anything about it, even in their minds, nor should their failure oblige us to accept those principles which they did not, even in intention, impose; whereas if they examined it with care, then they would have condemned it had they judged it to be erroneous; but there is no record of their having done this. Indeed, after some theologians began to examine it, one sees that they did not deem it to be erroneous, as one can read in Diego de Zuñiga's *Commentaries on Job*,[44] in regard to the words "Who shaketh the earth out of her place, etc." in chapter 9, verse 6; he discusses the Copernican position at length and concludes that the earth's motion is not against Scripture.

Furthermore, I would have doubts about the truth of this prescription, namely whether it is true that the Church obliges one to hold as articles of faith such conclusions about natural phenomena, which are characterized only by the unanimous interpretation of all the Fathers. I believe it may be that those who think in this manner may want to amplify the decrees of the Councils in favor of their own opinion. For I do not see that in this regard they prohibit anything but tampering, in ways contrary to the interpretation of the Holy Church or the collective consensus of the Fathers, with those propositions which are articles of faith or involve morals and pertain ⟨337⟩ to edification according to Christian doctrine; so speaks the Fourth Session of the Council of Trent. However, the motion or rest of the earth or the sun are not articles of faith and are not against morals; nor does anyone want to twist scriptural passages to contradict the Holy Church or the Fathers. Indeed, those who put forth this doctrine have never used scriptural passages, for it always remains the prerogative of serious and wise theologians to interpret these passages in accordance with their true meaning. Moreover, it is quite obvious that the decrees of the Councils agree with the Holy Fathers in regard to these details; for they are very far from wanting to accept as articles of faith similar physical conclusions or to reject as erroneous the contrary opinions, so much so that they prefer to pay attention to the primary intention of the Holy Church and consider it useless to spend time trying to ascertain those conclusions. Let me tell Your Most Serene Highness what St. Augustine (*On the Literal Intepretation of Genesis*, book 2, chapter 10) answers to those brethren who ask whether it is true that the heavens move or stand still: "To them I answer that these things should be examined with very subtle and demanding arguments to determine truly whether or not it is so; but I do not have the time to undertake and pursue these investigations, nor should such time be available to those whom we desire to instruct for their salvation and for the needs and benefit of the Holy Church."[45]

However, suppose one were to decide that, even in the case of propositions about natural phenomena, they should be condemned or accepted on the basis of scriptural passages which are unanimously interpreted in the same way by all the Fathers; even then I do not see that this rule would apply in our case, given that one can read in the Fathers different interpretations of the same passages. For example, Dionysius the Areopagite says that it was not the sun but the Prime Mobile which stopped;[46] St. Augustine thinks the same thing, namely that all heavenly bodies stopped; and the Bishop of Avila[47] is of the same opinion. Moreover, among the Jewish authors whom Josephus endorses, some thought that the sun did not really stop, but that it appeared so for the short time during which the Israelites defeated their enemies. Similarly, in the miracle at the time of Hezekiah,[48] Paul of Burgos thinks that it did not take place in the sun but in the clock. ⟨338⟩ At any rate, I shall demonstrate further below that, regardless of the world system one assumes, it is in fact necessary to gloss and interpret the words of the text in Joshua.

[5] Finally, let us grant these gentlemen more than they ask—namely, let us submit entirely to the opinion of wise theologians. Since this particular determination was not made by the ancient Fathers, it could be made by the wise ones of our age. The controversy concerns questions of natural phenomena and dilemmas whose answers are necessary and cannot be otherwise than in one of the two controversial ways; so they should first hear the experiments, observations, reasons, and demonstrations of philosophers and astronomers on both sides of the question, and then they would be able to determine with certainty whatever divine inspiration will communicate to them. No one should hope or fear that they would reach such an important decision without inspecting and discussing very minutely all the reasons for one side and for the other, and without ascertaining the facts: this cannot be hoped for by those who would pay no attention to risking the majesty and dignity of the Holy Writ to support their self-righteous creations; nor is this to be feared by those who seek nothing but the examination of the foundations of this doctrine with the greatest care, and who do this only out of zeal for the truth and for the majesty, dignity, and authority of the Holy Writ, which every Christian must strive to uphold. No one can fail to see that this dignity is desired and upheld with much greater zeal by one group than by the other—by those who submit in every way to the Holy Church and do not ask for the prohibition of this or that opinion, but only that they be allowed to present things whereby she could more reliably be sure of making the safest choice; and not by those who, blinded by their own interests or incited by malicious suggestions, preach that she immediately flash the sword since she has the power to do it, without considering that it is not always useful to do all that one can do. This opinion was not held by the holiest Fathers. Indeed, they knew how harmful and how contrary to the primary function of the Catholic church it would be to want to use scriptural passages to establish conclusions about nature, when by means of observation and necessary demonstrations one could at some point demonstrate the contrary of what ⟨339⟩ the words literally say; thus not only were they very circumspect, but they left precepts for the edification of others. From St. Augustine, *On the Literal Interpretation of Genesis*, book 1, chapters 18 and 19, we have the following: "In obscure subjects very far removed from our eyes, it may happen that even

in the divine writings we read things that can be interpreted in different ways by different people, all consistent with the faith we have; in such a case, let us not rush into any one of these interpretations with such precipitous commitment that we are ruined if it is rightly undermined by a more diligent and truthful investigation; such recklessness would mean that we were struggling for our opinions and not for those of Scripture, and that we wanted to make scriptural opinion conform to ours, when we ought to want to make ours conform to that of Scripture."[49] A little further, to teach us how no proposition can be against the faith unless it is first shown to be false, he adds: "It is not against the faith as long as it is not refuted by an unquestionable truth; if this happens, then it was not contained in the divine Scripture but originated from human ignorance."[50] From this one sees the falsehood of any meanings given to scriptural passages which do not agree with demonstrated truths; and so one must search for the correct meaning of Scripture with the help of demonstrated truth, rather than taking the literal meaning of the words, which may seem the truth to our weak understanding, and trying somehow to force nature and deny observations and necessary demonstrations.

Your Highness should also note with how much circumspection this very holy man proceeds before deciding to assert that some scriptural interpretation is so certain and sure that there is no fear of encountering disturbing difficulties. Not satisfied with just any scriptural meaning which might agree with some demonstration, he adds: "But if this were proved to be true by an unquestionable argument, it would still be uncertain whether by these words the writer of the holy books meant this or something else no less true; for if the rest of the context of the passage showed that he did not intend this, then what he did intend would not thereby be falsified but would still be true and more beneficial to know."[51] Now, what increases our amazement about the circumspection with which this author proceeds is the fact that he is still not completely sure upon seeing that demonstrative reasons, as well as the literal scriptural meaning and the preceding and subsequent text, ⟨340⟩ all point in the same direction, and so he adds the following words: "If the context of Scripture did not disprove that the writer meant this, one could still ask whether he might not have meant the other."[52] Still he does not decide to accept this meaning or exclude that one. Rather, he does not think he can ever be sufficiently cautious, and so he continues: "If we found that he could have meant the other, then it would be uncertain which of the two he intended; and if both interpretations were

supported by solid documentation, it would not be implausible to believe that he meant both."[53] Next he seems to want to give the rationale
for his procedure by showing us the dangers to which certain people
would expose themselves, the Scripture, and the Church; these are people who, concerned more with the preservation of their own errors than
with the dignity of Scripture, would want to extend its authority beyond the limits which it prescribes for itself. And so he adds the following words, which by themselves should suffice to repress and temper the
excessive license which some people arrogantly take: "In fact, it often
happens that even a non-Christian has views based on very conclusive
reasons or observations about the earth, heaven, the other elements of
this world, the motion and revolutions or the size and distances of the
stars, the eclipses of the sun and moon, the cycles of years and epochs,
the nature of animals, of plants, of rocks, and similar things. Now, it is
very scandalous, as well as harmful and to be avoided at all costs, that
any infidel should hear a Christian speak about these things as if he
were doing so in accordance with Christian Scripture and should see
him err so deliriously as to be forced into laughter. The distressing thing
is not so much that an erring man should be laughed at, but that our
authors should be thought by outsiders to believe such things, and
should be criticized and rejected as ignorant, to the great detriment of
those whose salvation we care about. For how can they believe our
books in regard to the resurrection of the dead, the hope of eternal life,
and the kingdom of heaven, when they catch a Christian committing an
error about something they know very well, when they declare false his
opinion taken from those books, and when they find these full of fallacies in regard to things they have already been able to observe or to
establish by unquestionable argument?"[54] Finally, we can see how of-
fended are the truly wise and prudent Fathers by these people who, in
order to support propositions they do not ⟨341⟩ understand, constrain
scriptural passages in certain ways and then compound their first error
by producing other passages which they understand even less than the
former ones. This is explained by the same saint with the following
words: "It is impossible to express sufficiently well how much harm and
sorrow those who are reckless and presumptuous cause to prudent
brethren. This happens when they begin to be rebuked and refuted for
their distorted and false opinions by those who do not accept the authority of our books, and so they put forth those same books to prove
and defend what they had said with very superficial recklessness and
very obvious falsity, and they even quote many of their passages from

memory, considering them supporting testimony, but without understanding either what they say or what they are talking about."[55]

To this type belong, I think, those who will not or cannot understand the demonstrations and observations with which the originator and the followers of this position confirm it, and who thus are concerned with putting forth Scripture. They do not notice that the more scriptural passages they produce, and the more they persist in claiming that these are very clear and not susceptible to other meanings besides what they advance, the greater the harm resulting to the dignity of Scripture if later the truth were known to be clearly contrary and were to cause confusion (especially if these people's judgment had much authority in the first place). There would be harm and confusion at least among those who are separated from the Holy Church, toward whom she is nevertheless very zealous like a mother who wants to be able to hold them on her lap. Your Highness can therefore see how inappropriate is the procedure of those who, in disputes about nature, as a first step advance arguments based on scriptural passages, especially when very often they do not adequately understand them.

However, if these people truly feel and fully believe they have the true meaning of some particular scriptural passage, it would have to follow necessarily that they are also sure of possessing the absolute truth about the physical conclusion they intend to discuss and, at the same time, that they know they have a very great advantage over the opponent, who has to defend the false side; for whoever is supporting the truth can have many sensory experiences and many necessary demonstrations on his side, ⟨342⟩ whereas the opponent cannot use anything but deceptive presentations, paralogisms, and fallacies. Now, if they know that by staying within the limits of the physical subject of discussion and using only philosophical weapons, they are in any case so superior to the opponent, why is it that when they come to the debate they immediately seize an irresistible and fearful weapon, so that their opponent is frightened at its mere sight? To tell the truth, I believe they are the ones who are frightened and are trying to find a new way of repelling the enemy because they are unable to resist his assaults. That is why they forbid him to use the reason he received through the Divine Goodness and why they abuse the very proper authority of the Holy Scripture, which (when adequately understood and used) can never conflict with clear observation and necessary demonstrations, as all theologians agree. However, the fact that these people take refuge in Scripture, to cover up their inability to understand and to answer the contrary arguments, should be of no advantage to them, if I am not mistaken, since till now such an opinion has never been condemned by the Holy Church. Therefore, if they wanted to proceed with sincerity, they could remain silent and admit their inability to discuss similar subjects; or else they could first reflect that it is not within their power, nor within that of anyone but the Supreme Pontiff and the sacred Councils, to declare a proposition erroneous, but they are free to discuss whether it is false; then, understanding that it is impossible for a proposition to be both true and heretical, they should focus on the issue which more concerns them, namely on demonstrating its falsity; if they were to discover this falsity, then either it would no longer be necessary to prohibit it because no one would follow it, or its prohibition would be safe and without the risk of any scandal.

Thus let these people apply themselves to refuting the arguments of Copernicus and the others, and let them leave its condemnation as erroneous and heretical to the proper authorities; but let them not hope that the very cautious and very wise Fathers and the Infallible One with his absolute wisdom are about to make rash decisions like those into which they would be rushed by their special interests and feelings. ⟨343⟩ For in regard to these and other similar propositions which do not directly involve the faith, no one can doubt that the Supreme Pontiff always has the absolute power of permitting or condemning them; however, no creature has the power of making them be true or false, contrary to what they happen to be by nature and de facto. So it seems more advisable to first become sure about the necessary and immutable truth of the matter, over which no one has control, than to condemn one side when such certainty is lacking; this would imply a loss of freedom of decision and choice insofar as it would give necessity to things which are presently indifferent, free, and dependent on the will of the supreme authority. In short, if it is inconceivable that a proposition should be declared heretical when one thinks it may be true, it should be futile for someone to try to bring about the condemnation of the earth's motion and sun's rest unless he first shows it to be impossible and false.

[6] There remains one last thing for us to examine: to what extent it is true that the Joshua passage[56] can be taken without altering the literal meaning of the words and how it can be that, when the sun obeyed Joshua's order to stop, from this it followed that the day was prolonged by a large amount.

Given the heavenly motions in accordance with the Ptolemaic system, this is something which in no way can happen. For the sun's mo-

tion along the ecliptic[57] takes place in the order of the signs of the zodiac, which is from west to east; this is contrary to the motion of the
Prime Mobile from east to west, which is what causes day and night;
therefore, it is clear that if the sun stops its own true motion, the day
becomes shorter and not longer and that, on the contrary, the way to
prolong it would be to speed up the sun's motion; thus, to make the sun
stay for some time at the same place above the horizon, without going
down toward the west, ⟨344⟩ it would be necessary to accelerate its
motion so as to equal the motion of the Prime Mobile, which would be
to accelerate it to about three hundred and sixty times its usual motion.
Hence if Joshua had wanted his words taken in their literal and most
proper meaning, he would have told the sun to accelerate its motion by
an amount such that, when carried along by the Prime Mobile, it would
not be made to set; but his words were being heard by people who
perhaps had no other knowledge of heavenly motions except for the
greatest and most common one from east to west; thus he adapted himself to their knowledge and spoke in accordance with their understanding because he did not want to teach them about the structure of the
spheres but to make them understand the greatness of the miracle of the
prolongation of the day.

Perhaps it was this consideration that first led Dionysius the Areopagite (in the Letter to Polycarpus) to say that in this miracle the Prime
Mobile stopped and, as a consequence of its stopping, all other celestial
spheres stopped. The same opinion is held by St. Augustine himself (in
book 2 of *On the Miracles of the Holy Scripture*),[58] and the Bishop of
Avila supports it at length (in questions 22 and 24 of his commentary
on chapter 10 of Joshua). Indeed one sees that Joshua himself intended
to stop the whole system of celestial spheres, from his giving the order
also to the moon, even though it has nothing to do with the prolongation of the day; in the injunction given to the moon one must include
the orbs of the other planets, which are not mentioned here, as they are
not in the rest of the Holy Scripture, since its intention has never been
to teach us the astronomical sciences.

I think therefore, if I am not mistaken, that one can clearly see that,
given the Ptolemaic system, it is necessary to interpret the words in a
way different from their literal meaning. Guided by St. Augustine's very
useful prescriptions, I should say that the best nonliteral interpretation
is not necessarily this, if anyone can find another which is perhaps better
and more suitable. So now I want to examine whether the same miracle
could be understood in a way more in accordance with what we read in
Joshua, if to the Copernican system we add ⟨345⟩ another discovery

which I recently made about the solar body. However, I continue to speak with the same reservations—to the effect that I am not so enamored with my own opinions as to want to place them ahead of others'; nor do I believe it is impossible to put forth interpretations which are better and more in accordance with the Holy Writ.

Let us first assume, in accordance with the opinion of the above-mentioned authors, that in the Joshua miracle the whole system of heavenly motions was stopped, so that the stopping of only one would not introduce unnecessarily universal confusion and great turmoil in the whole order of nature. Second, I think that although the solar body does not move from the same place, it turns on itself, completing an entire rotation in about one month, as I feel I have conclusively demonstrated in my *Sunspot Letters*;[59] this motion is sensibly seen to be inclined southward in the upper part of the globe, and thus to tilt northward in the lower part, precisely in the same manner as the revolutions of all planetary orbs. Third, the sun may be regarded as a noble body, and it is the source of light illuminating not only the moon and the earth but also all the other planets, which are in themselves equally dark; having conclusively demonstrated this, I do not think it would be far from correct philosophizing to say that, insofar as it is the greatest minister of nature and, in a way, the heart and soul of the world, it transmits to the surrounding bodies not only light but also (by turning on itself) motion; thus, just as all motion of an animal's limbs would cease if the motion of its heart were to cease, in the same way if the sun's rotation stopped then all planetary revolutions would also stop. Now, concerning the admirable power and strength of the sun I could quote the supporting statements of many serious writers, but I want to restrict myself to just one passage from the book *The Divine Names* by the Blessed Dionysius the Areopagite. He writes this about the sun: "Light also gathers and attracts to itself all things that are seen, that move, that are illuminated, that are heated, and in a word that are surrounded by its splendor. Thus the sun is called Helios because ⟨346⟩ it collects and gathers all things that are dispersed." And a little below that he again writes about the sun: "If in fact this sun, which we see and which (despite the multitude and dissimilarity of the essences and qualities of observed things) is nevertheless one, spreads its light equally and renews, nourishes, preserves, perfects, divides, joins, warms up, fertilizes, increases, changes, strengthens, produces, moves, and vitalizes all things; and if everything in this universe in accordance with its own power partakes of one and the same sun and contains within itself an equal anticipation of the causes of the many things which are shared; then certainly all the more reason, etc." Therefore, given that the sun is both the source of light and the origin of motion, and given that God wanted the whole world system to remain motionless for several hours as a result of Joshua's order, it was sufficient to stop the sun, and then its immobility stopped all the other turnings, so that the earth as well as the moon and the sun (and all the other planets) remained in the same arrangement; and during that whole time the night did not approach, and the day miraculously got longer. In this manner, by stopping the sun, and without changing or upsetting at all the way the other stars appear or their mutual arrangement, the day on the earth could have been lengthened in perfect accord with the literal meaning of the sacred text.

Furthermore, what deserves special appreciation, if I am not mistaken, is that with the Copernican system one can very clearly and very easily give a literal meaning to another detail which one reads about the same miracle; that is, that the sun stopped in the middle of heaven. Serious theologians have raised a difficulty about this passage: it seems very probable that, when Joshua asked for the prolongation of the day, the sun was close to setting and not at the meridian; for it was then about the time of the summer solstice, and consequently the days were very long, so that if the sun had been at the meridian then it does not seem likely that it would have been necessary to pray for a lengthening of the day in order to win a battle, since the still remaining time of seven hours or more could very well have been sufficient. Motivated by this argument, very serious theologians have held that the sun really was close to setting; ⟨347⟩ this is also what the words "Sun, stand thou still"[60] seem to say, because if it had been at the meridian, then either there would have been no need to seek a miracle or it would have been sufficient to pray merely for some slowing down. This opinion is held by the Bishop of Gaeta,[61] and it is also accepted by Magalhaens, who confirms it by saying that on the same day, before the order to the sun, Joshua had done so many other things that it was impossible to complete them in half a day; thus they really resort to interpreting the words "in the midst of heaven"[62] somewhat implausibly, saying they mean the same as that the sun stopped while it was in our hemisphere, namely above the horizon. We can remove this and every other implausibility, if I am not mistaken, by placing the sun, as the Copernican system does and as it is most necessary to do, in the middle, namely at the center of the heavenly orbs and the planetary revolutions; for at any hour of the day, whether at noon or in the afternoon, the day would have been lengthened and all heavenly turnings stopped by the sun stopping in the middle of the heavens, namely at the center of the heavens, where it is located. Furthermore, this interpretation agrees all the more with the literal meaning inasmuch as, if one wanted to claim that the sun's stopping occurred at the noon hour, then the proper expression to use would have been to say that it "stood still at the meridian point," or "at the meridian circle," and not "in the midst of heaven"; in fact, for a spherical body such as heaven, the middle is really and only the center.

As for other scriptural passages which seem to contradict this position, I have no doubt that, if it were known to be true and demonstrated, those same theologians who consider such passages incapable of being interpreted consistently with it (as long as they regard it as false) would find highly congenial interpretations for them; this would be especially true if they were to add some knowledge of the astronomical sciences to their expertise about Holy Writ. Just as now, when they consider it false, they think that whenever they read Scripture they only find statements repugnant to it, so if they thought otherwise they would perchance find an equal number of passages agreeing with it. Then perhaps they would judge ⟨348⟩ it very appropriate for the Holy Church to tell us that God placed the sun at the center of heaven and that therefore He brings about the ordered motions of the moon and the other wandering stars by making it turn around itself like a wheel, given that she sings:

> Most holy Lord and God of heaven,
> Who to the glowing sky hast given
> The fires that in the east are born
> With gradual splendours of the morn;
> Who, on the fourth day, didst reveal
> The sun's enkindled flaming wheel,
> Didst set the moon her ordered ways,
> And stars their ever-winding maze.[63]

They could also say that the word firmament is *literally* very appropriate for the stellar sphere and everything above the planetary orbs, which is totally still and motionless according to this arrangement. Similarly, if the earth were rotating, then, where one reads "He had not yet made the earth, nor the rivers, nor the poles of the terrestrial globe,"[64] one could understand its poles literally; for there would be no point in attributing these poles to the terrestrial globe if it did not have to turn around them.

CHAPTER IV

Galileo's Discourse on the Tides (1616)

⟨377⟩ To the Most Illustrious and Most Reverend Lord Cardinal Orsini: [1]

[1][2] I am honored that Your Most Eminent and Most Reverend Lordship has asked me to put into writing what I explained to you orally ten days ago, and the honor is much greater than I and my insubstantial discussions deserve. My only means of reciprocating, at least in part, being prompt obedience, here I am ready to serve and obey you in accordance with your request. I shall try to do so in the briefest and most concise way possible for such an amazing problem as the investigation of the true cause of the tides; this is all the more recondite and difficult to ascertain since we clearly see that everything written so far by serious authors is very far from quieting the mind of those who, in the contemplation of nature, desire to fathom beyond the surface. This tranquillity is attained only when the reason advanced as the true cause of the effect accounts easily and clearly for all symptoms and features that are seen in the effect. Since, as we saw in our private discussions, this peace of mind is not given to us by the reasons adduced so far by other writers on this question, I shall set them aside as ineffective. After all, your Most Eminent and Most Reverend Lordship was quite ⟨378⟩ satisfied with the refutations of them which I presented orally, though even earlier you yourself had not been overly impressed by them; and you agreed, indeed ordered me, that the extended discussion of such refutations for the general public wait until I treat this subject at greater length in my System of the World.[3]

Sensory experience shows us that the tides are not a process of expansion and contraction of seawater, similar to what happens with water heated by fire, which becomes more rarefied and rises with great
force due to heat and contracts and subsides when returning to its natural coldness. Instead they are a process of true local motion in the sea,
a displacement, so to speak, now toward one end and now toward the
opposite end of a sea basin, without any alteration of the element water
other than that deriving from change of place. Now, while we are discussing on the basis of sensory experience (a safe guide in true philosophizing), let us note that we see that motion can be imparted to water
in various ways; we shall be examining these in turn to see whether any
of them can reasonably be assigned as the primary cause of tides. I speak
of primary cause because, while we shall examine the many different
phenomena that are seen in connection with the tides, we shall consider
it necessary that many other secondary and, as they say, concomitant
causes should combine with the primary one to produce such various
phenomena; for from a single and simple cause can derive only a simple
and determinate effect. We shall thus begin our discussion with the investigation of the primary cause, which is general and without which
this regular movement of seawater would not exist. (I say regular even
though different seas experience different periods in their tides.)

One of the causes of motion is the inclination of the place or bed
where the fluid body is contained; this is why torrents rush into rivers,
and rivers flow into seas. However, because this flow is always toward
the same part of the incline, along which the water never goes back, this
reason has nothing to do with ⟨379⟩ our cause; it cannot be present in
the case of reciprocal motions toward opposite ends, as we see happening with seawater.

There is another way in which water can be agitated, namely by the
motion of the surroundings or some other external body impinging on
it; thus we see the water of lakes and seas agitated and driven where the
wind pushes it. However, such an agitation cannot be assigned as the
cause of our problem. For such agitations are very irregular, whereas
tides have their periods fixed; besides, tides occur even when winds have
ceased and there is the greatest calmness in the air; moreover, they
maintain their course in a given direction even when the wind blows in
the opposite direction at the same time.

Motions can also be given to water when the containing vessel is
somehow moved. This can occur in two ways, one of which would be
to raise and to lower alternately the two ends of the vessel; this motion

of oscillation would be followed by the contained water flowing toward the inclined end and, in turn, going back and forth along the length of the vessel. However, such a phenomenon of oscillation cannot take place in our case. In fact, even if the earth had some reciprocal oscillation, this would not provide the water with means for flowing back and forth; for it flows in an oscillating vessel insofar as the oscillation lowers now one and now the other end of the vessel, that is, insofar as this lowered end approaches the common center of heavy objects, toward which the weight of the water makes it flow; whereas if the earth had an oscillatory motion, even this oscillation would not bring any part of its surface closer to or farther from its center, which is the point toward which heavy bodies tend to move, and so the water would have no way of flowing toward it. Furthermore, the oscillation which can be attributed to the terrestrial globe involves a transverse inclination, that is, from north to south; but, on the contrary, tides are all in an east-west direction. Finally, the oscillation which some have attributed to the earth has a period of many thousands of years;[4] ⟨380⟩ whereas in the oscillatory and reciprocal motions of tides the times are very short, namely a matter of hours.

The other way of transmitting motion to water through the motion of the containing vessel is by moving the vessel forward, without tilting it in any way, but merely moving it with motion alternately accelerated and retarded; with this variation the water, besides moving with the container, moves also in different and sometimes even contrary ways. To be explicit, let us consider a large vessel full of water, such as, for example, a large boat like those carrying fresh water from one place to another over salt water. First we see that as long as the containing vessel, namely the boat, stands still, the water contained in it stays equally calm. Then let the boat begin moving, not slowly and gradually, but suddenly with great speed; I say that we see the water being left behind, rising toward the stern and subsiding toward the bow, because unlike the solid parts, the water is not firmly attached to the vessel, but rather its fluidity makes it somewhat detached and not constrained to follow every sudden move of the vessel; then little by little the water is reduced to obeying the motion of its container without any disturbance as long as it moves placidly and uniformly. On the other hand, when the boat is significantly restrained in its course, either by running aground or by some other intervening obstacle, the water inside it is not restrained in the same way by the impetus received, but rather, being somewhat detached from the container, it retains the impetus and flows and overflows toward the bow, subsiding near the stern. The more suddenly the vessel leaves the state of rest or is stopped while speeding, the more clearly this can be seen. For if the transition from the state of rest to accelerated motion is made gradually and by small degrees, or if one returns from fast motion to rest equally slowly, then there is an imperceptible or very small disobedience (so to speak) in the contained water; without resistance and equally slowly it acquires the same changes as the whole vessel.

⟨381⟩ Now, Most Eminent Lord, when I examine these and other facts pertaining to this cause of motion considered last, I would be greatly inclined to agree that the cause of tides could reside in some motion of the basins containing seawater; thus, attributing some motion to the terrestrial globe, the movements of the sea might originate from it. If this did not account for all particular things we sensibly see in the tides, it would thus be giving a sign of not being an adequate cause of the effect; similarly, if it does account for everything, it may give us an indication of being its proper cause, or at least of being more probable than any other one advanced till now.

[2] Let us, then, take the motion of the earth hypothetically,[5] in particular those same motions which many ancients and other recent philosophers have attributed to it on account of other sensible effects; and let us consider what consequence and relevance they may have for the present subject. For greater clarity let us briefly explain these motions attributed to the terrestrial globe.

The first and largest one is the annual motion under the ecliptic, from west to east, in an orbit or circle whose radius is the distance from the sun to the earth. The second is a turning by the terrestrial globe on itself and around its own center, taking place in a period of twenty-four hours, in the same direction, that is, from west to east, though around an axis somewhat inclined to the axis of the annual motion. I leave aside the third motion,[6] as of little or no relevance to this effect, due to its extreme slowness in comparison to the other two, which are very fast; for the speed of the rotation on itself already mentioned is about 365 times greater than this third motion (if indeed it is to be so called), while the speed of the annual motion is more than triple that of the diurnal speed, even if taken at the terrestrial equator.

For easier comprehension, let the circumference of the earth's orbit be *AFG*, around the center *E*; let *BCDL* be the terrestrial globe with center A; let the annual motion take place from point *A* toward part *F*, with the center of the terrestrial globe tracing the circumference *AFGI*

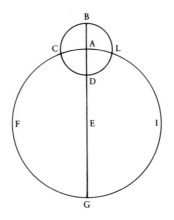

in about 365 days; and ⟨382⟩ let the turning on itself by the terrestrial globe be in the direction from *B* to *C* to *D*, etc. Let us understand that each of these two motions is in and of itself equable and uniform; that is, the center of the earth *A* traverses equal parts of the circumference *AFG* in equal times, and, similarly, point *B* and any other point of the circumference *BCDL* also traverse equal distances in equal times. Assuming this, we must first of all carefully notice that from the combination of the two motions there results a very unequal motion for the parts of the earth's surface, so that each of these parts moves with different speeds at different times of the day; this is indeed so, despite the fact that both of these two motions, that is, the annual by the earth's center along the earth's orbit *AFG* and the diurnal by the circumference *BCDL* on itself and around its own center *A*, are in and of themselves equable and uniform. I proceed to explain this more clearly.

Notice, then, that while the circle *BCDL* turns on itself in the direction *BCD*, there are in its circumference mutually contrary movements: for, while the parts near *C* go down, the opposite ones near *L* go up; and while the parts near *B* move toward the left, the parts on the opposite side near *D* move toward the right. Thus, in a complete rotation the point marked *B* first moves down and toward the left; when it is near *C*, it descends the most and begins to move toward the right; at *D* it no longer goes down, but moves most toward the right and begins to go up; and at *L* it ascends the most, begins to move slowly toward the left, and goes up till *B*. Now let us combine these specific motions of the parts of the earth with the general movement by the whole globe through the circumference *AFG*. We shall find that the absolute motion of the upper parts (near *B*) is always fastest, resulting from the composition of the annual motion along the circumference AF and the specific motion of part B, which two motions ⟨383⟩ reinforce each other and add up toward the left; on the other hand, the absolute motion of the lower parts near D is always slowest, since the specific motion of D, which here is fastest toward the right, must be subtracted from the annual motion along the circumference AF, which is toward the left; whereas the absolute motion resulting from the combination of the two motions (annual and diurnal) of the earth's parts at points C and L is intermediate and equal simply to the annual motion, since the turning of the circle $BCDL$ on itself does not make points C and L move either to the right or to the left, but only down and up, and it does not add to or subtract from the speed of the simple motion along the arc AF.

Thus I believe that it is now clear how, though each part of the earth's surface moves with two very uniform motions, nevertheless within a period of twenty-four hours it moves sometimes very fast, sometimes slowly, and twice at intermediate speeds; and this change results from the combination of these two uniform motions, diurnal and annual.

So far, then, we see that any body of water (be it a sea, a lake, or a pond) has a continuous but nonuniform motion, since it is retarded during some hours of the day and much accelerated during others; we also have the principle and the cause why the water contained in it, being fluid and not firmly attached to the container, flows and moves now in this and now in the opposite direction. Now we may begin to examine the particular details, so numerous and so diverse, which are observed in various seas and other bodies of water, our aim being to identify the specific and adequate causes. To do this, we must examine some other particular phenomena which take place in connection with these movements given to the water by the acceleration or retardation of the vessel containing it.

[3] The first is that whenever water is made to flow in this or that direction by a noticeable retardation or acceleration of its containing vessel, and it rises here and subsides there, it does not however remain in such a state. Rather, by virtue of its own weight and natural inclination to balance ⟨384⟩ and level itself out, it goes back with speed and seeks the equilibrium of its parts; and, being heavy and fluid, not only does it move toward equilibrium but, carried by its own impetus, it goes beyond and rises at the end where earlier it was lower; not resting here either, it again goes back, and with many and repeated oscillations it indicates that it does not want to change suddenly from the acquired speed to the absence of motion and state of rest, but wishes to do so

gradually and slowly. This is similar to the way in which a pendulum, after being displaced from the perpendicular, spontaneously returns to it and to rest, but not before having gone beyond it many times with a back and forth motion.

The second phenomenon to notice is that the just mentioned reciprocal motions take place and are repeated with greater or lesser frequency, namely in shorter or longer times, depending on the length of the vessels containing the water, that is, depending on the shorter or longer distance from one end of the vessel to the other; thus, the oscillations are more frequent for the shorter distances and more rare for the longer. And this is exactly what happens in the same example of pendulums, where we see that the oscillations of those hanging from a longer string are less frequent than those of pendulums with shorter strings.

And here is a third important point to know: it is not only the greater or lesser length of the vessel that causes the water to make its oscillations in different times; rather, the greater or lesser depth of the vessel and the water brings about the same difference. Thus, for water contained in vessels of equal length but unequal depth, the one which is deeper makes its oscillations in shorter times, and the vibrations of less deep water are less frequent.

Fourth, worthy of notice and diligent observation are two effects of water in such vibrations: one is the alternate rising and falling at both ends; the other is the flowing back and forth, horizontally so to speak. ⟨385⟩ These two different motions affect different parts of the water differently. For its ends are the parts that rise and fall the most; those at the middle do not move up or down at all; and for the rest, gradually those that are nearer the ends rise and fall proportionately more than the farther parts. On the contrary, in regard to the lateral motion back and forth, the middle parts go forth and come back the most, while the water at the ends does not flow at all except insofar as by rising it goes over the embankment and overflows its original bed; but where the embankment stands in the way and can hold it, it only rises and falls; finally, the water in the middle is not the only part that flows back and forth with speed and for large distances, for this is also done proportionately by its other parts, as they flow more or less depending on how near or far they are from the middle.

The fifth detail must be considered much more carefully, insofar as it is at least very difficult, if not impossible, to reproduce it experimentally and practically. The effect is this. In artificial vessels which, like the boats mentioned above, move now more and now less swiftly, the acceleration or retardation is shared to the same extent by the whole vessel and all its parts: thus, for example, as the boat slows down the forward part is not retarded any more than the back, but they all share the same retardation equally; the same happens in acceleration, so that as the boat acquires greater speed the parts in front are not accelerated any more than those in back, but the same speed is acquired fore and aft. All this is the result of the vessel being made of hard and solid substance, which does not deform or flow. However, in very large vessels like the very long basins of the seas, though they are nothing but certain hollows carved out of the solid terrestrial globe, nevertheless amazingly their extremities do not increase or diminish their motion together, equally, and simultaneously; rather it happens that, when one ⟨386⟩ extremity is greatly retarded, in virtue of the composition of the diurnal and annual motions, the other extremity finds itself still experiencing very fast motion. For easier comprehension, let us explain this by referring to the preceding diagram. In it, let us consider, for example, a portion of water spanning a quarter of the globe, such as the arc *BC*; here, as we explained above, the parts at *B* are in very fast motion due to the combination of the diurnal and annual motions in the same direction, whereas the parts at *C* are retarded and without the forward motion deriving from the diurnal rotation. If, then, we take a sea basin whose length equals the arc *BC*, we see how its extremities move simultaneously with great inequality. The differences would be greatest for the speeds of an ocean a hemisphere long, for end *B* would be in very fast motion, the other end *D* would be in very slow motion, and the middle parts at *C* would have an intermediate speed. Further, the shorter a given sea is, the less will it experience this curious effect of having its parts moving at different speeds during certain hours of the day. Thus if, as in one case, we observe acceleration and retardation causing the contained water to flow back and forth, despite the fact that they are shared equally by all parts of the vessel, what must we think will happen in a vessel placed so curiously that its parts acquire retardation and acceleration very unequally? It seems certain we can only say that here we have a greater and more amazing cause of stronger movements in the water. Though many will consider it impossible that we could experiment with the effects of such an arrangement by means of machines and artificial vessels, nevertheless it is not entirely impossible; I have under construction a machine, which I shall explain at the proper time, and in which one can observe in detail the effects of these amazing combinations of

motions. However, regarding the present subject, ⟨387⟩ let us be satisfied with what each can conceive with his imagination so far.

[4] Let us now go on to examine the properties of tides observed in
experience. First, there will be no difficulty saying why it happens that
there are no noticeable tides in ponds, lakes, and even small seas; this
has two corresponding causes. One is that, as the basin acquires different degrees of speed at different hours of the day, because of its smallness they are acquired with very little difference by all its parts, and the
forward as well as the backward parts (namely the eastern and the western) are accelerated and retarded almost at the same time and in the
same way; moreover, since this change occurs gently and gradually, and
not by a sudden obstacle and retardation or an immediate and very
large acceleration in the motion of the containing basin, it as well as all
its parts receive equally and slowly the same degrees of speed; from this
uniformity it follows that the contained water too receives the same
action with little resistance, and consequently it gives hardly any sign of
rising and falling and of flowing hither and thither. The second cause is
the reciprocal vibration of the water stemming from the impetus it receives from the container, which vibration has very frequent oscillations
in small vessels, as we have seen: for the earth's motions can cause agitation in the waters only at twelve-hour intervals, since the motion
of the containing basins is retarded and is accelerated the maximum
amount only once a day, respectively; but the second cause depends on
the weight of the water while in the process of reaching equilibrium,
and it has its oscillations at intervals of one hour, or two, or three, etc.,
depending on the size of the basin; now, mixed with the former cause,
which is very small in small vessels, the latter completely perturbs it and
renders it imperceptible; for, before the end of the operation of the primary cause with the twelve-hour period, ⟨388⟩ the secondary cause due
to the weight of the water emerges, and with its period of one hour,
two, three, or four, etc. (depending on the size and depth of the basin),
it opposes, perturbs, and removes the first, without allowing it to reach
the maximum or the middle of its effect. From this contraposition any
sign of tides remains completely obliterated or much obscured. I say
nothing of the constant accidental alterations due to air; disturbing the
water, they would not allow us to ascertain a very small rise or fall of
half an inch or less, which might actually be taking place in water basins
that are no longer than a degree[7] or two.

Second, I come to resolving the difficulty of how tidal periods can
commonly appear to be six hours, even though the primary cause of the

tides embodies a principle for moving the water only at twelve-hour intervals, that is, once for the maximum speed of motion and the other for maximum slowness. To this I answer, first, that the determination of the periods which actually occur cannot in any way result from the primary cause alone; instead we must add the secondary cause, depending (as we said) on the specific inclinations of water which, once raised at one end of the vessel by virtue of its own weight, flows to reach equilibrium and undergoes many oscillations and vibrations, more or less frequent depending on the smaller or greater length of the vessel and on the greater or smaller depth of the water. My second answer is that the approximately six-hour period commonly observed is no more natural or significant than any other; rather, it is the one which has been observed and described more than others, since it takes place in the Mediterranean Sea around which all our ancient writers and a large part of the moderns have lived. The length of this Mediterranean basin is the secondary cause that gives its oscillations a six-hour period; whereas on the eastern shores of the Atlantic Ocean, which extends to the West Indies, the oscillations have a period of about twelve hours, as one observes daily in Lisbon, located on the far ⟨389⟩ side of Spain;[8] now, this sea, which extends toward the Americas as far as the Gulf of Mexico, is twice as long as the stretch of the Mediterranean from the Strait of Gibraltar to the shores of Syria, that is, 120 degrees for the former and 56 degrees for the latter, approximately. Thus, to believe that tidal periods are six hours is a deceptive opinion and it has led writers to make up many fictional stories.

Third, from what has been said it will not be hard to investigate the reasons for so many inequalities observed in smaller seas, such as the Sea of Marmara and the Hellespont[9] and others, in some of which the waters oscillate every three hours, or every two, or every four, etc.; these differences have much troubled observers of nature as long as, ignorant of the true causes, they have resorted to such useless chimeras as the motions of the moon and other fictions, without ever thinking of considering the different lengths and depths of the seas. As stated above, these are a very powerful factor in determining the periods of oscillation of the waters; indeed, if first we were to ascertain the empirical truth of the matter and what occurs in various seas, and if moreover we had the demonstrations of how oscillations behave in proportion to the length and depth of basins, then it would be very easy to solve all difficulties quickly, especially by combining these secondary causes with the primary and general one stemming from the earth's motion.

Fourth, we can quickly explain why it happens that although some seas are very long, for example the Red Sea, nevertheless they are almost entirely lacking in tides. This occurs because its length does not extend from east to west, but from southeast to northwest. For, the earth's motions being from west to east, the impulses received by the water always cross the meridians and do not move from one parallel to another; thus, in seas which extend transversely in the direction of the poles and which are narrow in the other direction, no cause of tides remains but the contribution of some other sea, with which they are connected and which is subject to large motions.

Fifth, we can very easily understand the reason why, ⟨390⟩ in regard to the rise and fall of the water, tides are greatest at the extremities of gulfs and smallest in the middle. For, as explained above, experience shows us that in its vibrations water does not rise at all in the middle of a containing vessel, and it rises and falls the most at the ends. Thus it happens that at the end of the Adriatic, namely around Venice, tides show a variation of about three feet in height; but in places of the Mediterranean far from the extremities, such as the islands of Corsica and Sardinia, the variation is very small, and at the beaches of Rome and Leghorn[10] it does not exceed half a foot.

Sixth, remembering what has been noted above and what experience places before our eyes, we shall have very readily available the cause of the occurrence that in very large seas, though the rise and fall of the water is very small at the middle, nevertheless the currents now toward the west and now toward the east are very strong. This derives from the very nature of the vibrations of water: the less it rises and falls at the middle, the more it flows back and forth, while at the ends it does just the opposite. Furthermore, considering how the same quantity of water which moves slowly through a wide canal must flow with great speed when passing through a narrow place, we shall have no difficulty in understanding the cause of the immense currents which flow in the narrow channel that separates Sicily from Calabria;[11] for, although all the water contained in the eastern Mediterranean and bound by the width of the island and the Ionian Gulf may slowly flow into it toward the west, nevertheless, when constricted into this strait between Scylla and Charybdis,[12] it undergoes very great agitation. Similar to this and much greater, we understand, are the currents between Africa and the very large island of Madagascar, as the waters of the Indian and South Atlantic oceans which surround it flow and become constricted in the smaller channel between it and the South African coast. Very great and immense must be the currents in the Strait of Magellan, which connects the extremely vast South Atlantic and South Pacific oceans.

Seventh, to account for some more obscure and implausible phenomena observed in this subject, we have to make another very important consideration about the two ⟨391⟩ principal causes of tides, and then mix them together. The first and simpler of these is the definite acceleration and retardation of the earth's parts, dependent on the union of the annual and diurnal motions; this variation has a very fixed period, insofar as there is maximum acceleration at a certain time, maximum retardation at another time, and a consequent speedy transition from one to the other, and insofar as these changes take place in twenty-four hours. The other reason is the one that depends on the mere weight of water: stirred at first by the primary cause, it then tries to reach equilibrium by repeated oscillations; these are not determined by a single period in advance, but they have as many temporal differences as the different lengths and depths of sea basins; so it follows that, insofar as it depends on this second principle, some seas might flow back and forth in one hour, others in two, in four, six, eight, ten, etc. Now let us begin to join together the primary cause, whose fixed period between flow in one direction and in the opposite is twelve hours, with one of the secondary causes whose period is, for example, five hours: sometimes it will happen that the primary and secondary causes agree by both producing impulses in the same direction, and with such a combination (a unanimous consent, so to speak) the tides are large; other times, the primary impulse being somehow opposite to that of the secondary cause, and thus one principle taking away what the other one gives, the watery motions weaken a great deal and we have the sea in a state of exhaustion; on still other occasions, when the same two causes neither oppose nor reinforce each other, there are other variations in the increase or decrease of the tides. It may also happen that of two very large seas connected by a narrow channel, due to the mixture of the two principles, one sea has tidal motions in one direction while the other has them in the opposite; in this case, in the channel where the two seas meet there are terrible agitations with contrary motions, ⟨392⟩ vortices, and very dangerous boilings, as it is constantly observed and reported. These conflicting motions, dependent on the different positions and lengths of interconnected seas and on their different depths, give rise sometimes to those irregular disturbances of the water whose causes have worried and continue to worry sailors, who experience them without seeing any winds or other atmospheric disturbances that might produce them. These atmospheric disturbances must be taken into account in other cases, and we must regard them as a tertiary accidental cause capable of significantly altering the occurrence of the effects produced by the primary and more important causes. For there is no doubt that, for example, very strong winds from the east can support the water and prevent it from ebbing; thus, when at the appropriate time there is a second wave of flow (and a third), it will rise a great deal (much more than usual), and, if kept up by the strength of the wind for a day or two, it will produce destructive flooding.

Eighth, we must also note another cause of motion dependent on the great quantity of river waters flowing into seas that are not very large. Here, in channels or straits connected with such seas, the water is seen flowing always in the same direction. For example, this happens at the Bosphorus near Constantinople,[13] where the water always flows from the Black Sea to the Sea of Marmara. For in the Black Sea, because of its smallness, the principal causes of tides are of little effect; on the contrary, many very large rivers flow into it, such as the Danube, the Dnieper, the Don (through the Sea of Azov), and others; so, with such a superabundance of water having to go through the strait, here the flow is very noticeable and always southward. Moreover, we must note that, though this strait is very narrow, it is not subject to perturbations like those in the Strait of Messina. For the former has the Black Sea on the north and the Sea of Marmara, the Aegean, and the Mediterranean on the south, and (as we have already said) when a sea is long in a north-south direction it is not subject ⟨393⟩ to tides; on the contrary, because the Strait of Messina is interposed between two parts of the Mediterranean that extend for long distances in an east-west direction, namely in the direction of tidal currents, in the latter the disturbances are very large. Similarly, they would be extremely large between the Pillars of Hercules[14] if the Strait of Gibraltar were less wide, and they are reported to be immeasurable in the Strait of Magellan.

[5] This was what I advanced as the cause of these motions of the sea in my discussion with you, Most Eminent Lord. It was an idea which seemed to harmonize mutually the earth's motion and the tides, taking the former as cause of the latter, and the latter as a sign of and an argument for the former. Now, I remember telling you in our discussion that, in regard to the same motion, besides many signs being provided by the movements of heavenly bodies, others were furnished us by the elements, namely water and air; thus I think you will not be displeased if, for your memory, I also briefly note what I explained to you about the other argument pertaining to the air.

As a fluid and tenuous body not firmly attached to the earth, air does not seem bound to obey its motion, except insofar as the roughness and unevenness of the earth's surface carries away and takes with it the part next to it. We may believe this is the part that does not extend much above the tallest mountains. This portion of the air is all the more inclined toward the earth's turning, insofar as it is full of vapors, fumes, and exhalations, and these are all elemental substances inclined by their nature toward the same terrestrial motions. However, wherever the causes of motion are lacking, that is, wherever the earth's surface has large flat spaces and the air contains less of a mixture of earthly vapors, the cause that would make the ambient air conform completely to the earth's turning is partially ineffective. Thus, in such places, while the earth turns toward the east, one should constantly feel a breeze blowing from east to west, and such a wind ⟨394⟩ should be more noticeable where the terrestrial turning is fastest; this is in places farthest from the poles and close to the equatorial circle. But in fact experience seems to approve highly this philosophical discourse. For, in those parts of large seas that are far from land and inside the torrid zone, namely within the tropics, one feels a perpetual breeze blowing from the east; this is so constant that with it ships easily and readily go to the West Indies, and from there, sailing from the Mexican shores, with the same ease they move through the Pacific Ocean toward the East Indies (which are to the east for us but to the west from there). On the contrary, navigation toward the east is difficult and uncertain; one cannot sail always by the same routes, but must stay closer to land in order to find other accidental and occasional winds, caused by other principles, with which we inhabitants of terra firma are quite familiar. The causes of such winds are numerous and diverse, and for now there is no need to discuss them; these accidental winds are the ones which blow indifferently in all directions, and which disturb the smaller seas and those enclosed within a continent, thus facilitating navigation therein. But let us come back to seas which are far from the equator and surrounded by the rough surface of the earth, namely seas subject to those atmospheric disturbances upsetting the primary wind that should be constantly felt if these accidental impediments did not exist. Although it may seem that in these seas of ours navigation can indifferently take place with equal ease toward the east and toward the west, nevertheless, whoever takes the appropriate care will find that navigation toward the west is generally much easier and shorter. For I know from the diligent records Venetian merchants keep of the days of departure and arrival of ships to and from Alexandria and Syria, that if you average out the figures for one or more years, the times for return trips are more than twenty-five percent shorter than those of the outgoing voyages; this is a manifest sign that on the whole the easterly winds prevail over the westerly ones. Thus, the existence of a constant blowing of an easterly breeze around the terrestrial globe, especially near the equator and where the surface is even (like at sea), seems to agree with the earth's motion no less probably ⟨395⟩ than the many phenomena relating to the tides; this is especially clear if we make a comparison with the inane explanations proffered thus far by other authors to account for these same effects.

I could propose many other considerations if I wanted to delve into finer details. Many, many more could be advanced if we had abundant, clear, and truthful empirical reports of observations made by competent and diligent men in various places of the earth; for by comparing and collating them with the assumed hypothesis we could decide more firmly and ascertain more correctly the things that pertain to this very obscure subject. At the moment I only claim to have given something of a sketch, suitable at least for stimulating students of nature to reflect on this new idea of mine. I hope, however, that it does not turn out to be delusive, like a dream which gives a brief image of truth followed by an immediate certainty of falsity. This I submit to the judgment of intelligent investigators.

Finally, one last conclusion and seal for this short discourse of mine. If the given hypothesis, previously corroborated only by philosophical and astronomical reasons and observations, were declared fallacious and erroneous by virtue of more eminent knowledge, then one should not only call into question what I have written but consider it completely empty and out of place. Regarding the problems discussed, we should then either wish that those who showed the fallacy of our account would present their own or the correct reasons, or else regard these topics as such that the blessed God wants to conceal their knowledge from the human mind, or, finally and more advisedly, remove ourselves from these and other fruitless inquiries which take a great deal of time that could or should be more usefully spent on more beneficial studies. And now I reverently kiss your robe, and I humbly implore you for your goodwill.

Written in Rome, at Villa Medici, on 8 January 1616.

The Earlier Inquisition Proceedings (1615–1616)

Lorini's[1] Complaint (7 February 1615)[2]

Most Illustrious and Most Reverend Lord:[3]

Besides the common duty of every good Christian, there is a limitless obligation that binds all Dominican friars, since they were designated by the Holy Father the black and white hounds of the Holy Office. This applies in particular to all theologians and preachers, and hence to me, lowest of all and most devoted to Your Most Illustrious Lordship. I have come across a letter[4] that is passing through everybody's hands here, originating among those known as "Galileists," who, following the views of Copernicus, affirm that the earth moves and the heavens stand still. In the judgment of all our Fathers at this very religious convent of St. Mark, it contains many propositions which to us seem either suspect or rash: for example, that certain ways of speaking in the Holy Scripture are inappropriate; that in disputes about natural effects the same Scripture holds the last place; that its expositors are often wrong in their interpretations; that the same Scripture must not meddle with anything else but articles concerning faith; and that, in questions about natural phenomena, philosophical or astronomical argument has more force than the sacred and the divine one. Your Most Illustrious Lordship can see these propositions underlined by me in the above-mentioned letter, of which I send you a faithful copy.[5] Finally, it claims that when Joshua ordered the sun to stop one must understand that the order was given to the Prime Mobile and not to the sun itself. Besides this letter passing through everybody's hands, without being stopped by any of the authorities, it seems to me that some want to expound Holy Scripture in their own way and against the common exposition of the Holy Fathers and to defend ⟨298⟩ an opinion apparently wholly contrary to Holy Scripture. Moreover, I hear that they speak disrespectfully of the ancient Holy Fathers and St. Thomas; that they trample underfoot all of Aristotle's philosophy, which is so useful to scholastic theology; and that to appear clever they utter and spread a thousand impertinences around our whole city, kept so Catholic by its own good nature and by the vigilance of our Most Serene Princes. For these reasons I resolved, as I said, to send it to Your Most Illustrious Lordship, who is filled with the most holy zeal and who, for the position that you occupy, is responsible, together with your most illustrious colleagues, for keeping your eyes open in such matters; thus if it seems to you that there is any need for correction, you may find those remedies that you judge necessary, in order that a small error at the beginning does not become great at the end. Though perhaps I could have sent you a copy of some notes on the said letter made at this convent, nevertheless, out of modesty I refrained since I was writing to you who know so much and to Rome where, as St. Bernard said, the holy faith has lynx eyes. I declare that I regard all those who are called Galileists as men of goodwill and good Christians, but a little conceited[6] and fixed in their opinions; similarly, I state that in taking this action I am moved by nothing but zeal.[7] I also beg Your Most Illustrious Lordship that this letter of mine (I am not referring to the other letter mentioned above)[8] be kept secret by you, as I am sure you will, and that it be regarded not as a judicial deposition but only as a friendly notice between you and me, as between a servant and a special patron. And I also inform you that the occasion of my writing was one or two public sermons given in our church of Santa Maria Novella by Father Tommaso Caccini,[9] commenting on the book of Joshua and chapter 10 of the said book. So I close by asking for your holy blessing, kissing your garment, and asking for a particle of your holy prayers.[10]

Consultant's Report on the Letter to Castelli (1615)[11]

In the composition shown me today,[12] except for these three following items, I found nothing else to question.

On the first page, where it says "that in the Holy Scripture one finds many propositions which are false if one goes by the literal meaning of the words,"[13] etc., granted that this sentence can be taken in a benign sense, nevertheless at first impression it sounds bad. Certainly it is not right to use the word falsehood, in whatever manner it be attributed to Holy Scripture; for it is the infallible truth in every way.

As for the other item, on the second page where it says that "Holy Scripture has not abstained from perverting its most basic dogmas,"[14] etc., since the verbs "to abstain" and "to pervert" are always employed with negative connotations (we abstain from evil, and one perverts when something just is made unjust), they sound bad when they are attributed to Holy Scripture.

Bad-sounding, too, are those words on the fourth page that read, "Let us then assume and concede,"[15] etc., for in this statement one sees only a desire to concede the story of the sun stopped by Joshua as truly in the text of Holy Scripture, although from what follows those words can be understood in the right sense.

For the rest, though it sometimes uses improper words, it does not diverge from the pathways of Catholic expression.

Caccini's Deposition (20 March 1615)

Friday, 20 March 1615.

There appeared personally and of his own accord at Rome in the great hall of examinations in the palace of the Holy Office, in the presence of the Reverend Father Michelangelo Segizzi, O.P., Master of Sacred Theology and Commissary General of the Holy Roman and Universal Inquisition, etc., the Reverend Father Tommaso Caccini, son of the late Giovanni Caccini, Florentine, a professed priest of the Order of Preachers, Master and Bachelor from the convent of Santa Maria sopra Minerva in Rome, about thirty-nine years of age. Having been administered the oath to tell the truth, he declared as follows:

I had spoken with the Most Illustrious Lord Cardinal Aracoeli[16] about some things taking place in Florence, and yesterday he sent for me and told me that I should come here and tell you everything. Since I was told that a legal deposition is needed, I am here for this purpose. I say then that on the fourth Sunday of Advent of this past year I was preaching at the church of Santa Maria Novella in Florence, where I had been assigned by superiors this year as a reader of Holy Scripture, and I continued with the story of Joshua begun earlier. Precisely on this

Sunday I happened to read the passage of the tenth chapter of that book where the sacred writer relates the great miracle which God made in answer to Joshua's prayers by stopping the sun, namely "Sun, stand thou still upon Gibeon,"[17] etc. After interpreting this passage first in a literal sense and then in accordance with its spiritual intention for the salvation of souls, I took the opportunity to criticize, with that modesty which befits the office I held, a certain view once proposed by Nicolaus Copernicus and nowadays held and taught by Mr. Galileo Galilei, mathematician, according to public opinion very widespread in the city of Florence. This is the view that the sun, being ⟨308⟩ for him the center of the world, is immovable as regards progressive local motion, that is, motion from one place to another. I said that such a view is regarded as discordant with Catholic faith by very serious writers since it contradicts many passages of the divine Scripture whose literal sense, as given unanimously by the Holy Fathers, sounds and means the opposite; for example, the passage of the 18th Psalm, of the first chapter of Ecclesiastes, of Isaiah 38, besides the Joshua passage cited. And in order to impress upon the audience that such a teaching of mine did not originate from my whim, I read them Nicolaus Serarius's doctrine (fourteenth question on chapter 10 of Joshua): after saying that such a position of Copernicus is contrary to the common account of almost all philosophers, all scholastic theologians, and all the Holy Fathers, he added that he could not see how such an opinion is not almost heretical, due to the above-mentioned passages of Scripture. After this discussion I cautioned them that no one was allowed to interpret divine Scripture in a way contrary to the sense on which all the Holy Fathers agree, since this was prohibited both by the Lateran Council under Leo X and by the Council of Trent.[18]

Although this charitable warning of mine greatly pleased many educated and devout gentlemen, it displeased certain disciples of the above-mentioned Galilei beyond measure; thus some of them approached the preacher at the cathedral so that he would preach on this topic against the doctrine I expounded. Having heard so many rumors, out of zeal for the truth, I reported to the very reverend Father Inquisitor of Florence what my conscience had led me to discuss concerning the Joshua passage; I also suggested to him that it would be good to restrain certain petulant minds, disciples of the said Galilei, of whom the reverend Father Fra Ferdinando Ximenes,[19] regent of Santa Maria Novella, had told me that from some of them he had heard these three propositions: "God is not otherwise a substance, but an accident"; "God is sensuous because there are in him divine senses"; and "in truth the miracles said to have been made by the saints are not real miracles."

After these events Father Master Fra Niccolò Lorini showed me a copy of a letter written by the above-mentioned Mr. Galileo Galilei to Father Benedetto Castelli, Benedictine monk and professor of mathematics at Pisa, in which it seemed to me are contained questionable doctrines in the domain of theology. Since a copy of it was sent to the Lord Cardinal of Santa Cecilia,[20] I have nothing else to add to that.

Thus I declare to this Holy Office that it is a widespread opinion that the above-mentioned Galilei holds these two propositions: the earth moves as a whole as well as with diurnal motion; the sun is motionless. These are propositions which, according to my ⟨309⟩ conscience and understanding, are repugnant to the divine Scripture expounded by the Holy Fathers and consequently to the faith, which teaches that we must believe as true what is contained in Scripture. And for now I have nothing else to say.

He was asked: How he knows that Galileo teaches and holds the sun to be motionless and the earth to move, and whether he learned this expressly from others.

He answered: Aside from public notoriety, as I said before, I also heard from Mons. Filippo de' Bardi, Bishop of Cortona, at the time I stayed there and then in Florence, that Galilei holds the above-mentioned propositions as true; he added that this seemed to him very strange, as not agreeing with Scripture. I also heard it from a certain Florentine gentleman[21] of the Attavanti family, a follower of the same Galilei, who said to me that Galilei interpreted Scripture in such a way as not to conflict with his opinion. I do not recall this gentleman's name, nor do I know where his house is in Florence; I am sure that he often comes to service at Santa Maria Novella in Florence, that he wears priest's clothes, and that he is twenty-eight or thirty years of age perhaps, of olive complexion, chestnut beard, average height, and sharply delineated face. He told it to me this past summer, about the month of August, in Father Ferdinando Ximenes' room, the occasion being that Father Ximenes was telling me that I should not take too long discussing the miracle of the stopping of the sun when he (Ximenes) was around. I have also read this doctrine in a book printed in Rome, dealing with sunspots,[22] published under the name of the said Galileo, and lent to me by the said Father Ximenes.

Q:[23] Who the preacher at the cathedral is, to whom Galileo's disciples went in order to have a public sermon against the doctrine taught equally publicly by the plaintiff himself, and who those disciples are who made such a request to the said preacher.

A: The preacher at the Florence cathedral whom Galileo's disciples approached about preaching against the doctrine I taught is a Jesuit Father from Naples, whose name I do not know. Nor have I learned these things from the said preacher, since I did not even speak with him. Rather they have been told me by Father Emanuele Ximenes,[24] a Jesuit, whom the said preacher had asked for advice, and who dissuaded him. Nor do I know who were the disciples of Galilei who contacted the preacher about the above-mentioned matters.

Q: Whether he has ever talked to the said Galileo.

A: I do not even know what he looks like.

Q: What the reputation of the said Galileo is in the city of Florence regarding matters of faith.

A: By many he is regarded as a good Catholic. By others he is regarded with suspicion in matters of faith because they say he is very close to Fra Paolo,[25] of the Servite order, so ⟨310⟩ famous in Venice for his impieties; and they say that letters are exchanged between them even now.

Q: Whether he remembers from which person or persons he learned about these matters.

A: I heard these things from Father Master Niccolò Lorini and from another Mr. Ximenes,[26] Prior of the Knights of Santo Stefano. They told me the above-mentioned things. That is, Father Niccolò Lorini has repeated to me several times and even written to me here in Rome that between Galileo and Master Paolo there is an exchange of letters and great friendship and that the latter is a suspect in matters of faith. And Prior Ximenes did not tell me anything different about the closeness between Master Paolo and Galileo, but only that Galilei is a suspect and that, while being in Rome once, he learned how the Holy Office was trying to seize him, on account of which he ran away. This was told me in the room of the above-mentioned Father Ferdinando, his cousin, though I do not remember exactly if the said Father was present there.

Q: Whether he learned from the above-mentioned Father Lorini and the Knight Ximenes why they regarded the said Galileo to be suspect in matters of faith.

A: They did not say anything else to me, except that they regarded

him as suspect on account of the propositions he held concerning the
immobility of the sun and the motion of the earth, and because this man
wants to interpret Holy Scripture against the common meaning of the
Holy Fathers.

He added on his own: This man, together with others, belongs to an
Academy—I do not know whether they organized it themselves—
which has the title of "Lincean."[27] And they correspond with others in
Germany, at least Galileo does, as one sees from that book of his on
sunspots.[28]

Q: Whether he had been told himself in detail by Father Ferdinando
Ximenes the persons from whom he learned about those propositions,
that God is not a substance but an accident, that God is sensuous, and
that the miracles of the Saints are not true miracles.

A: I seem to remember that he gave the name of Attavanti, whom I
have described as one of those who uttered the said propositions. I do
not remember any others.

Q: Where, when, in the presence of whom, and on what occasion
Father Ferdinando related that Galilei's disciples had mentioned to him
the said propositions.

A: It was on several occasions (sometimes in the cloister, sometimes
in the dormitory, sometimes in his cell) that Father Ferdinando told me
he had heard the said propositions from Galileo's disciples; he did this
after I had preached that sermon, the occasion being that of telling me
that he had defended me against these people. And I do not remember
that there ever was anyone else present.

Q: About his hostility toward the said Galileo, toward the Attavanti
character, and also toward other disciples of the said Galileo.

A: Not only do I not have any hostility toward the said Galileo, but
I do not even know him. Similarly, I do not have any hostility or hatred
toward Attavanti, or toward other disciples of Galileo. Rather I pray to
God for them. ⟨311⟩

Q: Whether the said Galileo teaches publicly in Florence, and what
discipline; and whether his disciples are numerous.

A: I do not know whether Galileo lectures publicly, nor whether he
has many disciples. I do know that in Florence he has many followers
who are called Galileists. They are the ones who extol and praise his
doctrine and opinions.

Q: What home town the said Galileo is from, what his profession is,
and where he studied.

A: He regards himself as a Florentine, but I have heard that he is a

Pisan. His profession is that of mathematician. As far as I have heard, he studied in Pisa and has lectured at Padua. He is past sixty years old.

With this he was dismissed, having been bound to silence by oath and his signature having been obtained.

I, Fra Tommaso Caccini, bear witness to the things said above.

Ximenes' Deposition (13 November 1615)

⟨316⟩ 13 November 1615.

With regard to the order in the letter by the Roman Congregation of the Holy Office, dated in Rome the seventh of this month of November of the year 1615,[29] in the presence of the very reverend Father Master Lelio Marzari of Faventia, Inquisitor-General of the city of Florence and its possessions, etc.

Summoned, there appeared personally the reverend Father Master Ferdinando Ximenes, a professed priest of the Order of Preachers, forty years of age, who was sworn to tell the truth and declared the following.

He was asked: Whether he knows the cause of his summons.

He answered: No, Father.

Q:[30] Whether he is acquainted with a certain doctor residing in Florence named Galileo, and what he hears about him.

A: I have never seen him in the two years that I have been in Florence. But according to what I have heard about the opinion of the earth's motion and heaven's immobility, and what I have heard from those who talk to him, I say that it is a doctrine diametrically opposed to true theology and philosophy.

He was told to explain his assertion more clearly.

A: I have heard some of his students say that the earth moves and heaven is motionless; they have added that God is an accident; that there is no such thing as the substance of things nor is there continuous quantity, but everything is a discrete quantity and contains empty spaces; that God is sensuous, that he laughs, that he cries. But I do not know whether this is their own opinion or the opinion of their teacher Galileo mentioned above.

Q: Whether he has heard either Galileo himself or one of his disciples say specifically that the miracles attributed to the Saints are not true miracles.

A: I have no recollection of this particular point.

Q: From which one or ones in particular of Galileo's disciples he has heard that the earth moves and heaven stands still; that God is an accident; that things do not have a substance or continuous quantity, but rather discrete quantity and empty spaces; and that God is sensuous and subject to laughter and crying.

A: I heard the above-mentioned things and discussed them with the rector of Castelfiorentino, named Giannozzo Attavanti, a Florentine, at which discussions a certain Mr. Ridolfi, a Florentine and a Knight of Santo Stefano, was present.

Q: What the place, the time, the witnesses, and the occasion were.

A: Regarding the place, it was in my room, here at the convent of Santa Maria Novella; the time was last year, many times, but I could not say what month or ⟨317⟩ day; witnesses present were the above-mentioned knight, sometimes, and some of our friars whom I do not remember exactly.

Q: Whether from the words of that rector he was able to infer that said rector was speaking in earnest and thus was believing, asserting, and regarding such things to be true.

A: I do not think that the said rector Attavanti was saying and believing the above-mentioned things categorically since it seems to me he himself said that he deferred to the Church and was saying everything for the sake of the argument.

Q: Whether he has any specific information about the said rector Attavanti so as to be able to say that the latter is knowledgeable and was speaking argumentatively rather than categorically.

A: I know he has no grounding either in theology or in philosophy, and I think he is not a doctor. But I detected that he has, as it were, a smattering of the one and of the other; and I think that he was expressing more Galileo's opinion than his own. The occasion was that I was discussing with the said Attavanti problems of conscience, and we got involved arguing about some of Father Master Caccini's sermons. At that time, the latter was a preacher on the Holy Scripture here in our church of Santa Maria Novella and was explaining the story of Joshua and, among others, the words "the sun stood still." This was the occasion for our coming to discuss the above-mentioned things.

Q: Whether he rebuked the said rector Attavanti for his wrong beliefs and false arguments, and what the said rector may have answered.

A: I rebuked him most instantly and led him to see that the things being stated and discussed were false and heretical. For the truth is that the earth as a whole is motionless and founded upon its immobility, as the prophet says;[31] that heaven and the sun move; that God is a substance and not an accident, which indeed cannot be otherwise; and that it is nonsense to say that God is sensuous, laughs, and cries and that there is nothing but discrete quantity mixed with empty spaces.

Q: About any hostility either toward the said Galileo or toward the rector Attavanti.

A: I have never seen the said Galileo, as I stated above, nor have I ever had anything to do with him. Nor have I ever had any hostility toward the said rector Attavanti, but rather we have been friends. To be sure, I dislike the doctrine of the said Galileo since it is not in accordance with the orthodox Fathers of the Holy Church, or rather it goes against truth itself.

Q: Whether he wishes to declare anything else while testifying before the Holy Office.

A: I have nothing else to say, and what I have said above is all true.

With this the said Father was dismissed, under an oath of silence, after obtaining his signature.

Fra Ferdinando Ximenes, Master etc.

⟨318⟩ Recorded in Florence in the hall of the very reverend Father Inquisitor, by me, Fra Lodovico Iacoponi of Interamna, Chancellor of the Florentine Holy Office.

Attavanti's Deposition (14 November 1615)

14 November 1615.

Summoned in accordance with the above-mentioned letter,[32] there appeared personally, in the presence of the same person and at the same place as above, etc., the Reverend Lord Giannozzo Attavanti, Florentine nobleman, rector of Castelfiorentino, initiated into minor orders, thirty-three years of age, named as a witness, who was sworn to tell the truth and declared the following.

He was asked: Whether he knows the cause of his summons.

He answered: I do not know anything about it.

Q:[33] Whether he has studied letters here in Florence, and under which teachers.

A: I have studied letters in years past, and my teachers were Father Vincenzo da Civitella and Father Vincenzo Populeschi, both Dominicans.

Q: Whether he has had other teachers, especially laymen.

A: While I studied grammar and humanities, I was taught by Mr.

Simone dalla Rocca and Mr. Gio. Batta, nowadays a teacher of our Princes. And it is a year ago that Father Ximenes, O.P., gave me lessons on the casuistry of conscience.

Q: Whether he has any knowledge of a certain doctor, living here in Florence, whose name is Galileo Galilei, and learned letters from him.

A: I have never studied under him as his pupil. I have indeed discussed scholarly topics with him, as I ordinarily do with men of letters; in particular, I have discussed philosophical matters with him.

Q: Whether he has ever heard from Galileo, either in lecture or in dialogue, anything opposed to and not agreeing with Holy Scripture, philosophy, or our faith.

A: I have never heard Mr. Galileo say things that conflict with Holy Scripture or with our holy Catholic faith. But in regard to philosophical and mathematical matters, I have heard Mr. Galileo say, in accordance with Copernicus's doctrine, that the earth moves both around its center and as a whole, and the sun likewise moves around its center but (viewed from outside) does not have progressive motion, according to some letters published by him in Rome under the title *On Sunspots*, to which I refer in all this.

Q: Whether he has ever heard the above-mentioned Galileo interpret Holy Scripture, perhaps wrongly, in accordance with his opinion of the earth's motion and the sun's standing still. ⟨319⟩

A: I have heard him argue about Joshua's passage "the sun stood still upon Gibeon,"[34] in regard to which he admits that the sun stopped miraculously, but that nonetheless (viewed from the outside) it does not move with a progressive motion.

Q: Whether he has heard the above-mentioned Galileo assert that God is not a substance but an accident; also that God is sensuous, laughs, and cries, and in what way; and that the miracles attributed to the Saints are not true miracles.

A: About these particular things you should know the following, Father. One day I happened to be talking to Father Ferdinando Ximenes, O.P., in his room at Santa Maria Novella here in Florence, the discussion being in the manner of a disputation and for the sake of learning, and the topic being St. Thomas's ideas on whether God is a substance or an accident, and what St. Thomas argued in the *Contra Gentes*, whether God is sensuous, laughs, cries, etc. As I said, it was in the manner of a disputation, and not otherwise. Now, a certain Father Caccini, also a Dominican, at the time a preacher at Santa Maria Novella, was staying in a room near the one of the said Father Ximenes;

hearing us argue with each other in the manner of a disputation, perhaps he imagined that I was reporting the above-mentioned things as asserted or believed by the said Mr. Galileo, but this is not true. In regard then to the miracles of the Saints, we did not discuss them in any way, and I do not know anything on the matter. Thus we determined, in accordance with St. Thomas's doctrine, that God is not sensuous, nor does he laugh, nor cry, since otherwise he would be an organic body, which is false; rather he is a most simple substance.

Q: What made him think of this and of mentioning the said Father Caccini, as above, that is, as being wrong about the argument between the witness and the said Father Ferdinando Ximenes.

A: I mentioned the said Father Caccini as above because of another previous incident: I was discussing with the said Father Ximenes again in his room the topic of the sun's motion, and when the said Father Caccini heard us he left his room and came to us; he said that it was a heretical proposition to assert that the sun stands still and its center does not move, in accordance with Copernicus's opinion, and that he wanted to preach about it from the pulpit, as it later happened.

Q: What the degree of certainty, the place, the time, other witnesses, and the occasion were.

A: As I said above, I know it with certainty and through my own hearing. The place was Father Ximenes' room. The time was the month of August or July of the year 1613, but I do not recall the exact day. There was no one else present besides the said Father Ximenes and myself. The occasion was that I was learning the casuistry of conscience from the said Father Ximenes, and so we came upon the above-mentioned arguments, in the manner of a disputation and for the sake of learning, and not otherwise.

Q: What he hears about the same above-mentioned Galileo concerning the faith.

A: I regard him as a very good Catholic; otherwise he would not be so close to these Most Serene Princes.

Q: About any hostility, ill-will, or hatred toward the said Father Caccini. ⟨320⟩

A: I have never spoken with him before or after those times; I have no dealings with him; and I do not even know his first name.

Q: Whether he wishes to declare anything else while testifying before the Holy Office.

A: I have nothing else to say, and what I said above is the pure and simple truth.

With this the above-mentioned witness was dismissed, under an oath of silence, after obtaining his signature.

I, Giannozzo Attavanti, confirm the above.

Recorded in Florence in the hall of the very reverend Father Inquisitor, by me, Fra Lodovico Iacoponi of Interamna, Chancellor of the Florentine Holy Office.

The present copy agrees word for word with the original.

Fra Lodovico Iacoponi,
Chancellor of the Florentine Holy Office.

Consultants' Report on Copernicanism
(24 February 1616)

Assessment made at the Holy Office, Rome, Wednesday, 24 February 1616, in the presence of the Father Theologians signed below.

Propositions to be assessed: ⟨321⟩

(1) The sun is the center of the world and completely devoid of local motion.

Assessment: All said that this proposition is foolish and absurd in philosophy,[35] and formally heretical since it explicitly contradicts in many places the sense of Holy Scripture, according to the literal meaning of the words and according to the common interpretation and understanding of the Holy Fathers and the doctors of theology.

(2) The earth is not the center of the world, nor motionless, but it moves as a whole and also with diurnal motion.

Assessment: All said that this proposition receives the same judgment in philosophy and that in regard to theological truth it is at least erroneous in faith.

Petrus Lombardus, Archbishop of Armagh.[36]

Fra Hyacintus Petronius, Master of the Sacred Apostolic Palace.

Fra Raphael Riphoz, Master of Theology and Vicar-General of the Dominican Order.

Fra Michelangelo Segizzi, Master of Sacred Theology and Commissary of the Holy Office.

Fra Hieronimus de Casalimaiori, Consultant to the Holy Office.

Fra Thomas de Lemos.

Fra Gregorius Nunnius Coronel.

Benedictus Justinianus,[37] Society of Jesus.

Father Raphael Rastellius, Clerk Regular, Doctor of Theology.

Father Michael of Naples, of the Cassinese Congregation.

Fra Iacobus Tintus, assistant of the Most Reverend Father Commissary of the Holy Office.

Inquisition Minutes (25 February 1616) [38]

Thursday, 25 February 1616.

The Most Illustrious Lord Cardinal Millini [39] notified the Reverend Fathers Lord Assessor and Lord Commissary of the Holy Office that, after the reporting of the judgment by the Father Theologians against the propositions of the mathematician Galileo (to the effect that the sun stands still at the center of the world and the earth moves even with the diurnal motion), His Holiness ordered the Most Illustrious Lord Cardinal Bellarmine to call Galileo before himself and warn him to abandon these opinions; and if he should refuse to obey, the Father Commissary, in the presence of a notary and witnesses, is to issue him an injunction to abstain completely from teaching or defending this doctrine and opinion or from discussing it; and further, if he should not acquiesce, he is to be imprisoned.

Special Injunction (26 February 1616)

Friday, the 26th of the same month.

At the palace of the usual residence of the said Most Illustrious Lord Cardinal Bellarmine and in the chambers of His Most Illustrious Lordship, ⟨322⟩ and fully in the presence of the Reverend Father Michelangelo Segizzi of Lodi, O.P. and Commissary General of the Holy Office, having summoned the above-mentioned Galileo before himself, the same Most Illustrious Lord Cardinal warned Galileo that the above-mentioned opinion was erroneous and that he should abandon it; and thereafter, indeed immediately,[40] before me and witnesses, the Most Illustrious Lord Cardinal himself being also present still, the aforesaid Father Commissary, in the name of His Holiness the Pope and the whole Congregation of the Holy Office, ordered and enjoined the said Galileo, who was himself still present, to abandon completely the above-mentioned opinion that the sun stands still at the center of the world and the earth moves, and henceforth not to hold, teach, or defend it in any way whatever, either orally or in writing; otherwise the Holy Office would start proceedings against him. The same Galileo acquiesced in this injunction and promised to obey.

Done in Rome at the place mentioned above, in the presence, as witnesses, of the Reverend Badino Nores of Nicosia in the kingdom of Cyprus and Agostino Mongardo from the Abbey of Rose in the diocese of Montepulciano, both belonging to the household of the said Most Illustrious Lord Cardinal.

Inquisition Minutes (3 March 1616)

The Most Illustrious Lord Cardinal Bellarmine having given the report that the mathematician Galileo Galilei had acquiesced when warned of the order by the Holy Congregation to abandon the opinion which he held till then, to the effect that the sun stands still at the center of the spheres but the earth is in motion, and the Decree of the Congregation of the Index[41] having been presented, in which were prohibited and suspended, respectively, the writings of Nicolaus Copernicus *On the Revolutions of the Heavenly Spheres*, of Diego de Zuñiga *On Job*, and of the Carmelite Father Paolo Antonio Foscarini, His Holiness ordered that the edict of this suspension and prohibition, respectively, be published by the Master of the Sacred Palace.

Decree of the Index (5 March 1616)

Decree of the Holy Congregation of the Most Illustrious Lord Cardinals especially charged by His Holiness Pope Paul V and by the Holy Apostolic See with the Index of books and their licensing, prohibition, correction, and printing in all of Christendom. To be published everywhere.

In regard to several books containing various heresies and errors, to prevent the emergence of more serious harm throughout Christendom, the Holy Congregation of the Most Illustrious Lord Cardinals in charge of the Index has decided that they should be altogether condemned and prohibited, as indeed with the present decree it condemns and prohibits them, wherever and in whatever language they are printed or about to be printed. It orders that henceforth no one, of whatever station or condition, should dare print them, or have them printed, or read them, or have them in one's possession in any way, under penalty specified in the Holy Council of Trent and in the Index of prohibited books; and under the same penalty, whoever is now or will be in the future in possession of them is required to surrender them to ordinaries[42] or to inquisitors, immediately after learning of the present decree. The books are listed below:

Calvinist Theology (in three parts) by Conradus Schlusserburgius.

Scotanus Redivivus, or Erotic Commentary in Three Parts, etc. ⟨323⟩

Historical Explanation of the Most Serious Question in the Christian Churches Especially in the West, from the Time of the Apostles All the Way to Our Age by Jacobus Usserius,[43] professor of sacred theology at the Dublin Academy in Ireland.

Inquiry Concerning the Pre-eminence Among European Provinces, Conducted at the Illustrious College of Tübingen, in 1613 A.D., by Fridericus Achillis, Duke of Wittenberg.

Donellus's Principles, or Commentaries on Civil Law, Abridged so as . . . , etc. by Hugo Donellus.

This Holy Congregation has also learned about the spreading and acceptance by many of the false Pythagorean doctrine, altogether contrary to the Holy Scripture, that the earth moves and the sun is motionless, which is also taught by Nicolaus Copernicus's *On the Revolutions of the Heavenly Spheres* and by Diego de Zuñiga's *On Job*. This may be seen from a certain letter published by a certain Carmelite Father, whose title is *Letter of the Reverend Father Paolo Antonio Foscarini, on the Pythagorean and Copernican Opinion of the Earth's Motion and Sun's Rest and on the New Pythagorean World System* (Naples: Lazzaro Scoriggio, 1615), in which the said Father tries to show that the above-mentioned doctrine of the sun's rest at the center of the world and the earth's motion is consonant with the truth and does not contradict Holy Scripture. Therefore, in order that this opinion may not creep[44] any further to the prejudice of Catholic truth, the Congregation has decided that the books by Nicolaus Copernicus (*On the Revolutions of Spheres*) and Diego de Zuñiga (*On Job*) be suspended until corrected; but that the book of the Carmelite Father Paolo Antonio Foscarini be completely prohibited and condemned; and that all other books which teach the same be likewise prohibited, according to whether with the present decree it prohibits, condemns, and suspends them respectively. In witness thereof, this decree has been signed by the hand and stamped with the seal of the Most Illustrious and Reverend Lord Cardinal of St. Cecilia, Bishop of Albano, on 5 March 1616.

P.,[45] Bishop of Albano, Cardinal of St. Cecilia.
Fra Franciscus Magdalenus Capiferreus, O.P., Secretary.
Rome, Press of the Apostolic Palace, 1616.

Galileo to the Tuscan Secretary of State
(6 March 1616)[46]

Most Illustrious Lord and Most Honorable Patron:
I did not write to Your Most Illustrious Lordship at the time of the
last postal delivery because there was nothing new to communicate to
you, insofar as they were about to make a decision on that business
which I mentioned to you as being of public interest, and which concerned me only because my enemies wanted to drag me into it at any
cost. I am referring to the ⟨244⟩ deliberations by the Holy Church about
Copernicus's book and his opinion on the earth's motion and sun's stability, which about a year ago was the subject of a complaint in Santa
Maria Novella and then here in Rome on the part of the same friar,[47]
who called it heretical and against the faith. Together with his followers,
he tried orally and in writing to make this idea prevail, but events have
shown that his effort did not find approval with the Holy Church.[48] She
has only decided that that opinion does not agree with Holy Scripture,
and thus only those books are prohibited which have explicitly maintained that it does not conflict with Scripture. There is only one such
book, in the form of a letter by a Carmelite Father, printed last year,
and it alone is prohibited. Diego de Zuñiga, an Augustinian hermit who
printed a book on Job thirty years ago and held that that opinion is not
repugnant to Scripture, is suspended until corrected; and the correction
is to remove from it a page in the commentary on the words "Who
shaketh the earth out of her place, etc."[49] As for the book of Copernicus
himself, ten lines will be removed from the Preface to Paul III, where he
mentions that he does not think such a doctrine is repugnant to Scripture; as I understand it, they could remove a word here and there, where
two or three times he calls the earth a star.[50] The correction of these
two books has been assigned to Lord Cardinal Caetani. There is no
mention of other authors.
     As one can see from the very nature of the business, I am not men-
tioned, nor would I have gotten involved in it if, as I said, my enemies
had not dragged me into it. What I have done on the matter can always
be seen from my writings pertaining to it, which I save in order to be
able to shut the mouth of malicious gossipers at any time, and because

I can show that my behavior in this affair has been such that a saint would not have handled it either with greater reverence or with greater zeal toward the Holy Church. This perhaps has not been done by my enemies, who have not refrained from any machination, calumny, and diabolical suggestion, as their Most Serene Highnesses and also Your Lordship will hear at length in due course. Furthermore, experience has in many ways made me grasp how much reason I had to fear some persons' dislike of me, ⟨245⟩ which I think I mentioned to you, and so I can believe that the same feeling will make them alter the story when they relate it; therefore, I beg Your Lordship, if need be, to keep for me until my return the regard which my sincerity deserves. At any rate, I am very sure that the coming here of the Most Illustrious and Most Reverend Lord Cardinal[51] will relieve me of the need to say even a word, such is the reputation I enjoy everywhere at this Court. Above all, Your Lordship will learn with how much calmness and moderation I have behaved, and how respectful I have been of the reputation of those who, on the contrary, very sharply and without any reservation have always sought the destruction of mine; you will be surprised. I say this to Your Very Illustrious Lordship in case you hear something which would make me look worse, for it would be absolutely most false, as I hope you will learn from other more reliable sources.

As for my trip to Naples,[52] so far the weather and the roads have been abominable; if they improve, I shall see what I can do, for I want to give top priority, over any other business of mine, to my being here when the Lord Cardinal arrives. In the meantime, I am grateful for the benevolence of their Most Serene Highnesses; I am also infinitely obliged to you, as my irreplaceable patron and protector, and I reverently kiss your hands.

Rome, 6 March 1616.

To Your Most Illustrious Lordship.

> Your Most Devout and Most Obliged Servant,
> Galileo Galilei.

Galileo to the Tuscan Secretary of State
(12 March 1616)

Most Illustrious Lord and Most Honorable Patron:

I have already reported to Your Most Illustrious Lordship the decision taken by the Congregation of the Index about Copernicus's book, namely that his opinion does not agree with Holy Scripture and therefore the book is suspended until corrected. The correction will be made soon.[53] The only passage involved is in the Preface to Pope Paul III, where he mentioned that his opinion does not contradict Scripture; and some words will be removed from the ending of chapter 10 of book 1, where, after explaining the arrangement of his system, he writes: "Such truly is the size of this structure of the Almighty's."[54]

⟨248⟩ Yesterday I went to kiss the feet of His Holiness,[55] with whom I strolled and reasoned for three-quarters of an hour during a very warm audience. I first paid my respects to him in the name of our rulers the Most Serene Highnesses; he accepted them warmly and ordered me to return them with equal warmth. Then I related to His Holiness the reason for my coming here. I told him how, just before leaving, I had renounced any favor which their Most Serene Highnesses could have done me, as long as it was a question of religion and integrity of life and morals, and he approved of my decision with much repeated praise. I pointed out to His Holiness the maliciousness of my persecutors and some of their false calumnies, and here he answered that he was aware of my integrity and sincerity. Finally, since I appeared somewhat insecure because of the thought that I would be always persecuted by their implacable malice, he consoled me by saying that I could live with my mind at peace, for I was so regarded by His Holiness and the whole Congregation that they would not easily listen to the slanderers, and that I could feel safe as long as he lived. Before I left he told me many times that he was very ready at every occasion to show me also with actions his strong inclination to favor me. I have been glad to report this to Your Most Illustrious Lordship, thinking that you would be pleased with it, as also would their Most Serene Highnesses, in view of their humaneness.

I am continually supported by the Most Illustrious and Most Distinguished Lord Prince of Sant'Angelo,[56] son of the Duke of Acquasparta and very devout servant of their Most Serene Highnesses. He is very well aware of how much his family owes to the Medici family, to whom he desires very much to give greater proof of his loyalty; a good occasion for this would be for him to marry into the family of the Most Illustrious Lord Marquis Salviati, as they are in the process of negotiating.[57] If a holy life, a sublime mind, and an unspeakable gentleness of very noble manners deserve to be taken into account together with the nobility of blood and with wealth, this Lord is very well endowed with them. I know it as a result of a long and very close relationship, and I wanted Your Lordship to learn it from me too; for the deal cannot be concluded without the endorsement of their Most Serene Highnesses, and so if the occasion presents itself for Your Most Illustrious Lordship to support this Lord, you should know that ⟨249⟩ he will undertake every effort to ensure a happy life for the woman he will take as his wife. I know that the low status of my rank should restrain me from mentioning these affairs, but, since the benevolence of this Lord makes him have a high regard for me in this matter, I could not rebuff the confidence he has in me without being disrespectful. So Your Lordship should excuse me and accept the affection with which I should like to serve my patrons. Finally, I remind you that I am your very devout servant, I humbly kiss your hands, and I pray the Lord God to give you the greatest happiness.

Rome, 12 March 1616.

To Your Most Illustrious Lordship.

> Your Most Devout and Most Obliged Servant,
> Galileo Galilei.

Cardinal Bellarmine's Certificate (26 May 1616)[58]

We, Robert Cardinal Bellarmine, have heard that Mr. Galileo Galilei is being slandered or alleged to have abjured in our hands and also to have been given salutary penances for this.[59] Having been sought about the truth of the matter, we say that the above-mentioned Galileo has not abjured in our hands, or in the hands of others here in Rome, or anywhere else that we know, any opinion or doctrine of his; nor has he received any penances, salutary or otherwise. On the contrary, he has only been notified of the declaration made by the Holy Father and published by the Sacred Congregation of the Index, whose content is that the doctrine attributed to Copernicus (that the earth moves around the sun and the sun stands at the center of the world without moving from east to west) is contrary to Holy Scripture and therefore cannot be defended or held.[60] In witness whereof we have written and signed this with our own hands, on this 26th day of May 1616.

> The same mentioned above,
> Robert Cardinal Bellarmine.

CHAPTER VI

Galileo's Reply to Ingoli (1624)

P. 156 - 200
202 - 204
209 - 211
281 - 293

⟨509⟩ To the Very Illustrious and Very Distinguished Mr. Francesco Ingoli of Ravenna:

[1][1] Eight years have already passed, Mr. Ingoli, since while in Rome I received from you an essay written almost in the form of a letter addressed to me.[2] In it you tried to demonstrate the falsity of the Copernican hypothesis, concerning which there was much turmoil at that time. In particular, you dealt with the location and motion of the sun and the earth, maintaining that the latter is at the center of the universe and completely motionless and the former in motion and as far from the said center as from the earth itself. To confirm this you advanced three types of arguments: the first astronomical, the second philosophical, the third theological. Then, very courteously, you urged me to reply if I had noticed any fallacy or incorrect reasoning in them. Moved by your sincerity and other courteous feelings I had detected in the past, and completely sure that you had offered your thoughts to me with a candid heart and no envy, I examined them more than once and was desirous of reciprocating as best I could the sincerity of your soul. I concluded inwardly there was no more appropriate means of putting my desire into practice than to remain silent.[3] For in this way I would not be spoiling the satisfaction which I imagine you felt in coming to think that you had convinced a man like ⟨510⟩ Copernicus; at the same time I would be leaving intact your reputation with those who had read your essay, insofar as it was within my power. I will not say that

the regard for your reputation made me disregard my own, for I never believed mine is that tenuous; indeed I felt no one who has carefully examined your objections to that opinion which I then considered true could infer from my silence that I have less intelligence than is sufficient to refute them all. I say all, except the theological ones, concerning which it seems to me one must proceed differently from the others; in fact, they are not subject to refutations but are only liable to interpretations. However, I have now discovered very tangibly that I was completely wrong in this belief of mine: having recently gone to Rome to pay my respects to His Holiness Pope Urban VIII, to whom I am tied by an old acquaintance and by many favors I have received, I found it is firmly and generally believed that I have been silent because I was convinced by your demonstrations; some even regard them as necessary and unanswerable. Although their being so regarded is of some relief for my reputation, nevertheless experts as well as nonexperts have formed a very low opinion of my competence: the former because they understand the ineffectiveness of the objections and yet they see me being silent, the latter because they are capable of judging only by the results and so from my silence have inferred the same conclusion. Thus I have found myself being forced to answer your essay, though, as you see, very late and against my will.

Note, Mr. Ingoli, that I do not undertake this task with the thought or aim of supporting as true a proposition which has already been declared suspect and repugnant to a doctrine higher than physical and astronomical disciplines in dignity and authority. Instead I do it to show that earlier I was not so blind and stupid that, because I had not seen or understood your objections, I was thereby led to think ⟨511⟩ the Copernican hypothesis could and should be true and the common Ptolemaic one false. To this should be added another reason; it is the following. The arguments you advanced were not lightly regarded by persons of authority who may have spurred the rejection of the Copernican opinion by the Congregation of the Index; further, as I understand it, your essay circulated in various foreign nations, and perhaps even among heretics. Now, for the sake of my reputation and that of others, it seems proper to deprive them of the opportunity to think less of our doctrine than they should; for it is not the case that among Catholics no one knew that your essay leaves much to be desired, or that the rejection of Copernicus's opinion was based on a belief that you were right, or that no one feared it could ever happen that our adver-

saries might provide some sure and conclusive demonstration or decisive observational evidence in favor of the truth of that opinion. Moreover, I am thinking of treating this topic very extensively,[4] in opposition
to heretics, the most influential of whom I hear accept Copernicus's
opinion; I would want to show them that we Catholics continue to be
certain of the old truth taught us by the sacred authors, not for lack of
scientific understanding, or for not having studied as many arguments,
experiments, observations, and demonstrations as they have, but rather
because of the reverence we have toward the writings of our Fathers and
because of our zeal in religion and faith. Thus, when they see that we
understand very well all their astronomical and physical reasons, and
indeed also others much more powerful than those advanced till now,
at most they will blame us as men who are steadfast in our beliefs, but
not as blind to and ignorant of the human disciplines; and this is something which in the final analysis should not concern a true Catholic
Christian—I mean that a heretic laughs at him because he gives priority
to the reverence and trust which is due to the sacred authors over all
the arguments and observations of astronomers and philosophers put
together. To this we should, finally, add another benefit, that is, to understand how little one should rely on human reason and human wisdom, and therefore how much one owes the higher sciences; ⟨512⟩ they
are the only ones capable of clearing up the blindness of our mind and
teaching us those things which we could never learn from our experience and reasoning.

These considerations can provide, if I am not mistaken, not only suitable excuses for common people but also urgent reasons why I resolved
to answer your essay. As far as you personally are concerned, I am not
sure whether I should beg your forgiveness for the excessive delay, since
you yourself repeatedly requested a reply, or for asking you to accept
kindly and quietly something that will perhaps uncover clearly those
fallacies for which you had received praise. Please do not deny me such
a pardon, for my silence of eight years can assure you that I have never
desired to belittle your reputation. Furthermore, from the nature of my
answers you can understand that it is not they, but your own objections,
which produced a fruit that may perhaps leave a bitter taste in your
mouth, a consequence which I am far from relishing. For, Mr. Ingoli, if
your philosophical sincerity and my old regard for you will allow me to
say so, you should in all honesty have known that Nicolaus Copernicus
had spent more years on these very difficult studies than you had spent
days on them; so you should have been more careful and not let yourself

be lightly persuaded that you could knock down such a man, especially with the sort of weapons you use, which are among the most common and trite objections advanced in this subject; and, though you add something new, this is no more effective than the rest. Thus, did you really think that Nicolaus Copernicus did not grasp the mysteries of the extremely shallow Sacrobosco? That he did not understand parallax? That he had not read and understood Ptolemy and Aristotle? I am not surprised that you believed you could convince him, given that you thought so little of him. However, if you had read him with the care required to understand him properly, at least the difficulty of the subject (if nothing else) would have confused your spirit of opposition, so that ⟨513⟩ you would have refrained or completely abstained from taking such a step.

Since what is done is done, let us try, as far as possible, to prevent you or others from multiplying errors. So I come to the arguments you give to prove that the earth, and not the sun, is located at the center of the universe. Because the first of these, taken from the parallax of the sun and the moon, is new and your own, I will examine it in more detail than the other common and old ones. Furthermore, since I notice from it that you lack more detailed and exact information, allow me to explain these more exact details.

[2] I know you are aware of the fact that our vision takes place in a straight line, and that if such a line is extended beyond a given object and if other visible objects lie along the line, all of them appear conjoined with each other; on the other hand, things lying off this line are seen separated from it, either to the right or to the left, depending on whether they are located on this or on that side of the line. Thus, if someone is looking at, for example, the planet Venus and imagines a straight line drawn from his eye to the center of the planet and extended into the starry heavens, he will see Venus joined with any star which happens to lie on that line; if such a line happens to hit the first degree of Aries, one will say that Venus appears joined to or lying in the first degree of Aries. Moreover, it happens very rarely that two observers of the same object are both situated in the same straight line with the object, but it almost always happens that they are separated and experience vision along different lines; these meet and intersect at the object, become separated more and more as they are extended, and finally end up at different points of the firmament; that is why the same object appears to the two observers joined to or lying at two different points of the heavens. Now, this difference of apparent position, caused by the

different locations of the two observers, is commonly called parallax or diversity of aspect.

Now I go on to apply this consideration to the two visible objects you mention, namely the sun and the moon; as they ⟨514⟩ are seen by various observers from different places on the earth very far from each other, there is no doubt whatever that they appear lying in different places of the highest heaven. Thus, for example, the moon, which to someone in the east may appear at the first degree of Taurus, to someone who looks at it from the west may simultaneously appear at the second or third degree; in short, it appears located in different parts of the firmament to those who observe it from different places on the earth's surface. Now, one of the primary aims of astronomers is to be able to determine at what place of the firmament the stars appear to be located at any given time to any observer; they soon discovered that this would be impossible if among the innumerable apparent places they did not select a fixed and stable one, to which the others would then be referred and from which they would be measured. So they agreed to the convention that the true and real place of the firmament in which or under which a given planet is said to be located is the point where the straight line ends that goes from the center of the earth and passes through the center of the planet; thus only the person whose eye is on such a line sees the sun and moon in their true places; since this line originates at the center of the terrestrial globe, it cuts the surface at a right angle and determines the point in the heavens which lies perpendicularly above the observer, called the *vertical point* or in Arabic language the *zenith*.

Thus there are two places where a planet is located in the firmament: the apparent place is the one determined by the straight line from the eye of the observer to the center of the planet; the true place is the one marked by the straight line drawn from the center of the earth to the center of the planet. These two places coincide and become the same only when the eye of the observer is in the line of the true place, which is to say when the planet is at the zenith or vertical point; otherwise the true place and the apparent one are always different. The interval between the two is called the parallax of the sun or the moon. Because parallax is nothing but the distance in the heavens ⟨515⟩ between the line of the true place and the line of the apparent one, it is clear that the more or less the two lines separate, the greater or smaller the parallax becomes; hence, in short, its magnitude is defined and determined by the size of the angle the two lines form at the center of the star; and since this angle is always equal to the one on the opposite side, we can equally correctly determine the magnitude of the parallax by the angle formed at the center of the star by the two lines (one from the center of the earth, the other from the eye of the observer).

This angle, and consequently the parallax, may increase or decrease for two reasons: one is the greater or smaller distance on the earth between the observer and the line of the true place of the star; the other is the smaller or greater height of the same star, namely its distance from the earth. For a better comprehension of all this, consider the following two diagrams.[5] In the first, let A be the center of the earth and let DFE be the equator; let the star be at B and the observer at D; then $AEBC$ is the line of the true place and DBG that of the apparent place; the angle of the parallax is CBG, or the other one on the opposite side but equal to it, namely DBA; but if the observer is closer to the line of the true place, for example at F, and if we draw the new line of the apparent place, namely FBH, the parallax will be smaller, that is, determined by the angle HBC or FBA. In the other diagram, let $AEBC$ be the line of the true place and let the angle CBG or DBA be the magnitude of the parallax while the star is at B; but if it were at S, namely nearer the earth, draw the line DSH, which will be the new line of the apparent place; then the angle CSH or DSA will be the magnitude of the parallax, and being an external angle of the triangle DSB, it will be larger than the angle DBA. Therefore, the greater vicinity of the star to the earth ⟨516⟩ makes the parallax greater; and to speculate whether the lines DB and AB, when extended toward G and C, end up in a sphere which is near, far, or very far, has nothing to do with making the parallax larger or smaller, since it does not change the angle CBG at all. This is the measure and the magnitude of the parallax studied by Copernicus and all other astronomers in regard to the sun and the moon.

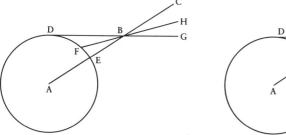
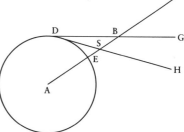

From this one can easily see the error which, if I am not mistaken, you commit when, to prove that the sun cannot be at the center of the firmament, you argue as follows: a center is the point most distant from the surface of a sphere out of all the points contained in the sphere; if, then, the sun were at the center, it would be farther from the firmament than the moon is; so the parallax of the sun should be greater than that of the moon; but it is much smaller, according to Copernicus and all astronomers; therefore, the sun cannot be at the said center. Here the error is very clear, given that it is not the distance of a star from the firmament (or from anything else which you may place as a boundary for the parallax) which makes it greater, but the vicinity of the star to the eye of the observer, namely to the earth. Now, if the parallax is to disturb Copernicus's position, it would be necessary for you to show that according to his position the sun is nearer the earth than the moon is; this is something which he never said or thought; instead, he takes the distances among the three bodies (sun, moon, and earth) to be exactly the same as other astronomers do; hence, this business of the parallaxes leaves things perfectly unchanged, and it does nothing to weaken the Copernican system.

As I understand it, this error originates from another paralogism, which is this. Always keeping firmly in your mind that the earth is located at the center of the firmament, you then inwardly inferred (as a necessary consequence) that, since the moon is nearest the earth, it is much farther from the firmament than the sun, which is that much farther from the earth than the moon is; this is equivalent to saying that the sun is ⟨517⟩ much closer to the firmament than the moon is. Then, hearing that astronomers observe a much greater parallax for the moon than for the sun, you came to the thought that the greater distance from the firmament causes the greater parallax; however, this argument is correct only as long as the earth, namely the eye of the observer, is at the center of the firmament and not otherwise. Now, that the earth rather than the sun is at the center of the firmament is the point in question; but you take it as known.

Let me now explain to you why it does not necessarily follow that the sun is closer to the firmament than the moon is, unless one first assumes that the earth is at the center; in so doing I shall be pointing out to you another error. With Ptolemy and Copernicus we speak of the firmament insofar as in it we measure the magnitude of the parallax of the sun and the moon, which is nothing but the distance between the line of the true place and the line of the apparent place. Furthermore, the primary use of the parallax is to calculate the eclipses of the sun, for the precision of which the parallax of the moon is of the greatest importance. As you know, such eclipses occur only when the sun and the moon are in conjunction. However, when the moon is in conjunction with the sun, in the Copernican system it is much farther from the firmament than the sun (I say firmament to mean that part of the firmament where you want to measure the parallax); for, drawing the straight line from the center of the earth through the centers of the moon and sun which determines their true places in the firmament, anyone can understand that the sun is as much closer than the moon to that part as the distance between the moon and the sun. Thus, according to your own thought, which is that a star farther from the firmament has a greater parallax than a nearer one, the parallax of the moon must be greater than that of the sun. So you see the error you commit when you say that the farthest point from the circumference of a circle is the center: for although any other point is closer to some part of the circumference, it is correspondingly farther from some other part; ⟨518⟩ this counts against you because the part of the circumference around which we measure the parallax is the one to which the center is closer than the other points are. I say this because, in calculations of lunar eclipses, when the moon may be said to be closer to the firmament than the sun, parallaxes are not taken into consideration and do not have any use.

Let us continue to assume that the firmament is enclosed within a spherical surface, although neither you nor any other man in the world knows or can humanly know what its shape is or whether it even has any shape. Now, to remove the error more effectively, I ask what reason persuades you that the center is farther from that surface than any other point. As for me, I do not believe any such thing. For when you assert that the center is the farthest point from the surface, either you mean the whole surface or some part; if the whole, I say that all points inside the sphere are equidistant from the whole surface, since between each of them and the whole surface is interposed the whole volume of the sphere; but if you do not mean the whole surface taken together but its parts taken separately, then the situation becomes worse for you, since there are more parts to which the center is the closest point than there are parts from which it is farthest. This can be easily proved. Let *ABCD* be a circle with *E* as its center; take any other point *F*, and through it and the center draw the diameter *FEA*; divide *EF* in half at point *O*,

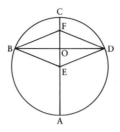

and through it draw the perpendicular *BOD* to the diameter; then draw the straight lines *BE*, *BF*, *ED*, and *DF*. Since the lines *EO* and *FO* are equal, *OB* is in common, and the angles at point *O* are right angles, the hypotenuses *EB* and *BF* are equal, and so are also *ED* and *DF*; thus, the lines drawn from point *F* to points *B* and *D* are equal to the radius. Now, it is clear, by the seventh proposition of the third book,[6] that all other lines from the same point *F* to any point whatsoever of the circumference *BCD* are ⟨519⟩ smaller than the radius, but that all the others drawn from the same point *F* to any points of the circumference *BAD* are greater than *FB* or *FD* (namely, than the same radius). Since the portion *BAD* of the circle is greater than the remaining *BCD* (for containing the center), therefore there are more parts of the circumference of the circle from which point *F* is farther than the center is, as compared to parts to which it is closer than the same center. What has just been demonstrated for a circle applies to a sphere.

It is false, therefore, to suppose that the center is farther from the circumference than is any other point whatsoever; instead, all other points are as distant as it is from the whole circumference taken together, and in general they are more distant from the parts taken separately. To escape the error, then, you should have said that the center is farther from some parts of the circumference than any other point from the same parts. However, this is not your only error, as I stated above, and as you yourself could have discovered from your own words, were it not that you had been carried away by a spirit of opposition into using technical terms with a meaning different from their proper one. You yourself write that at the apogee the sun has a smaller parallax than at the perigee; but then you interpret apogee and perigee as meaning near to or far from the firmament; yet these terms mean far from and near to the earth. Magini himself, who in this context and in the place you cite treats of the parallax extensively, never defines it in terms of the eighth sphere, but always in terms of the earth, as all other astronomers also do. But there is more. Tell me, Mr. Ingoli: do you believe it could ever happen that a star farther from the earth would have a parallax greater than a nearer one? You must necessarily answer no. Then I ask you a second question, which is whether in the Copernican system the moon is ever farther from the earth than is the sun. Again, you must answer no; these distances remain exactly the same ⟨520⟩ as in the Ptolemaic system. Now, if, as I believe, you have always heard these things, then I do not know how you managed to write that if the Copernican system were true the parallax of the sun would be greater than that of the moon. Whoever believes that the greater or smaller distance from the eighth sphere makes the parallax greater or smaller must likewise believe that the parallax and other stellar distances observed with quadrants, sextants, astrolabes, and other instruments are also greater or smaller depending on whether larger or smaller instruments are used; for the degrees of the circumference of a quadrant are defined in the same way as for the circumference of the zodiac or other circles imagined in the sky. But the truth is that these quantities are measured by the angles at the center of an instrument, which is regarded as the center of the celestial circles, and such angles do not increase or decrease by increasing or decreasing the circumferences which they subtend; therefore, the magnitude of parallax and of other intervals always remains the same, whether it is measured with large or small instruments and whether it is referred to celestial circles that are as near or as far as you like. If this is not sufficient to remove the misunderstanding, then I would have to say you believe that in a clock the hours shown by a longer hand around a larger circumference are longer than those shown by a shorter pointer around a smaller circle. Moreover, you mention Tycho and his parallax tables; but why did you not try to learn whether in calculating them he uses stellar distances from the earth or from the firmament? You would then have discovered your error, for you would have found that it is not a matter of distance from the firmament, and you would have ascertained that its being three or four or a thousand times nearer or farther does not change the parallax by a hairbreadth. However, even without reading Tycho or others, it should have dawned on you that in an exact calculation of parallaxes there could be no place at all for the distance from the firmament; for it is unknown to all, and what is unknown cannot serve as a foundation for what can be known.

⟨521⟩ Finally, in this first argument of yours, there remains to consider what you write against anyone who would free Copernicus from your objection by saying that it is enough that the moon is nearer the earth than is the sun. You retort, albeit very indirectly, that this solution is worthless because parallaxes must be to each other as the distances, but these are in the ratio of 18 to 1, whereas the parallaxes are in the ratio of 22 to 1 and therefore etc. Now, if you think you can refute me because parallaxes do not show the ratio you feel they should show, then (following your manner of reasoning) if it were the case that parallaxes did not have to show the ratio which you mention and which they do not in fact show, I could proceed in my way very comfortably. Now, the truth is that parallaxes do not have to show that ratio, but another one, which happens to be the one they actually show; thus the mistake is yours. Furthermore, how frivolous it is to say: "Parallaxes decrease by receding from the earth; therefore, since such recession causes the decrease, parallaxes must show the same ratio as the distances." Which geometry teaches that effects must correspond proportionately to their causes? I could show you a thousand contrary cases; but to be brief I shall adduce only one of them, which you must have frequently come across in your astronomical calculations and computations. Consider a circle with radius AB and tangent BD, and, moving one degree at a time from B to R, draw the secants AC, AD, and AR: it is clear that moving from B to R causes the tangents and the secants to get longer and that their increase must be proportionate in some way to the increase of the arcs; but these, increasing by one degree, increase equally; therefore, according to your view, these secants and tangents must also increase equally; now, this is most false since both of them continuously change ⟨522⟩ the proportions of their increments, and not only do they not increase equally, but the increments are 2, 3, 4, 10, 100, 1000, and 10,000 times greater than one another. So you see how far your discussion is from the right road. I will say more: if parallaxes

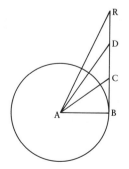

must show the ratio of distances, and if the parallax of the moon is twenty-two times greater than that of the sun, and if according to you parallaxes depend on the distance between observed bodies and the eighth sphere, then you must be thinking that the moon is twenty-two times farther from the eighth sphere than is the sun; this is the same as saying that the distance between the moon and the sun is twenty-one times greater than that between the sun and the eighth sphere; and this is a most absurd consequence, given that, taking a fixed star of average magnitude to be as large as the sun, the distance between the sun and the eighth sphere turns out to be four hundred times greater than that between the sun and the moon. Look what self-interest and pride can lead to! What I mean is this (so that you and others can understand better): in order to destroy Copernicus's doctrine, you take it as a very powerful criticism to object that his position cannot be true because a quantity of 22 should be 18; yet in your Ptolemaic position you are not at all worried that this same quantity, which should be 400, is rather $\frac{1}{21}$, namely that what should be 8400 is 1. Finally, Mr. Ingoli, let me take away from you every subterfuge, or rather let me free you from the temptation of adding errors upon errors, which you might do by making distinctions and qualifications to try to show that, if parallax is understood that way and not this way, then it may be that you spoke correctly in this sense and not in that. I tell you that the parallax of which Copernicus and all other astronomers speak is the one in terms of the angle made by the intersection of the line of the true place and the line of the apparent place; this is always the same, in the Copernican as well as in the Ptolemaic system, and from it one cannot derive the least assistance either pro or con this or that hypothesis. For you to come back with some qualification, limitation, or any other invention, will have a result similar to the one produced in the story of a plaintiff ⟨523⟩ at the trial of a notary public charged with fraud; having heard this, and knowing that conviction would have carried the punishment of cutting off the right hand, an enemy of the notary introduced some witnesses who without exception testified that they had seen the defendant wearing a mask, and he argued that this constituted fraud; the judge laughed with great amusement and dismissed this plaintiff, telling him that the right hand was cut off for persons who falsified contracts and wills, not for those who falsified themselves with a mask, and that therefore his charge did not damage at all the poor notary. In the same way, yours has nothing to do with Copernicus. Let this be sufficient in regard to your first argument.

[3] As regards the second one, following Sacrobosco you claim to be able to demonstrate that the earth is at the center of the firmament because a fixed star appears of the same magnitude when found in any part of the heavens you like. Here I tell you that this lacks not just one but all conditions necessary for a good proof. To begin with, you assume that the stars of the firmament are all placed on the same sphere. This is so doubtful that neither you nor anyone else will ever prove it. Restricting myself to the conjectural and the probable, I say that even for four fixed stars, let alone all, there is not a single point you care to choose in the universe from which they are equally distant; it is up to you to prove the opposite. However, even if it were true that the firmament is a spherical body, what makes you certain that a given star always appears of the same magnitude, so that you can argue that our eye and the earth are at the center of such a sphere? This claim is full of difficulties, which make it very uncertain. First, very few fixed stars can be seen when they are near the horizon. Second, their apparent magnitude is always altered in various ways by vapors and other impediments. Third, even if there were not such alterations, what natural eye can ever detect a very small change that might take place in two or three or four hours? By what instrument can such details be detected? Instead, thus far eyes as well as instruments have been so unable to make similar discriminations that even in determining the apparent diameter of fixed stars observers have been deceived by more than one thousand percent; ⟨524⟩ now see whether they could not be deceived by one in a thousand, indeed by much less. Finally, the same authors who place the earth at the center claim that, because its radius is insignificantly small compared to the great distance of the stellar sphere, therefore stars do not appear larger near the middle of the heavens than near the horizon, even though at the former location they are actually closer to us than at the latter by a terrestrial radius; hence, you should admit that the earth would have to be extremely close to the stellar sphere in order that the approach and recession of a fixed star to the earth due to the diurnal motion (which is less than a radius) could cause a noticeable change in its apparent magnitude. However, for Copernicus the earth is not so far from the center or so near the stellar sphere that the difference of a radius could cause a sensible change in the apparent magnitude of a star, since the distance between the earth and the fixed stars is many hundreds of times the distance between the earth and the sun. Nor is he forced to admit any of those things which you, Tycho, and others regard as great difficulties; I shall extensively explain this in the proper place

and at the proper time,[7] but to free you and others from your errors I shall now briefly touch upon it, especially since here lies the answer to another one of your objections.

These opponents of Copernicus make certain calculations based on the premise that, although the earth's motion in its annual orbit produces some curious and extremely large changes in the case of the planets, it does not cause any similar effects in the case of the fixed stars; they calculate that the stellar sphere would have to be so far away that a fixed star would itself have to be many times larger than the whole annual orbit in order for it to be visible to us with the magnitude with which it appears to us; this in turn would mean a size many thousands of times bigger than the sun itself, which they regard as the greatest absurdity. However, my calculations show me that this business proceeds very differently, namely that taking an average fixed star to be as large as the sun and no larger is sufficient to solve all difficulties which, through their own errors, they have attributed to Copernicus. Their errors were to consider the ⟨525⟩ apparent magnitudes of the stars (fixed as well as wandering) much greater than what they are; this false view has made them err so much that, for example, whereas they thought they could correctly say that Jupiter is eighty times bigger than the earth, the truth is that the earth is thirty times bigger than Jupiter is, which involves an error of 240,000 percent. Returning to our task, I say that if you measure Jupiter's diameter exactly, it barely comes to 40 seconds, so that the sun's diameter becomes 50 times greater; but Jupiter's diameter is no less than ten times larger than that of an average fixed star (as a good telescope will show us), so that the sun's diameter is five hundred times that of an average fixed star; from this it immediately follows that the distance to the stellar region is five hundred times greater than that between us and the sun. Now, what would you expect if the earth is displaced from the center of the stellar sphere by one or two parts out of five hundred, in regard to whether stars appear smaller at the horizon than at the meridian? Who will be so simpleminded as to believe that ordinary astronomers can detect such a small increase or decrease in the diameter of a star, when we can grasp with our hands that in similar observations they have been deceived so seriously, as I mentioned above? Thus, the objections of opponents are removed, as you see, simply by taking fixed stars, for example those of the third magnitude, to be equal in size to the sun. However, since with the telescope we see countless others that are much smaller than even those of the sixth magnitude, since we can reasonably believe there are many others which are not observable with the telescopes built so far, and since there is no difficulty in believing that they are equal in size and occasionally larger than the sun, at what remote distance do you feel we can say without difficulty that they must be located? Fixed stars, Mr. Ingoli, give off their own light, as I have proved elsewhere; so they lack nothing to be entitled ⟨526⟩ to be called and considered suns; and if it is true (as commonly believed) that the remotest parts of the universe are the shelters and habitations of purer and more perfect substances, then they should be no less bright and luminous than the sun itself. However, the light of all of them together, as well as the apparent magnitude of all of them together, does not reach one-tenth of the apparent magnitude and light of the sun; now, the sole cause of both these effects is their distance; so what should we believe it to be?

[4] Now I come to your third argument, taken from Ptolemy. Here I should first of all call to your attention that, of all the reasons which are advanced in the problem under discussion, some are true while others are false; and among the false ones there are occasionally some which have a semblance of truth, by contrast to others which any ordinary examination immediately reveals for what they are, namely false and out of place. Now, it so happens that in your attempt to refute the Copernican position, apart from the theological arguments, you produce things which are all really false and for the most part of that kind of falsity which is very obvious. To those which prima facie have some semblance of truth belongs this argument you take from Ptolemy. Indeed there are also others advanced by him in his *Almagest* which not only have a semblance of truth, but which I say are even conclusive from the viewpoint of the entire Ptolemaic position, though quite inconclusive from the viewpoint of the entire Copernican system. So, you will say, can the same propositions be both conclusive and inconclusive at will? Certainly not, when taken absolutely and from the universal viewpoint of all nature; but when combined with some other false proposition, they can be conclusive on that assumption. An example of this is the discussion we are presently handling.

With Ptolemy you say: if the earth were not at the center of the stellar sphere, we should not be able to always see half of this sphere; but we do see it; therefore, etc. Then you prove in various ways that what we see is half, no more or less. The first of these is taken from the observation of two fixed stars opposite to each other, such as ⟨527⟩ the Eye of Taurus and the Heart of Scorpio, which are such that when one rises the other sets and, vice versa, when one sets the other rises; this shows necessarily that the part of the sky above the earth is equal to the part below, and consequently that each is a hemisphere, and, since this phenomenon is observable for all horizons, that the earth is at its center. The argument is beautiful and worthy of Ptolemy; combined with another assumption of his it is necessarily conclusive, but if you deny that, the argument proves nothing. Indeed I am surprised that many astronomers of great renown and followers of Copernicus have had no little difficulty in answering this objection, and that they did not think of the true and very easy reply—that is, to deny Ptolemy's other assumption, from which this argument acquires its strength. Notice, Mr. Ingoli, it is true that, given the simultaneous rising and setting of two fixed stars at all horizons in turn, it is necessary to say the earth is at the center of the stellar sphere, but only as long as the earth is motionless and the rising and setting derives from the motion and turning of the stellar sphere; but if we let this sphere be still and let the terrestrial globe turn on itself (as Copernicus does), then regardless of where it is placed the same thing will always happen to the two fixed stars, namely their simultaneous rising and setting. For a clearer comprehension, let the center of the stellar sphere be D; let the earth A be as far as you wish from this center; and let the horizon be along the straight line BC. Now, if the earth and the horizon stand still, if the stellar sphere moves around its center D, and if a star rises at C while another sets at B, it is clear that when the C star will be at B, the B star will not have returned to C, since the arc CEB above the earth is smaller than the remaining one below the earth; rather it will be at S, assuming arc BS to be equal to arc CEB; therefore, the B star will be late in rising, after the setting of the C star, by a period corresponding to the arc SC. But let us now assume that the stellar sphere is motionless, and that the earth turns on itself, thus carrying along the horizon CB; then there is no doubt ⟨528⟩ that when the B end of the horizon will be at C, the C end will be at B;

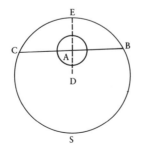

moreover, if the two stars *B* and *C* were previously one at the eastern end and the other at the western end, after the earth has completed one turn they will both return to the same places in the same period; thus, as you see, this periodic rising and setting proves nothing about the location of the earth. Nor can one infer that we see half of the sky from what you add, namely that one can always see 90 degrees from the zenith to the horizon along the meridian. For, letting the line *BC* in the same diagram represent any horizon, if from the center *A* we draw a perpendicular to *BC* that ends at the vertical point, then on both sides of this line we shall have two right angles, each of which is 90 degrees; as far as what the two arcs *BE* and *EC* really are, that is neither seen, nor known, nor knowable, nor of any use to know. What you add thereafter is equally false, when you say that if the earth were not at the center then one could not see half of the heavens; for, assuming heaven to be spherical and the earth far from the center, half of the former would still be seen by all those inhabitants of the earth whose horizon would go through the center of the heavens.

Let us now examine what you add to fault the answer of those who might say that the amount by which the visible part of the heavens exceeds or falls short of an exact hemisphere is imperceptible, because the orbit along which the earth revolves is insignificantly small relative to the immensity of the stellar sphere. It would not be necessary for me to consider it further, since I give a different explanation of the effect in question, namely that the diurnal motion belongs to the earth and not to the heavens; nevertheless, I want to call your attention to some details worth noticing. To begin with, you claim with great determination, and with the support of Tycho's authority, that to want to make Copernicus's annual orbit imperceptible relative to the immense size of the stellar sphere would require the distance to the stars to be 14,000 radii of this annual orbit, which is equivalent to $16\frac{1}{2}$ million earth radii; you really say this with too much confidence, and you rely too much on the mere authority of a man, when you use it to criticize conclusions so far-reaching about nature. If the present context and level of discussion would permit it, I could show you ⟨529⟩ how wrong is Tycho in this regard, and how he does not produce anything of any weight against Copernicus; indeed he shows not to have understood the basic idea of the Copernican system and not to know what phenomena should be visible or not visible in the fixed stars through the annual motion attributed to the earth. However, you will hear of this some other time. For now, so that I do not seem to evade the strength of your point, let us assume it is true that the annual orbit appears imperceptible relative to the stellar sphere, and that to appear such requires the fixed stars to be 16,506,000 terrestrial radii away. What impossibility or inconvenience do you find in that, Mr. Ingoli? It seems to me that all impropriety lies in men's imagination, and not at all in nature itself. To see why this is true, let us examine the absurdities which you bring forth.

You first say that, given such immensity, the universe would be asymmetric. If as a mathematician you take this term *asymmetric* in its true meaning, it means incommensurable; and then you cannot escape one of two errors. For, although incommensurability is a relation between two terms, you are considering only one of them, since you do not say with respect to what this immense size turns out to be incommensurable. However, if you inwardly meant that the stellar sphere is incommensurable in relation to the earth's orbit, you still err, since by saying the former contains the latter so many times you are treating their radii as numbers, and hence making them commensurable; and if the radii are commensurable, their spheres will be such all the more. Next, if you were using the term *asymmetric* improperly, and you wanted to mean what we might label *disproportionate*, your claim is still arbitrary and not necessarily valid. Do you not know it is still undetermined (and I believe it will always be from the viewpoint of human sciences) whether the universe is finite or infinite? If it were in fact infinite, how could you then say that the size of the stellar sphere was disproportionate to the annual orbit, since in relation to the universe this sphere would itself be much less than a grain of sand compared to it? On the other hand, if the universe were in fact finite and bounded, what reason would you have for saying ⟨530⟩ that the stellar sphere was disproportionate to the earth's annual orbit, other than that the former would hold the latter too many times because the ratio of their diameters was 14,000? If this is a valid reason, disproportionate are all those things of the same kind which are such that one is this many or more times greater than another. Thus, since there are fish so small that a whale can hold a greater number of them, and since an elephant can hold a greater number of ants, therefore whales and ants would be disproportionate animals; hence also, in your opinion they do not exist because such disproportions are not allowed by nature. Furthermore, as I have already stated, the sun has no characteristic for which we can remove it from the herd of fixed stars, so it is most reasonable to say each fixed star is a sun;

now, start to think how much space in the universe you assign to the
sun for its own habitation, where it can be alone and free from its other
companion stars; consider then the innumerable multitude of stars, and
assign to each as its property an equal amount of space; then you will
find yourself absolutely forced to deal with a total sphere much greater
than the one you now deem to be too vast. As for me, when I think
about the world perceived by our senses, I cannot say at all whether it
is large or small. I will say it is very large compared to the world of
earthworms and the like, which have only the sense of touch to measure
it, and so cannot regard it as greater than the space they occupy; and I
do not find it repugnant that the world perceived by our senses, compared to the universe, may be as small as the world of worms compared
to ours. As regards what may be apprehended by the intellect, besides
the senses, my reason and my mind are not comfortable either in conceiving it finite or in conceiving it infinite; so on this I defer to what is
established in the higher disciplines. Therefore, to consider such an immensity excessive is an effect of our imagination, and there is no flaw in
nature.

Then you go on to write that if such a distance to the fixed stars were
real, this would destroy their ability to affect ⟨531⟩ things here on earth;
and then you confirm this point with the example of the sun's influence,
which diminishes so much in the winter due to its moving away from
our zenith, even though this movement is very small compared to the
distance of the fixed stars. If I may be allowed to speak freely, with your
reputation in mind, I wish you had not written this, especially the confirmation by the example of the sun. For either this example is relevant
or it is not; if not, you already admit your error; but if you regard it as
relevant, then you become liable to other and greater faults. To begin
with, for you to be able to say legitimately that the action of the winter
and distant sun is weak presupposes having experienced that of the
summer, when it is near; for if its action were always of the same
strength, you could not say whether this or that one was weak. Thus
your example presupposes necessarily that you have experienced the
effect of the sun at two different distances; so, to be able to argue similarly about the stars, you need to have them placed at two different
distances. These two different distances might be, respectively, yours
and that of Copernicus; because you say that the Copernican distance
is not appropriate, you must be assuming that their effects are appropriate at your distance. But this is what is at issue, and so your argument
begs the question; for with equal correctness I can say that the distance

to the fixed stars is what Copernicus claims, and it is precisely what is required for the stars to affect us the way they do; and if you say that from such a great distance they could not affect us, with no less justice I shall tell you that, if the distance were less, they would affect us so violently as to destroy the world. What must have happened is the following: when you first heard about this new Copernican hypothesis, you got the idea that to accommodate it in nature it would be necessary to enlarge the stellar sphere immeasurably; since this feat is not within the power of Copernicus or any other man, you felt reassured in your original opinion and you still remain fixated upon it. These discussions being based on an idle imagination, they should not, therefore, be put forth in the study of serious and important questions; nor should one claim at the end to have firmly demonstrated and proved anything of importance.

⟨532⟩ Regarding the point about the sun being hotter in the summer than in the winter due to its being nearer our zenith, which you mention to license your conclusion, either it does not relate properly to the concept being exemplified or else it is directly against you, if I am not mistaken. For if you attribute the greater or lesser influence to the greater or lesser elevation toward the zenith, this is completely and absolutely out of place, since the enlargement of the stellar sphere does not increase or decrease the declination of stars toward the zenith, but leaves it at the same value. On the other hand, if you want to explain the action of the sun by its smaller or greater distance from the earth, then it turns out that the sun is much farther in the summer than in the winter, being at the apogee in the former; hence, if you wanted to say something about the fixed stars that would correspond to the evidence from and example of the sun, you should have said that placing them as far as Copernicus says would have made them too influential rather than less effective; indeed, like stones or hail falling from higher regions, their effects would have been too forceful—in short more fit for the destruction than for the conservation of earthly things. Here, Mr. Ingoli, are the fruits produced by arguments based on useless fictions and incoherent and unfounded concepts.

There remains to consider how well you supported the other part of your objection, namely that, given so great a distance, fixed stars would have to be larger than the annual orbit. I have already stated above that in the derivation by Tycho and others to support this difficulty there are many fallacies, which I explain elsewhere. For now I tell you this: given that, as you say, fixed stars placed at so great a distance would have to be as large as the annual orbit in order to appear of a magnitude of 2 or 3 minutes, it does not follow that they are actually that large, since their apparent diameter is not even one-sixtieth of 3 minutes; so from this it is already clear that you and Tycho make the stellar sphere sixty times greater than required to save Copernicus's position, and that you do so somewhat arbitrarily and not because you have examined carefully the apparent magnitude of fixed stars. ⟨533⟩ This is a pruning or subtraction of no small moment, I mean to cut by ninety-eight percent the size you condemned. Nor can it be true, if you do not mind, that I ever said a fixed star subtends 2 minutes, as you claim I have; for it was many years ago that I learned by sensory experience that no fixed star subtends even 5 seconds, many not even 4, and innumerable others not even 2.

[5] Moving on to the fourth argument, you criticize the Copernican system by saying, in accordance with Tycho's authority, that the eccentricities of Mars and Venus are different from what Copernicus assumed, and likewise that Venus's apogee is not stable, as he believed. Here it seems to me you want to imitate the man who wanted to tear down his house to the foundations because the chimney made too much smoke, saying that it was uninhabitable and architecturally flawed; and he would have done it had not a friend of his pointed out that it was enough to fix the chimney without tearing down the rest. So I say to you, Mr. Ingoli: given that Copernicus went astray in regard to that eccentricity and apogee, let us revise this, which has nothing to do with the foundations and the essential structure of the whole system. If other ancient astronomers had had your attitude, namely to tear down all that had been built every time one found a particular detail which did not correspond to a previous hypothesis, not only Ptolemy's great edifice would not have been built, but one would have remained always without a roof and in the dark about heavenly matters. Thus, after Ptolemy assumed that the earth is motionless at the center and of insensible size relative to the heavens, that the sun and the firmament are in motion, etc., and after he then said, for example, that the year is a constant, if you had discovered the inequality of the annual cycles, you would have turned the sun, the earth, and the heavens upside down and denied all that had been taken as true about them till then. If painters were to bleach a whole canvas whenever they were shown a small imperfection in the finger or eye of a figure, it would take a long time indeed to represent a whole scene. Copernicus was induced to reject the Ptolemaic system, not because he discovered some small fallacy in some particular

motion of a planet, ⟨534⟩ but because of an essential and inadmissible incongruity in the structure of all planetary orbits and because of very many great difficulties which are then all removed in his system. So I reply that if one had to change the whole structure of the world at the least problematic novelty which is discovered in some part of the heavens, one would never get anywhere; for, I assure you, never does one so accurately observe the motions, sizes, distances, and arrangements of the orbs and stars that there is no need of constant correction, even if everyone were a Tycho or a hundred times better than Tycho. Nor should you think that there do not remain in the heavens various motions, changes, anomalies, and other occurrences which have not yet been discovered or observed and which are perhaps unobservable and inexplicable by their own nature. Who can guarantee to us that planetary motions are not all mutually incommensurable and hence liable, indeed bound, to eternal emendations since we can only handle them as commensurable? However, we are dealing with essential dilemmas, which necessarily must be either this way or that way and do not admit of a third alternative—that is, dilemmas such as whether the sun moves or stands still, whether the earth moves or not, whether it is at the center or off, and whether the stellar sphere turns or is motionless; of these, therefore, we can speak with some confidence, and the conclusions asserted about them are not subject to every novel detail which is discovered and observed about the proper motions of planets. So, leave the foundations of the Copernican edifice alone, and repair as you wish the eccentricities of Mars and Venus, and move the latter's apogee, for these are quantities that have nothing to do with the stability or location of the sun or the earth.

[6] I now come to the two arguments which you call physical. They seem to me to abound in that kind of paralogism which assumes as true what is at issue. I believe that such fallacies originate in you from your inability to free your mind of some notions and propositions to which you have been accustomed for a long time.

Your first argument takes this form: we see the denser and heavier of simple bodies occupy the lower places, ⟨535⟩ as we see happening to earth in relation to water, and to water in relation to air; but the Earth[8] is a body denser than the sun, and the lowest place in the universe is the center; therefore, the Earth, and not the sun, occupies the center.

Here I note first that when you say heavier bodies occupy lower places (giving the examples of water, air, and earth), by these two terms *lower* and *higher* you must be meaning nothing but under our feet toward the center of the terrestrial globe and above our heads toward the heavens; for if by *lower* you meant the center of the universe, then the paralogism would already be out in the open, since you would be taking as obvious what is in question, namely that the Earth is located at the center of the universe. Besides, this being lower is something finite which ends at the center of the Earth, but it does not extend to infinity as being higher does; for a straight line perpendicular to the Earth's surface and passing through our feet and our head can be extended to infinity and will always have higher parts; on the other hand, the same cannot be done toward the center because the line admittedly has lower parts until one arrives at the center, but by proceeding further one begins to go toward higher parts. A similar arrangement may likewise be attributed to the moon, sun, Venus, Jupiter, and every other star. Being spherical, they each have a center, and the parts around it are equally inclined to move toward it whenever they are displaced from it; thus in the moon, in the sun, and in the other stars, the lowest place is the center of each, and higher places are toward the surface and, beyond that, toward the surrounding heavens. Not only can we define higher and lower places of such a kind for these solid bodies, but also for any orbs and spheres which are turning around some point; thus for the orbs of the four Medicean[9] planets which turn around Jupiter, the center of the latter will be their lowest place, what is beyond their orbs will be a higher place, and what is lowest for the earth (namely its center) will be higher for the Medicean planets. The orbits of the other planets will also have such a lowest place, and it will be the center ⟨536⟩ of their revolutions, while higher places will be those beyond their orbits toward the rest of the surrounding heavens. Next, it is doubtful whether for the universe of fixed stars it is appropriate to have a lowest place, namely the center, and higher places, namely toward the outer parts. Still, given the uncertainty, a No is much more reasonable than a Yes, since (as I also stated above) I do not think the fixed stars are all placed on a spherical surface, so as to be equidistant from a particular point, such as the center of their sphere; indeed only God knows whether for any group larger than three there is a single point from which they are equidistant. Finally, let us assume, to your advantage, that all fixed stars are placed at equal distance from a single center; then in the whole universe we would have as many centers and as many lower and higher places as there are globes like the Earth and orbs turning around various points.

Let us now resume your argument. My next point is that you must be committing either an error of form or else the material fallacy of not proving what you want to prove. In fact, to avoid the formal error, the argument must be expressed as follows: simple bodies (such as air, water, and earth) are such that the denser and heavier occupy the lower places, namely those nearer the center of the Earth; this is shown by experience, inasmuch as water is above earth, and air above water; but the Earth is denser and heavier than the sun; therefore, the Earth, and not the sun, occupies the lower places, namely those clearly occupied by earth in relation to water and air. Thus the argument proves nothing but that the Earth, and not the sun, occupies a lower place and one nearer the center of the Earth; I admit this, and I would have admitted it even without the syllogism. On the other hand, if by lower place in your conclusion you did not mean the center of the Earth, but the center of the universe, then either you have a syllogism with four terms [10] (equivocating between the center of the Earth and that of the universe) or else you are assuming as known what is in question, namely that being a very heavy body the Earth occupies the center of the universe. Then, if it is legitimate for you to switch from the center of the Earth to that of the stellar sphere, ⟨537⟩ equally correctly I shall be able to conclude that the Earth occupies the center of Jupiter or of the moon, because these are also lower places in the universe, no less than the center of the Earth.

However, you will say, in supposing in the premises that denser and heavier bodies occupy the lower places in the universe, you have not supposed it as self-evident, but as demonstrated by the example of air, water, and earth, among which earth occupies the lowest place. If such was your intention, you are then still erring very seriously in many other details. It will be necessary, first of all, for you to give bodies two inclinations: one would be that their parts have gravity, namely an inclination toward the center of their globe; the other would be an inclination of a whole globe toward the center of the universe. Only in this way, and not otherwise, will the parts of earth and water seek to form their own globe, and then the latter would try to occupy the center of the universe. However, you have no reason for denying that the same thing would also apply to the moon, the sun, and the other heavenly bodies. For you cannot say that their parts lack that same inclination of seeking to form their own globe, which you know the parts of the Earth to have; but if this inclination is sufficient to make the Earth seek the center of the universe, the same inclination will have the same effect in the other globes; hence, taking this doctrine as true, one would have to say that all heavenly bodies, being dense and heavy, incline toward the lower places of the universe, namely the center. To give you as much advantage as possible, one could say that because the Earth is denser and heavier than the moon, the sun, and the other stars, it occupies the said center; but why do the others not fall on top of the Earth in order to approach their desired center as much as possible? You fail to notice, and this is another error, that to reason correctly you would have to expand your minor premise asserting "but the Earth is a denser and heavier body than the sun"; you would have to add that not only earth, but also water and air, are denser and heavier bodies than the sun, because for you they too are in a lower place. However, I do not think you will ever convince anyone of that, not even your inward self. ⟨538⟩ But what am I saying? You do seem to be convinced of it, and similarly you would like to convince me, on the basis of the authority of Aristotle and all the Peripatetics,[11] who say that heavenly bodies have no gravity. Here, before I go any further, I must tell you that in natural phenomena human authority is worthless. Like a lawyer, you seem to capitalize on it; but nature, Dear Sir, makes fun of constitutions and decrees of princes, emperors, and monarchs, and at their request it would not change one iota of its laws and statutes. Aristotle was a man, saw with his eyes, heard with his ears, and reasoned with his brain. I am a man and see with my eyes much more than he did; as regards reasoning, I believe he reasoned about more things than I, but whether he reasoned better than I about those things which we have both examined will be shown by our arguments and not by our authorities. You will say: "Such a great man, who has had so many followers!" But that means nothing because it is the antiquity and the number of years elapsed that give him so many followers. A father may have twenty children, but this does not necessarily imply that he is more prolific than a son of his who has only one child, if the father is sixty and the son twenty years old. But let us return to our subject.

To Aristotle's errors you add a greater one, which is to suppose as true that which is in dispute. In his philosophizing Aristotle first concluded that the Earth, being heaviest, occupies the center of the celestial sphere; then from this, seeing that the moon, sun, and the other heavenly bodies do not fall toward this point, which he regarded as being desired by all heavy bodies, he concluded that those bodies are devoid of gravity. Now you, arguing in a circle, assume as known that heavenly bodies are devoid of gravity in order to prove that which was used as proof of this lack of gravity, namely that the Earth is at the lowest place of the universe and is located there due to its being heavy. The error which you and Aristotle share is the following. When you say "heavy bodies are such that their proper and natural inclination is to go to the center," by center either you mean the central point of the whole collection of heavy bodies, which for terrestrial bodies is the center of the Earth, or you mean the center of the whole celestial sphere. If you mean ⟨539⟩ the first, then I say that the moon, sun, and all other heavenly globes are no less heavy than the Earth, and that their parts all seek to form their own globe; thus, if such a part were separated from such a globe, it would return to the whole in the same way we see the parts of the Earth doing, and you will never be able to prove the opposite. On the other hand, if you mean the second alternative, I say that the Earth too is devoid of any gravity and does not seek the center of the universe, but that it stays in its place, as the moon stays in its own place.

In addition to these issues, Mr. Ingoli, I see you together with your Peripatetics entangled in a strange labyrinth in order to find and determine where is this delicious center of the universe. Aristotle deemed it to be the point around which all heavenly orbs turn; that is, not the stellar sphere, but the orbs of Saturn, Jupiter, Mars, and all other planets. Indeed, believing all these orbs to be concentric, he believed he could locate the center of the stellar sphere insofar as he thought he could assert the centers of the former and that of the latter are the same; for, considering the stellar sphere by itself, it was difficult, indeed impossible, to be able to find its center, due to its immeasurable vastness. Thus Aristotle was absolutely clear that he took the center of the universe to be the center of the planetary orbs, and at the latter he placed the Earth. Now, in our time, it is as clear and glaring as the sun itself that the sun, and not the Earth, is located at such a center; indeed I believe you understand this. Nevertheless, although you grasp that Aristotle was highly mistaken about the facts, motivated by excessive attachment you still pay lip service to his words; you would rather turn the world upside down to find a center for the universe to replace the lost one of Aristotle than confess the error; and, instead of fearing the supreme force of nature and truth, full of trust you expect help for your cause from the useless authority of a man. If any place in the world is to be called its center, that is the center of celestial revolutions; and everyone who is competent in this subject knows that it is the sun rather than the Earth which is found therein.

⟨540⟩ Once these things are understood, it becomes completely immaterial to consider whether the sun is more or less dense and heavy than the Earth, which is something neither I nor you know or can know with certainty. On the other hand, speculatively speaking, I would rather believe that the sun is denser and heavier. This would be so even for the Peripatetic doctrine, which considers heavenly bodies inalterable and incorruptible and the Earth the opposite, and which seems to make density and solidity more consistent with very long durability than is rarity; for we see gold (the heaviest of all terrestrial elements), the extremely hard diamonds, and other gems come closer to incorruptibility than do other bodies that are less heavy and less rigid. To be sure, you regard fires as similar to the sun, because of their luminosity, and consequently you like to infer that, just as they consist of a tenuous, rare, and light substance, likewise so must be the sun. It seems to me you do not reason correctly here. For, on the contrary and with much greater verisimilitude, I can say that, since our fires are of such extremely short (indeed momentary) duration due to their consisting of such rare substances, given that you together with Aristotle consider the sun eternal and inexhaustible, it must consist of an extremely dense and solid substance; besides, I believe its brilliance is very different from the brilliance of our burning substances. At the end of your argument you produce your usual philosophical authorities to prove that the center must be called the lowest part and the surface or circumference the highest part. I answer that these are words and names which prove nothing and have nothing to do with how things really are; for I deny that the Earth is at the lowest place, as well as that it is at the center. Further, if in your view this name *center* has the power of drawing the Earth to it, why do you not place us in the firmament, where there are centers by the thousands, since every star is a perfect globe and every globe has its own center?

[7] Let us finally hear the argument based on the sieve, concerning which, if (as I believe) you have confidence in it, I beg you to keep your confidence even after I show that it proves the complete opposite of what you now think it proves. Please do not act the way ⟨541⟩ most modern polemicists act: they first impress on their minds the conclusion without hearing other reasons and demonstrations; then, once having acquired the impression, they agree totally and very readily with every stupid and crude reason put forth in support of it; on the other hand, when faced with all kinds of evident and conclusive demonstrations to the contrary, they are motionless and immovable, having adopted the principle that perfect and true philosophizing consists of not letting oneself ever be convinced by any reason or experiment, however unequivocal it may be. You say that, as a result of the circular motion of a sieve,

the tiny pieces of soil mixed with the wheat end up at the center of the sieve, and therefore that in a similar way the earth was sieved (as it were) by the turning of the heavens, and so must have already been pushed to the center of the heavens. Let us accept the analogy. But notice, Mr. Ingoli, that in sieving wheat one does not turn the sieve around its center at all; this is obvious because one always keeps his hands on the same parts of the sieve, and so it is impossible for the sieve to be able to turn around its center without the siever's hands and arms becoming detached from his body. In such an operation the motion of the sieve is the following: it is shaken and moved in such a way that its center moves along the circumference of an imaginary circle parallel to the ground, whose center may be conceived as remaining suspended in air between the arms and stomach of the siever. By this shaking, the impurities of the wheat collect at the center of the sieve; but such a motion has nothing to do with the motion of heaven, which is motion around its own fixed and stable center. However, to adapt this kind of experiment for our purpose, you would have to keep the center of the sieve always in the same place and make the sieve speedily turn around it; then, while it is turning in this manner, throw some small rocks or pieces of soil and observe what happens; unquestionably you will see them move toward the circumference until they touch the rim of the sieve, where they will stop. Now, since for you the sieve experiment is valid, ⟨542⟩ change your mind and say that the earth must necessarily be far from the center. Indeed, if you consider more exactly the behavior of the small rocks in the original experiment you introduced yourself, you will notice that their collecting at the center of the sieve is nothing but their coming to the circumference of the motion in process, since the center of the sieve moves along the circumference of this circular motion. I could also tell you that the effect you attribute to the sieve happens when it moves, but not when it is standing still; now, the sieve which we know with certainty moves is the one found within the orbit of Saturn, namely the planetary orbs, and at their center is not the earth but the sun. Therefore, either your example is not true, or it is irrelevant to our purpose, or the sun is heavier than the earth.

[8] There follow, in your essay, the arguments with which you claim to be able to show the stability of the earth and to strip it of all the motions assigned to it by Copernicus—that is, the diurnal motion of rotation and two annual motions, one around the sun along the ecliptic and the other also of rotation but almost opposite to the diurnal. Although the annual motion around the sun would be condemned if you

had demonstrated that the earth is located at the center of the universe, nevertheless you also produce other reasons against the former (I believe for extra security).

As regards the diurnal motion, namely the motion of rotation every twenty-four hours from west to east, there are many reasons and experiments produced by Aristotle, Ptolemy, Tycho, and others. However, you get by very lightly by mentioning only two, namely the much used one about heavy bodies falling vertically on the surface of the earth and the other one about projectiles, which without any difference move through equal distances toward the east as well as toward the west and toward the south as well as toward the north. Maybe you get by so quickly, I believe, because of the great obviousness and ⟨543⟩ necessity with which they seem to convince you. However, these and others were well known to Copernicus, and they were examined by him and much more attentively by myself;[12] and I know they are all such that either nothing can be proved on the affirmative or negative side, or else if any consequence can be deduced it favors the Copernican opinion. Moreover, I say I have other evidences not previously observed by anyone, which are necessarily convincing about the certainty of the Copernican system (as long as we remain within the limits of human and scientific inquiry). As all these things need more extended analyses to explain them, I reserve them to some other time. In the meantime, to answer what is sufficient for the things you touched upon, I repeat that you and all the others begin by having the earth's stability firmly impressed upon your minds, and because of this you fall into two errors: one is always to commit equivocations, assuming as known what is in question; the other is that when you think of experiments which could be made to discover the truth, without having made them you take them as made and report them as resulting in favor of your conclusion. I shall try as concisely as possible to make you grasp these two errors; some other time you will be able to see this point treated at great length, with the answers to all the objections which seem prima facie to have some probability, but have none at all.

[9] With Aristotle and others, you say the following. If the earth were turning on itself every twenty-four hours, then rocks and other heavy bodies falling from on high downward, for example, from the top of a tower, would not hit the earth at the foot of the tower; for, during the time the rock spends in the air going down toward the center of the earth, the earth would proceed at great speed toward the east, carrying along with it the foot of the tower, and so the rock would necessarily

be left behind by a distance equal to how much the earth's turning would have advanced forward in the same time, which would be many hundreds of feet. They confirm this reasoning with an example taken from another experiment,[13] where they say the same thing can be clearly seen: if one lets a rock fall freely from the top of the mast of a ship standing still in harbor, it falls vertically and hits the foot of the mast precisely at that point which ⟨544⟩ is perpendicularly below the place from which the rock was dropped; this effect does not happen (they say) when the ship moves at a fast speed; for in the time the rock takes to go from top to bottom, and during which it falls freely and perpendicularly, the ship moves forward and leaves the rock behind toward the stern many feet from the foot of the mast; the corresponding thing should happen to the rock falling from the top of a tower, if the earth were turning with such a great speed. This is the argument, in which I see very clearly the mistakes I am talking about.

My reasons are as follows. When you and Aristotle infer that the rock falling from the top of the tower moves in a straight line perpendicular to the earth's surface, you do this, and can do this, only from seeing how during its fall it goes on licking (as it were) the surface of the tower, erected perpendicularly on the earth's surface; this makes one perceive the line traced by the rock as being also straight and perpendicular. However, here I say that from this appearance one cannot infer that conclusion unless one assumes the earth is motionless while the rock falls, which is the question at issue. For, if with Copernicus I say the earth turns and consequently carries with it the tower and also us who are observing the rock, then we can say the rock moves with a motion composed of the general circular motion toward the east and the accidental straight motion toward the whole to which it belongs, and the result of these is a motion inclined toward the east; in this case, the motion which is common to me, the rock, and the tower is for me imperceptible, as if it did not exist, and only the other remains observable, which the tower and I do not share, namely the motion of getting nearer the earth. Here, then, is the misunderstanding exposed, if I have been able to explain myself sufficiently. Let me add another point. Arguing from parts to whole, you and Aristotle elsewhere say that, since the parts of the earth are seen to move naturally straight downward, one can infer that such is the natural inclination of the whole earth, namely to desire the center and, having already reached it, to stop therein. Therefore, arguing much more correctly from whole to parts, I shall say that, since the natural inclination of the terrestrial globe is

⟨545⟩ a circular motion around its center every twenty-four hours, this is also the inclination of the parts, and so by their nature they circle the earth's center every twenty-four hours, and this is their congenital, proper, and completely natural behavior; if by some violence they are removed from their whole, then as an accident the other motion of fall is added to that. Here I reason all the more correctly than Aristotle and you, inasmuch as you attribute to the earth a natural motion which it has never had and will not have for all eternity, namely straight motion toward the center, whereas I give it and all its parts a natural motion which perpetually belongs to them and in which they are perpetually engaged.

The other error involves your putting forth experiments as having been made and corresponding to your needs, without ever having carried them out or made the observations. To begin with, if you and Tycho wanted sincerely to confess the truth, you should say that you have never tested whether or not any difference is observable in shooting artillery toward the east and toward the west, or toward the north and toward the south (especially in regions near the pole, where according to you the effect should be more noticeable); I am moved to believe this, indeed to be sure of it, by the fact that you put forth as certain and unequivocal other experimental observations which are much easier to make and which I am so sure you did not make as to be able to say that whoever makes them will find the effect to be contrary to what you claimed with excessive confidence. One of these experiments is precisely that of the rock falling from the top of the mast on a ship; the rock always ends up hitting the same spot, whether the ship is standing still or moving forward fast; it does not strike away from the foot toward the stern, as you believed (on account of the ship moving forward while the rock comes down through the air). Here I have been a better philosopher than you in two ways: for, besides asserting something which is the opposite of what actually happens, you have also added a lie by saying that it was an experimental observation; whereas I have made the experiment, and even before that, natural reason had firmly persuaded me that the effect had to happen the way it indeed does. ⟨546⟩ Nor was it difficult for me to discover your deception. For you first imagine someone on a motionless ship at the top of the mast who drops a rock while all is still, and then you fail to notice that when the ship is in motion the rock no longer starts from rest, since the mast, the man at the top, his hand, and the rock are also moving with the same speed as the whole boat; I still frequently meet people with such a thick skull that I cannot put it into their head that, because the man on the mast keeps his arm still, the rock does not start from rest. Thus, I say to you, Mr. Ingoli, that when the ship is moving forward, the rock also moves with the same impetus; this is not lost just because the person who held it opens his hand and drops it, but is permanently conserved in the rock, and so is sufficient to allow it to follow the ship; no longer held by the person, the rock falls down on account of gravity, thus combining the two motions into a transversal and inclined (and perhaps also circular) motion toward where the ship is going; and so it strikes the same spot on the ship where it fell when all was still. From this you can understand how the same experiments advanced against Copernicus by his opponents speak much more in his favor than in theirs; for the motion transmitted by the ship is certainly accidental to the rock, and so, if it is conserved to such an extent that the effect is observed to be exactly the same whether the ship is still or moving, then what doubt can remain that the rock brought to the top of the tower and sharing the same speed as the terrestrial globe would conserve it while falling? Note, in fact, that such a speed would be a natural, primary, and eternal inclination, unlike that of the ship, which is accidental.

[10] As regards the projectile motions of artillery, I have no doubt that things happen precisely as Tycho and, following him, you say, although I have not made the experiments; that is, no difference is seen, and the shots range the same ⟨547⟩ in any direction you wish. However, I add (a point Tycho did not understand) that this happens because it must happen this way whether the earth moves or stands still, and that no difference can be conceived, as you will understand with clear reasons at the proper time.[14] In the meantime, there is another way to remove these and all other difficulties of this kind, such as the flight of birds and how they could keep up with so much motion, and also the suspension of clouds in the air and the fact that they are not always running toward the west, as you think it should happen if the earth were in motion. To solve all these apparent difficulties I say that, as long as water, land, and air (which are their environment) do the same thing together, namely either jointly move or jointly stand still, then necessarily all phenomena appear to us exactly the same in the one as well as in the other state. I say all, meaning those involving the mentioned motions of falling bodies, projectiles upward or sideways in this or that direction, birds flying toward the east or the west, clouds, etc. But watch out, Mr. Ingoli, for some other effect which might be observed in the air, water, land, or heavens,[15] and which might enable us to discover the

truth with certainty; watch out, I say, because I am very much inclined to think it may occur clearly to your disadvantage. In regard to the phenomena mentioned, consider this single experiment, which is very suitable for pointing the right road to you, by showing you, as I said, the impossibility of getting from them anything that might serve one iota to clear up this question.

Shut yourself with a friend in the largest cabin under the deck of a large ship, and make sure you have with you some flies, butterflies, and similar small flying animals; bring also a large tank of water, with some small fish in it; arrange also some vase up high dripping into a lower one with a narrow neck. When the ship is standing still observe carefully how those small flying animals go with equal speed toward all parts of the room; you will see the fish swim indifferently toward any side of the tank whatever; the falling drops will all go into the vase below; if you throw any object to your friend you need not exert greater effort in one direction than in another, as long as ⟨548⟩ the distances are equal; and if you jump with your feet tied together, you will reach equal distances in all directions. After carefully observing all these things, let the ship move with as much speed as you wish; as long as the motion is uniform and does not fluctuate here and there, you will not notice the least change in all the things mentioned, nor from any of them or anything within yourself will you be able to ascertain whether the ship is advancing or standing still. If you jump on the floor, you will reach the same distances as before, nor, because the ship is moving very fast, will you jump farther toward the stern than toward the bow, even though while you are in the air the floor advances in a direction opposite to your jump; if you throw a fruit to your friend, you will not need to throw it harder to reach him when he is toward the bow and you astern than when your positions are reversed; the drops will fall into the lower vase without any being left behind astern, even though while the drop is in the air the ship advances many inches; the fish in their water will not have to work harder to swim toward the front than toward the rear side of the tank, but with equal ease they will go to catch the food you give them from any part of the rim of the tank; the butterflies and flies will be able to fly indifferently in all directions, nor will they ever be stuck to the wall toward the stern, as if they were tired of keeping up with the fast course of the ship, from which they are separated when they remain suspended in air; and if you produce some smoke by burning a little incense, you will see it rise on high and linger there, and like a small cloud move indifferently no more in one direction than in another. If you ask me for the cause of all these effects, at the moment I answer: "Because the general motion of the ship is indelibly conserved in all those things contained in it, being transmitted to the air and to them, and not being contrary to their natural inclination." Some other time you will hear detailed answers explained at length. Now, once you have made all these observations, and seen how these motions (though accidental ⟨549⟩ and temporary) appear exactly the same when the ship moves as well as when it stands still, will you not abandon all doubt that the same must happen in regard to the terrestrial globe, as long as the air moves together with it? This is especially true since that general motion, which for the case of the ship is accidental, is taken by us as natural and proper for the case of the earth and terrestrial things. Add to this that for the case of the ship we cannot determine from the things within it what it does, even though with ease we can make it move or make it stand still; how can it be possible to determine this for the earth, which in our experience has always been in the same state?

[11] I go on to the arguments which you, following Tycho, give to do away with the annual motion. In these I discern more clearly than ever that neither you nor he have an adequate idea of Copernicus's world system and of the phenomena and effects which it implies and which should appear to our eyes. Instead, confusing the old and customary notions with the new suppositions, your reasoning continues to be misconceived.

You give four arguments against the annual motion along the zodiac. The first is based on your not seeing any variation at all in the latitudes of the rising and setting of the fixed stars. You claim that as a consequence of the said motion they should vary noticeably every eight days; for the earth together with the horizon moves from south to north with a motion that becomes noticeable periodically after eight days, and given (as Copernicus says) that the fixed stars are motionless, it follows necessarily that in the same time the latitudes of their rising and setting would vary noticeably; this you say is not seen; therefore, etc. This argument is ineffective on many counts.

First, I do not know how much I should believe that you and Tycho have made diligent observations of the latitudes of the rising and setting of the fixed stars; I suspect that the postulated stability of the earth persuaded you of the immutability of those latitudes, rather than the observed immutability convincing you of the terrestrial stability. Secondly, I am strengthened in this opinion by the uncertainty of such observations, ⟨550⟩ which are very difficult, if not impossible, to make with the precision required; for very few stars are visible near the horizon, and here refractions hinder significantly our seeing them in their true and real position; in fact, the impediment is so great that it has happened several times one could see both the sun and moon above the horizon while the moon was already eclipsed, an event which makes us certain it is possible for a star to be actually below the horizon while it appears to us above it; hence a star's apparent rising and setting may be deceptive, as a result of this difficulty, by a much greater amount than the very small difference which should be noticed due to the earth's annual motion. Third, you claim that if this motion belongs to the earth, since it carries the horizon along, the variation would be noticeable every eight or ten days and thus would be perceived as such in the fixed stars; to this I answer that such a motion is noticeable and very much so where it must appear such, but not where it must not appear such. Does it not seem very noticeable in the sun, whose horizontal latitudes vary about 50 or 60 degrees? I want to facilitate your understanding of this business with a very appropriate example, which will remind you of a phenomenon I believe you have observed many times in going from Padua to Venice by boat. As you looked at the trees planted along the banks of the Brenta river, and at others farther away, as well as at many others much further, all the way to the foothills of the Alps, the nearest ones seemed to pass by quickly in a direction opposite to the motion of the boat; others somewhat farther also seemed to move in the opposite direction, though more slowly than the near ones; then, by contrast to the former and the latter, others even farther seemed both to move in the opposite direction and to follow the progression of the boat; finally, the farthest ones always appeared the same, as if they were following the boat, just as it happens with the moon, which due to its great distance at night appears to move over roofs and eaves when one is walking on a street, even though it really stays behind. Thus, as the boat which is our horizon moves along the annual orbit, it leaves our neighboring sun significantly behind, but, by ⟨551⟩ contrast with it, the very distant stars appear to us to be following us completely. I do not want you to assume the distance to the fixed stars as greater than three hundred diameters of the annual orbit (though one could easily assume it as more than a thousand). Now, imagine someone three hundred feet away drawing two lines one foot apart from each other, and then, looking only at the lines, see whether you can detect that they are not parallel to each other; undoubtedly their imperceptible difference will remove every difficulty. The difference in the latitudes of the rising of fixed stars which you are talking about is similar and, for other reasons, much smaller; being imperceptible, it should no longer bother you. But of this I shall say much more some other time.[16]

[12] Let us go on to your second reason, based on polar elevations. You consider it an impossibility that they should not vary, becoming higher when the earth approaches north and lower when it is in the south, if in the annual motion the earth were approaching and receding from the north by a distance equivalent to the whole diameter of the annual orbit, which is twice the distance from the earth to the sun. You strengthen this consequence by means of experience, which shows that when a man moves on earth only 60 miles toward the north, the pole rises by a degree; your reason then infers that if the same man is carried by the terrestrial globe toward the north not only the same 60 miles but many hundreds of thousands, then a much greater variation than the one mentioned should be perceived; nevertheless, no sensible variation is observed; from this you infer the stability of the earth. Here, Mr. Ingoli, is a very clear witness of what I stated above, namely that you have not adequately understood the Copernican opinion and are incapable of stripping yourself of the old ideas impressed in your mind, and so you confuse heaven and earth and utter great inanities.

Indeed I say to you that not only approaching toward or receding from the north by a diameter of the annual orbit is not supposed to produce any change in the polar elevations, but this would not come about even with a ⟨552⟩ displacement of one hundred or one thousand such diameters; and I am very surprised that you and especially Tycho were dazzled so childishly. But let us look for the cause of the dazzling. Having learned from Sacrobosco that the earth is motionless at the center of the stellar sphere, and that this sphere is the one that undergoes the diurnal rotation, you fixed in it the axis of this rotation, the poles, and the equator, namely the great circle through points on the stellar sphere equidistant from both poles. You then transferred to earth these things you believe are really in the heavens, defining the poles, axis, and equator of the earth to be perpendicularly under those of the heavens. On the other hand, Copernicus makes the firmament stand still and attributes the diurnal motion to the earth, and so he removes the axis, poles, equator, and other circles from heaven, and attributes everything to the earth, since such things are not found in a sphere that does not turn on itself. It is true then that with our imagination we can transfer them to the heavens, and call the "axis" of the world that of the earth extended to the stellar sphere, the "poles" the two points where the axis intersects it, and the "equator" the great circle produced by the plane of our terrestrial equator extended to there. Now, someone on earth who is standing over the great circle of the diurnal rotation, namely over the equator, has the horizon passing through both poles; and if he moves away from the equator by walking on the earth's surface toward one of the poles, then his horizon will shift by a corresponding amount, and consequently the said pole will rise. However, if the man stops at any place, and if the earth continues to turn around the same axis and the same poles, then, regardless of where in the universe the earth is transported, neither the equator nor the horizon nor the axis nor the poles will show the least variation to him. Let me illustrate with a very appropriate example Tycho's error and yours. Your misunderstanding is exactly like that of someone who is standing at the stern of a galley, looks at the top of the foremast by means of a quadrant, and finds it, for example, at an elevation of 30 degrees above his horizon; who then walks twenty or thirty steps on the deck toward the mast and finds its elevation increased by 10 degrees; and who finally is so ⟨553⟩ simple-minded as to expect that the same should happen if, instead of him moving on the galley to approach the mast, the whole galley were moving in the same direction while he remained standing at the stern, and so simpleminded as not to understand that even if the galley moved by as many miles or thousands of miles, let alone twenty or thirty steps, the elevation of the top of the mast would always remain the same. When you, Mr. Ingoli, let the earth move toward the north in accordance with Copernicus, you then forget that the poles of its diurnal motion are actually on earth, and in the heavens only by imagination, and you do not consider that when the earth recedes toward the north it brings with it our horizon and its real poles, and that as the latter move so do the imaginary ones in heaven; because this motion is shared by us and the poles, it does not produce any change, and it is as if it did not exist. Let me help you as much as possible. You should have said that with such a motion there would be a change, not in the elevation of the pole, but in the elevation of some fixed star, such as, for example, the nearby Polaris, and then add that, since this is not seen, one could thereby infer the stability of the earth. However, Copernicus already answered this by saying that, because of the immense distance of the fixed stars, such a change is imperceptible. Besides this I add some other things, which you will hear at the proper time.[17] For now I say that since you have not made such observations yourself, you should not trust Tycho and his instruments so blindly; it may happen that they are unable to detect such minute quantities, which perhaps some day could be observed by means of other much bigger, much better, and much different instruments.

[13] If you have followed what I have said so far, then you will understand without my help the fallacy of your third argument based on the inequality of the days, which fallacy is rooted in the same misunderstandings. I repeat to you that the horizons, zenith, axis, poles, and diurnal rotation (which defines the daylight and nighttime arcs, namely the parallels to the equator) are all features of the earth, and they have nothing to do with the firmament or its stars, which might as well not exist in this case. ⟨554⟩ Then the fact that the equator and its axis always keep the same inclination and direction relative to the zodiac (namely the circle of the annual motion) makes the solar rays (which produce the daylight) cut those parallels sometimes all in equal parts (which happens when those rays simultaneously reach both poles), and sometimes in unequal parts, the longer arcs being those of daylight when the earth is toward the south, and those of nighttime when it is toward the north; the only exception is the equator, which is always cut equally since it is a great circle. But I know very well that these subjects are so abstract that some other longer explanation is needed to make oneself understood. You will have it at the proper time.

[14] The fourth argument is an arbitrary invention of Tycho based on something which, in my opinion, he never observed and could not have observed. I am referring to the motion of comets when they are in opposition to the sun. Now, if it is true, as I most certainly believe, that their tail always points away from the sun, then it is impossible for us to see any of them when they are in opposition to the sun since in this case their tail would be invisible. Furthermore, what does Tycho know with certainty about a comet's own motion, as to be able confidently to assert that, when mixed with the earth's motion, it should produce some phenomenon different from what is observed? Regarding himself as the arbiter and ruler of all astronomical affairs, he finds true and right only those things which correspond to his observations or imaginations; thus, having invented a very implausible theory of comets, he saw nothing in comets that could support the Copernican hypothesis, and he preferred to deny and reject the latter than to abandon his own conceited whim.

[15] There remain for me to consider the objections which you and Tycho make against the third motion of the earth,[18] the annual one around its own center, in the opposite direction from the annual orbital

motion. Here you say, first, that once the orbital motion is taken away, so is this motion. I admit this for now, but the former has not yet been removed, and so the latter likewise stays. In your second objection you consider it impossible that the earth's axis moves or can move with such a great correspondence to the ⟨555⟩ annual orbital motion that it is as if it were motionless. I answer that this is not only not impossible but necessary, and that such an effect is seen very clearly happening in every body which is freely suspended, as I have shown to many people. You can do the experiment yourself: place a floating wooden ball in a glass of water, take the latter in your hands and extend your arms, and then turn on your feet; you will see the said ball turn on itself with a motion contrary to yours and complete a rotation in the same time you complete yours. This is what you will necessarily see happening; some other time[19] you will understand how the ball really does not turn at all, but rather always keeps the same direction relative to any stationary point outside your rotating body, which is then the same property that Copernicus attributes to the earth. From this one can also take care of the third argument: for you say it is impossible that in the same body the center and the axis should have contrary motions, and this is not only not impossible, but necessary (as long as we understand the motions as Copernicus does). And do not say that the difficulty becomes greater by also adding the diurnal motion, given that you regard it as a great absurdity that the same body at the same time should move with so many different motions. For I do not consider it at all absurd to move with ten or one hundred motions let alone with three, as you will understand some other time; to be sure, ultimately from the composition of all there results but one motion, so that if the moving body by means of some point on it left a trace of all its motions, it would leave only a very simple line.

[16] I now go on to the three physical arguments you put forth to prove the stillness of the earth. Stripped of the ornaments with which you embellish it, the first of these is this. Heavy bodies are less suitable for motion than bodies which are not heavy, because experience shows this; but the earth is the heaviest of all bodies known to us; therefore, we must say that nature has not given it so many motions, especially the diurnal, which is so fast that it traverses 19 miles in one minute. ⟨556⟩ I should have to tell a very long story if I had to note all the fallacies which are present in this and similar discussions; I shall touch upon what suffices to show its complete ineffectiveness.

First, to my eyes everything appears the opposite of yours. You see heavy bodies being very reluctant to engage in any motions, natural as well as violent, and light bodies being correspondingly more so inclined; whereas, beginning with natural motions, I see a piece of cork move more speedily and readily than a feather, a piece of wood more than a piece of cork, a stone more than wood, and a piece of lead more than a rock. I see the same thing for the case of violent motions. For I see that if in the artillery one uses balls of different substances and fires them with the same charge, a lead ball moves much faster and longer than a wooden ball, and a bundle of straw or tow fiber much less than the latter. I see that if balls of cotton, wood, and lead are suspended from equal strings, and they are all equally given cause to move, the cotton ball will stop in the shortest time, the wooden one will last much longer in its back and forth motion, and the leaden one even longer. On the other hand, if at the bottom of a vase full of water one attaches a string shorter than the depth of the water, if at the other end of the string one ties a float or some other light body, and if the latter is displaced from the perpendicular and then released, it will quickly stop when it moves back to the perpendicular, without undergoing any oscillations like those of heavy pendulums in air or even in the same water. I see bowl makers and tin-plate turners adding very heavy wooden wheels to their machines in order to make them retain longer the impetus they acquire; and the same is done with flywheels in many other machines. I see that the air in a room immediately comes to rest after it has been agitated; on the other hand, this does not happen to the water in an aquarium, which retains the impetus and flutters for a long time after the cause of the agitation has ceased. I would have gladly heard the observations that convinced you otherwise, but you do not mention any.

Second, where did you get the idea that the terrestrial globe is so heavy? As for me, either I do not know what heaviness is, or the terrestrial globe is neither heavy nor light, just like all the other globes in the universe. ⟨557⟩ For me (and, I believe, for nature) heaviness is that innate inclination by which a body resists its being removed from its natural place, and by which it spontaneously returns there when it has been removed by force; thus a pail of water, lifted up high and dropped, returns to the sea; but who will say that in the sea the same water is heavy, since it is free in it and yet does not move? In saying that bodies which are not heavy are more suitable to motion than the heavy ones, you utter, in my opinion, a proposition which is diametrically opposite to the truth, because the truth is that bodies which are not heavy are most unsuitable above all others. For, since motion can only occur in some medium, and since heaviness or lightness can only be experienced relative to the medium, nonheavy bodies are just those which are specifically as heavy or as light as the medium in which they find themselves. Thus a body is neither heavy nor light in water if it is specifically as heavy as water; but such a body will not move at all with natural motion in that medium, since it is there neither heavy nor light; nor will it move with violent motion there, except insofar as it is attached to a mover, and it will immediately cease moving when abandoned by the latter. On the other hand, a body which is heavy in the same medium will naturally go down in it, and will move in it conserving the virtue received from the projecting agent; and the heavier such a body is, the more it will do both things.

What you add at the end goes to show how much your emotions dominate your reason. For you charge it is a great absurdity to want the earth turning on itself every twenty-four hours, and such a speed seems to you exorbitant; yet you praise and admit as a very easy thing the motion of one hundred thousand bodies bigger than the earth at a speed one hundred thousand times higher than that, such being the fixed stars and the diurnal rotation attributed to their sphere. However, if you resort to admitting similar enormities in order to persist in your opinion, or better, in your first assertion, what hope do you leave to anyone that all ⟨558⟩ the evidence in the world could ever persuade you of a very tangible truth, which you might have once denied?

[17] Your second argument is based on a physical principle which claims that every natural body can undergo only one, and no more than one, natural motion; since the natural motion of the earth is to move toward the center, so many circular motions cannot in any way naturally belong to it; and if they are not natural for it, how could it move for such a long time? A very revealing answer to this objection would be the one you would give to someone who was asking you: Mr. Ingoli, you say that the natural motion of the terrestrial globe is to move toward the center; but how can this be natural for it if it has never moved nor will it ever move with such a motion? According to your own philosophical views circular motion has no contrary motion, whereas rest is contrary to any movement. Now, why are you bothered so much that the earth keeps on moving circularly for such a long time, given that this is not a movement contrary to what you call its natural motion, and you do not have the least worry in the world in saying that it has been and will be eternally motionless, which is against its natural inclination to move? How less wrong would it have been to say that it is natural

for the earth to stand still, since for you it has always been so! What I have said would more than answer your objection. But I want to add something more, and I say that if natural bodies move by nature with any motion, this can only be circular motion, nor is it possible that nature should have given any of its integral bodies the propensity to move with rectilinear motion. I have many confirmations of this proposition, but for now only one must suffice. It is this. I assume that the parts of the universe are set up in the best arrangement, so that none is outside its own place, which is to say that nature and God have perfectly ordered their edifice. Given this, it is impossible that any of these parts should by nature move with straight motion or motion other than circular, because whatever moves with straight motion changes place, and if this change is natural then its earlier place was ⟨559⟩ preternatural for it, which goes against the assumption. Therefore, if the parts of the world are well ordered, straight motion is superfluous and nonnatural, and it can only occur when a body is by violence removed from its natural place; then perhaps it will return there by a straight line, for it appears to us that this is what the parts of the earth do when separated from their whole. I said "it appears to us" because I am inclined to believe that nature does not employ straight motion even for such an effect.[20] These inconveniences do not follow in the case of circular motion, which can be used in nature without in any way disordering the best arrangement of the parts; for whatever turns on itself does not change place, and whatever moves on a circumference presents no impediment to others and always goes toward the places it leaves, so that its motion is a perpetual leaving and a perpetual returning. Moreover, straight motion is a movement toward a place where it is impossible to arrive, since a straight line is by nature extensible to infinity; but a circular line is necessarily bounded and finite, although Peripatetics believe the opposite, namely that circular lines and motions are infinite, and straight lines and motions finite and bounded. And do not tell me that there are the center and the circumference to act as ends of straight lines, first because no circumference terminates a straight line in such a way that it could not be extended beyond there to infinity; furthermore, to assume this center and this circumference is an instance of human arbitrariness, and it is like wanting to adjust architecture to the edifice, rather than building in accordance with the precepts of architecture. So I conclude[21] that if the earth has a natural inclination to motion, this can only be toward circular motion, and I leave straight motion to be used by the parts, not only of the earth, but of the moon, sun, and all

other integral bodies of the universe; if these parts are by violence separated from their whole, and consequently brought to a wrong and disordered arrangement, they will return to their whole by the shortest path.

[18] There remains your third and last argument. However, before I
examine it, I want to bring forth a certain fitness which I once was in
the custom of pointing out to nonprofessionals who were incapable of
esoteric demonstrations, in order to persuade them to consider it much
more probable that the sun, and not the earth, is motionless ⟨560⟩ and
located at the center of celestial revolutions. What I used to say was
this. There are eight bodies (namely the earth and the seven planets),
seven of which absolutely and incontrovertibly move, and only one of
which (and no more) can be motionless; this single one must necessarily
be either the earth or the sun. Now let us see whether we can find out
which one of them moves on the basis of some very probable conjecture.
Because motion and rest are very significant natural phenomena (indeed
nature is defined in terms of them),[22] and because they are extremely
different from each other, it is necessary that there be a great difference
between the condition of those which incessantly move and the condition of any other that eternally stands still. Being then in doubt as to
whether the earth or the sun is motionless, and being certain that the
other six move, if by some strong correspondence we were sure about
which one of the two (earth or sun) conforms more to the nature of the
other six moving bodies, then very reasonably we could attribute motion to that one. But nature graciously points the way in this search for
truth by means of two other phenomena, which are no less basic and
important than rest and motion, namely light and darkness; for enor-
mous should be the difference in nature between a body that shines
brilliantly with an eternal light and another one that is very dark and
completely devoid of light. Now, of the six indubitably moving bodies,
we are sure that they are by their nature totally devoid of light; and we
are equally certain that exactly so is also the earth. Therefore, we can
very resolutely assert that the earth's conformity to the other six planets
is very great, and that on the contrary the discrepancy between the sun
and these bodies is equally great. Now, if the nature of the earth is very
similar to that of moving bodies, and the essence of the sun very different, will it not be much more probable (other things being equal) that
the earth rather than the sun imitates with motion its other six consorts?
Add to this another no less notable harmony, which is that in the Copernican system all fixed stars, also intrinsically luminous bodies like
the sun, are eternally at rest. You twist out of order this extremely well-

ordered story in order to conclude the opposite, and merely mentioning this should suffice to make you avoid the error and discover its faults. You say the following: ⟨561⟩ Copernicus attributes motion to all the luminous parts of the sky, namely to the planets, and he denies it to the sun, which is the most luminous of all, in order to attribute it to the earth, which is an opaque and dark body; but nature, which is judicious in all her works, does not do these things. You should rearrange this, Mr. Ingoli, and say: Copernicus attributes rest to all the luminous parts of the world, which are the fixed stars and the sun, and he gives motion to all the opaque and dark ones, which are the planets and the earth, the latter being like the former; this is what one would expect from nature, which is very judicious in all her works.

This is what occurs to me now, in answer to your physical and astronomical objections against the system of Nicolaus Copernicus. You will be able to see this topic treated at much greater length if I shall have the time and strength to finish my Discourse on the Tides,[23] which takes the motions attributed to the earth as hypotheses and consequently gives me great leeway to examine at length all that has been written on this subject. Finally, I beg you to receive these answers of mine graciously, and I hope you do that both because of your innate courtesy and also because it is the appropriate thing to do for every lover of truth. For, if I have correctly resolved your objections, your gain will not be small, since you will be exchanging falsehoods for truths; on the contrary, if I am wrong, then your point of view will show itself all the more clearly.

CHAPTER VII

Miscellaneous Documents (1618–1633)

Galileo to Archduke Leopold of Austria
(23 May 1618)

Most Serene Lordship and Most Honorable Patron:

I am still afflicted with the same ills which Your Most Serene Highness found me to have when I was very undeservedly favored and honored by your infinite kindness. To my suffering from the corporal afflictions has been added a more painful mental one, namely not to have been able nor to be able still (at least in part) to satisfy the suggestions of Your Highness, by putting together in accordance with my inclination some discussions on problems that I think would not displease you. Thus I am forced to implore you very humbly to attribute to my infirmity the delay ⟨390⟩ in obeying your requests more fully, and to accept the few little things I am sending with the present letter. They include two telescopes, a longer and a shorter one. The bigger one can be used by Your Highness and the other members of your family for the observation of celestial phenomena; it is in fact the same spyglass with which I have been making observations for the past three years, and, if I am not mistaken, you will find it excellent. The other smaller one will be easier to handle and very good for observations on earth; for the latter the longer one will show objects bigger and sharper, but they will be a little harder to find.

I am also sending you another even smaller spyglass attached to a brass headpiece. This is made without any decoration since Your Highness can only use it as a model or sample from which to build another which better fits the shape and size of your head or the head of whoever would be wearing it. It is impossible to adjust this instrument and device without examining the actual head and eyes of the individual who must use it since the adjustment consists of changing the position higher or lower, and more to the right or to the left, by extremely small amounts. Your Highness will not lack artisans who can do an accurate job on the basis of this model. However, I beg you to keep it secret as much as you can, on account of some interests of mine.

Furthermore, I send you a copy of my printed *Sunspot Letters*, as well as my short discourse on the cause of the tides,[1] which I happened to compile a little more than two years ago in Rome at the request of the Most Illustrious and Reverend Lord Cardinal Orsini. That was when those theologians were thinking of prohibiting Nicolaus Copernicus's book and the opinion of the earth's motion, put forth in that book and held by me as true at that time, until it pleased those Lordships to suspend the book and declare this opinion false and repugnant to Holy Scripture.[2] Now, I know how suitable it is to obey and believe the decisions of superiors, as they can discern higher truths where my lowly mind ⟨391⟩ by itself cannot arrive; and so, since the present essay I am sending you is based on the earth's motion and contains one of the physical arguments I advanced in support of this motion, I consider it a poem or a dream, and Your Highness should take it as such. Nevertheless, because even poets sometimes appreciate some of their own images, similarly I have some regard for this conceit of mine. Moreover, since I had already written and shown it to the above-mentioned Lord Cardinal and to a few others, I later let other copies pass through the hands of other gentlemen in high places; my purpose was that if perhaps someone else separated from our Church were to attribute to himself this fancy of mine, as it has happened with many others of my inventions, then there could be the testimony of persons above all suspicion that I had been the first to dream up this chimera. What I send you is really only a sketch because it was hastily written while I was hoping that Copernicus would not be judged to be in error eighty years after the publication of his work; thus I was planning to elaborate it at leisure, adding much more on the same topic, including other evidence, rearranging it, and expressing it in better form. However, a single voice from heaven awakened me and dissolved like a fog all my confused and contorted images. May Your Most Serene Highness accept it kindly, unseemly as it is; if ever divine mercy will grant me to be in a state of being able to work harder, you may expect from me something else more real and solid. In the meantime, rest assured that I feel so much obliged for your infinite courtesy that I not only consider it impossible to be able to be released from such an obligation but am always ready at the least sign from you to try to show you how grateful a servant I am.

⟨392⟩ Finally, I bow down most humbly, I kiss your garments with all due reverence, and I beg you when the occasion comes to remind your Most Serene Sister, my Ladyship, with how much devotion I respect both Your Highnesses. May God the Lord grant you the greatest happiness.

Florence, 23 May 1618.

To Your Most Serene Highness.

Your Most Humble and Obliged Servant,
Galileo Galilei.

Correction of Copernicus's On the Revolutions
(15 May 1620)

⟨400⟩ The Fathers of the Holy Congregation of the Index decreed that the writings of the distinguished astronomer Nicolaus Copernicus, *On the Revolutions of the World*, were to be absolutely prohibited,[3] because he does not treat as hypotheses, but advances as completely true, principles about the location and the motion of the terrestrial globe that are repugnant to the true and Catholic interpretation of Holy Scripture; this is hardly to be tolerated in a Christian. Nevertheless, since Copernicus's work contains many things that are very useful generally, in that decision they were pleased by unanimous consent to allow it to be printed with certain corrections according to the emendation below, in places where he discusses the location and motion of the earth not as a hypothesis but as an assertion. In fact, copies to be subsequently printed are permitted only with the above-mentioned places emended as follows and with this correction added to Copernicus's preface.

Emendation of the passages in Copernicus's book which are deemed suitable for correction:

In the preface, toward the end, delete everything from the beginning of the last paragraph up to the words "this work of mine," and substitute: "For the rest, this work of mine . . ."[4]

In book 1, chapter 1, page 6,[5] where it says "However, if we consider the matter more closely . . . ," substitute: "However, if we consider the question more closely, we think it is immaterial whether the earth is placed at the center of the world or away from the center, so long as one saves the appearances of celestial motions."[6]

In chapter 8 of the same book, this whole chapter could be expunged since it explicitly treats of the earth's motion while it refutes the ancient arguments proving its rest; however, since it is preferable to speak problematically, so as to satisfy scholars and to keep integral the book's sequential order, it may be emended as follows.

First, on page 6, delete the sentence from "Why therefore" to the words "We sail out," and correct the passage in this manner: "Why therefore can we not grant it the motion suitable to its shape, rather than rendering unstable the whole universe, whose limits are unknown and cannot be known, and why not grant that the things which appear in heaven happen in the same manner as expressed by Virgil's Aeneas?"[7]

Second, on page 7, the sentence beginning with "I also add" should be corrected this way: "I also add that it is no more difficult to attribute motion to that which is in a place in a container, namely to the earth, than to the container."[8]

⟨401⟩ Third, on the same page, at the end of the chapter, the passage from the words "You see" till the end of the chapter is to be deleted.[9]

In chapter 9, page 7, correct the beginning of this chapter up to the sentence "For the fact that . . ." thus: "If, then, I assume that the earth moves, I think that we now have to see also whether several motions can belong to it. For the fact that . . ."[10]

In chapter 10, page 9, correct the sentence beginning with "Consequently" thus: "Consequently we should not be ashamed to assume . . ."[11] And a little below that, where it says "is correctly attributed to the motion of the Earth," substitute: "is consequently attributed to the motion of the Earth."[12]

Page 10, at the end of the chapter, delete the very last words: "Such truly is the size of this structure of the Almighty's."[13]

In chapter 11, the title of the chapter is to be changed in this manner: "On the Hypothesis of the Triple Motion of the Earth, and Its Demonstration."[14]

In book 4, chapter 20, page 122, in the title of the chapter, delete the words "these three stars," since the earth is not a star, as Copernicus would have it.[15]

Fra Franciscus Magdalenus Capiferreus, O.P., Secretary of the Holy Congregation of the Index.

Rome, Press of the Apostolic Palace, 1620.

Anonymous Complaint About The Assayer
(1624 or 1625)[16]

Having gone through in the past few days the book by Mr. Galileo Galilei entitled *The Assayer*, I came to consider a doctrine once taught by some ancient philosophers, but effectively refuted by Aristotle, though refurbished by the same Mr. Galilei. I then decided to compare it with the true and certain rule of revealed doctrines, which enlighten us more surely and more certainly than any physical evidence, though they shine in dark surroundings as a test and a reward for our faith. In the light of this lamp, I found that it appears false, or else (for I do not wish to be a judge) very difficult and dangerous, in the sense that whoever takes it to be true may then falter in reasoning and judging about more serious things. Thus I thought of letting Your Most Reverend Paternity know about it, and of asking you (as I hereby do) to tell me your feelings about it, for my own information.

On page 196, line 29,[17] of the cited book the above-mentioned author wants to explain a proposition put forth by Aristotle in many places (that "motion is the cause of heat"),[18] and to modify it for his purposes; so he tries to prove that the qualities commonly called the color, odor, taste, etc., of the thing to which they are commonly attributed are nothing but mere ⟨246⟩ words and exist only in the sensitive body of the animal who perceives them.[19] He explains this with the example of the tickling or titillation caused by something touching an animal in certain areas and how, if the animal's senses are taken away, the phenomenon of tickling is no different from touching and rubbing a marble statue and is entirely a matter of our perception; and he concludes that similarly the qualities which are perceived by our senses and are called tastes, odors, colors, etc., are not in the objects where they are commonly believed to be, but merely in our senses, just as tickling is not in the hand or feather which, for example, is touching the sole of a foot, but merely in the animal's sensitive organ.

However, this reasoning seems to me to be faulty for assuming as proved what should be proved, namely that in every case the thing perceived is in us because the act of perceiving is in us; nor would it be correct to argue that, since the vision with which I see the light of the sun is in me, therefore the light of the sun is in me. Nevertheless, be that as it may, I do not stop to examine that.

The author goes on to explain this doctrine of his, and he tries to demonstrate what these qualities are from the point of view of the object and the subject. As one can see on page 198, line 12,[20] he begins to explain them in terms of the atoms of Anaxagoras or Democritus, which he calls minima or minimum particles, and into which he says bodies are constantly separated. When applied to our senses, they penetrate our bodies, and, depending on the varieties of contact and on the different shapes of these minima (smooth or rough, hard or soft) and on whether they are many or few in number, they pierce us differently and go through us with more or less difficulty or with a beneficial effect on our breathing, and consequently with our annoyance or pleasure. He says that the minima of the element earth correspond to the sense of touch, which is more material and corporeal; to taste correspond those of water, which he calls fluid minima; to smell those of fire, which he calls igneous; to hearing those of air; and to vision he attributes light, concerning which he says he has very little to say. Then on page 199, line 25,[21] he concludes that to make us perceive tastes, odors, etc., nothing is needed in the bodies that have taste, odor, etc., but sizes, shapes, and number, and that odors, tastes, colors, etc., are nowhere else but in the eyes, tongue, nose, etc.; thus, if these organs are taken away, the just mentioned qualities are not distinguished from the atoms except in name.

⟨247⟩ Now if this philosophy of qualities is admitted to be true, it seems to me there follows a great difficulty in regard to the existence of the qualities of bread and wine which in the Holy Sacrament are separated from their own substance;[22] for we find therein the qualities of touch, vision, taste, etc., and so according to this doctrine one would have to say that there are in it the minimum particles with which the substance of bread affects our senses. Now, if these particles were substantial, as Anaxagoras said, and as this author seems to agree on page 200, line 28,[23] it would follow that in the Sacrament there are substantial elements of the bread and the wine, which is an error condemned by the Sacred Council of Trent, Session 13, Canon 2.[24]

On the other hand, if they were merely sizes, shapes, number, etc. (as more likely he seems to confess, in agreement with Democritus), it would follow that (since all these qualities are modalities, or forms of quantity, as others say) when the Sacred Councils, especially the Council of Trent at the cited reference, determine that in the Sacrament after the consecration only the accidents of bread and wine remain, there would remain only quantity in the form of triangles, acuteness, obtuseness, etc., and the existence of those accidents would be explained only on the basis of these quantities. This consequence seems to me to go against the common view of theologians, who teach that in the Sacrament there remain all the sensible accidents of bread and wine (color, odor, taste), not mere words; whereas the quantity of the substance does not remain, as is well known from an authoritative opinion (Suarez, *Metaphysicarum Disputationum*, volume 2, section 2, number 2).[25] Not only is this so, but also the same consequence is formally repugnant to the truths of the Sacred Councils since, whether these minima are explained according to Anaxagoras or Democritus, if they remain after the consecration, then a consecrated host will have the substance of bread no less than a nonconsecrated host, given that according to him a corporeal substance consists of an aggregate of atoms arranged in this or that manner, with this or that shape, etc. On the other hand, if these minima do not remain, it follows that none of the accidents of bread remain in a consecrated host since, as this author says on page 197, line 1,[26] there are no other accidents but shapes, sizes, motion, etc., and these are forms of quantity or quantitative substance, ⟨248⟩ so that it is impossible to separate them in such a way that they exist without the substance or quantity of which they are accidents, as all philosophers and theologians teach.

This is the difficulty which came into my mind about this doctrine, and which I propose and submit to Your Most Reverend Paternity for whatever you will wish to tell me in regard to my assessment mentioned above. With my best regards.

Guiducci to Galileo (18 April 1625)

⟨265⟩ My Most Illustrious, Distinguished, and Honorable Sir:

It is several weeks since I have written to you or received a letter from you, though I have always had news of you, your good health, and your continuing to write your Dialogues.

I have had occasion to meet several times with the Lord Prince of Sant'Angelo[27] to discuss you and your works already written and in the process of being written. On advice from His Excellency I have postponed giving to Ingoli the letter you wrote him, and I shall continue to postpone it until you tell me to disregard the considerations of the Prince. His considerations are these. First, some months ago a pious person proposed to the Congregation of the Holy Office that *The Assayer* be prohibited or corrected, charging that it contains praises for Copernicus's doctrine regarding the earth's motion.[28] As a result a cardinal undertook to find out more information and to report about the case. It fortunately happened that he put in charge of the matter Father Guevara, who is General of an order of Theatines[29] called (I believe) Minors, and who then went to France with the Lord Cardinal Ambassador.[30] He read the work diligently, and, since he liked it very much, he praised and extolled it very much to that cardinal; furthermore, he put in writing some defenses to the effect that, even if that doctrine of motion were held, it did not seem to him to deserve condemnation; so the matter quieted down for the time being. Now, since we lack this support which could guard our flank in regard to that cardinal, it does not seem advisable to run the risk of a beating; for in the letter to Ingoli Copernicus's opinion is explicitly defended, and though it is clearly stated that this opinion is found false by means of a superior light, nevertheless those who are not too sincere will not believe that and will be up in arms again. Furthermore, we lack the protection of the Lord Cardinal Barberini, who is absent; we are opposed here by another powerful man, who once was one of your chief defenders; and His Holiness is very disturbed by the mess of the war,[31] so that one could not speak to him; thus the matter would surely be left to the discretion and intelligence of the friars. For all these reasons it seemed right, as I said, to wait and let the matter rest rather ⟨266⟩ than keeping it alive by having oneself persecuted and by having to shield oneself from those who can deal blows with impunity. In the meanwhile time may benefit the cause.

As I wrote to you, Sarsi's work[32] is still not being printed. I believe that, in light of the difficulties the Genoese are having, he is worried about his fatherland.[33]

I hope to be back there before mid-May. When I leave I shall give to Mr. Filippo Magalotti the letter you wrote to Ingoli, so that he can save it and present it whenever you want.

The Lord Prince Cesi has told me that the Lincean Academy has bestowed on me a singular favor, namely admitting me to membership as an Academician. I know this has happened principally on account of you, and so I acknowledge your support most of all, and in due course I will show you my gratitude. In the meantime, let this serve as a beginning or, if you will, as thanks for the news I received. Finally, I kiss your hands, and I pray the Lord God to give you every happiness.

Rome, 18 April 1625.

To Your Very Illustrious and Most Distinguished Lordship.

Yesterday I spoke for a long time with the Most Illustrious Lord Cardinal Orsini,[34] who asked me what kind of a man Cosimo Lotti was in regard to the construction of fountains, for he had been proposed to His Most Illustrious Lordship as singularly qualified in this trade. I answered that I knew he was a painter, but did not know anything else. Then he asked me whether I knew any good person for such a job. I said that I did not know anyone, but that I had heard you say that there was in Rome someone who was very good and did original work, though I did not know whether he was still alive. If you want to propose someone for this, let me know; also let me have some information on Cosimo Lotti, about whom I later learned that he has worked at the Castle. The Lord Cardinal remains very affectionate toward you, but Apelles[35] has a great influence on His Most Illustrious Lordship.

> Your Most Affectionate and Most Obliged Servant,
> Mario Guiducci.

Galileo to the Tuscan Secretary of State[36]
(7 March 1631)

Most Illustrious Lord and Most Honorable Patron:

As Your Most Illustrious Lordship knows, I was in Rome to get a license for my Dialogue[37] and have it printed and published, and so I delivered it into the hands of the Most Reverend Father Master of the Sacred Palace.[38] He commissioned his associate Father Fra Raffaello Visconti[39] to examine it with the greatest care and to note whether he had any misgivings about it or whether it contained any ideas to be corrected. The latter did that very strictly, as I too asked him to do. ⟨216⟩ While I was applying for the license and for the personal endorsement of the same Father Master, His Most Reverend Paternity decided that he wanted to read it again himself. He did this and returned the book licensed and endorsed by his own signature, so that after a two months' stay in Rome I returned to Florence. However, I was thinking of sending the book back there after I had completed the table of contents, the dedication, and other things; it would have been handed over to the Most Illustrious and Distinguished Lord Prince Cesi, head of the Lincean Academy, to take care of the printing as he had been in the habit of doing for other works of mine and by other Academicians. There followed the death of the same Prince, as well as the suspension of commerce,[40] so that printing the work in Rome was hindered. Thus I made the decision to have it printed here; I found a suitable publisher and printer; and we made the necessary agreements. For this I also got a license here from the Most Reverend Vicar, the Most Reverend Inquisitor, and the Most Illustrious Lord Niccolò Antella.[41] Since I thought it proper to inform the Father Master in Rome about what was happening, and about the obstacles hindering its printing in Rome in accordance with the intention I had expressed to him, I wrote to His Most Reverend Paternity that I was thinking of printing it here. By way of the Most Distinguished Ladyship wife[42] of the ambassador, he let me know that he wanted another look at the work and that therefore I should send him a copy. Then, as you know, I came to Your Most Illustrious Lordship to find out whether at that time such a large volume could have been sent to Rome safely, and you told me clearly No, and that simple letters were barely going through safely. So I wrote again, informing him of such an obstacle and offering to send the book's preface and ending; there, authorities could add or remove at will and insert all the qualifications they wanted, since I would not refuse giving these thoughts of mine the labels of chimeras, dreams, paralogisms, and empty images, and since I would always defer and submit everything to the absolute wisdom and indubitable doctrines of the higher sciences, etc.; as for reviewing the work again, this could be done here by a person acceptable to His Most Reverend Paternity. He agreed to this, and I sent the work's preface and ending. He settled on the very Reverend Father Fra Giacinto Stefani, Consultant to the Inquisition, as the new reviewer; the latter reviewed the entire work with the greatest accuracy and severity (as I also asked him to do), noticing even some minutiae ⟨217⟩ which should not cause a shadow of a scruple even to my most malicious enemy, let alone to himself. Indeed, His Paternity has stated that at more than one place in my book tears came to his eyes when seeing with how much humility and reverent submission I defer to the authority of superiors, and he acknowledges (as also do all those who have read the book) that I should be begged to publish such a work, rather than being hindered in many ways which I need not list for now. Some months ago Father Benedetto Castelli[43] wrote to me that he had met the Most Reverend Father Master several times and heard from him that he was about to return the above-mentioned preface and ending, revised to his complete satisfaction; but this has never happened, and I no longer feel like mentioning the matter. In the meantime the work stays in a corner, and my life is wasted as I continue living in constant ill health.

So yesterday I came to Florence, first on orders from the Most Serene Patron[44] to see the drawings of the Cathedral's facade and then to appeal to his kindness. I wanted to let him hear about the state of this business and to ask him to arrange, with the advice of Your Most Illustrious Lordship, that at least we may understand clearly the feelings of the Most Reverend Father Master; furthermore, if you both agreed, Your Most Illustrious Lordship with the approval of His Highness could have written to the Most Distinguished Lord Ambassador to speak with the Father Master, conveying to him His Highness's desire that this business be brought to an end, in part to know what kind of man His Highness employs in his service. However, not only was I unable to speak with His Highness, but also I could not stay to look at the drawings, because I became very sick. Just now a messenger from Court came here to find out about my condition, which is such that I really should not have gotten out of bed were it not for the opportunity and desire to mention this business of mine to Your Most Illustrious Lordship. So I beg you to do me the favor of accomplishing what I was unable to do yesterday, by following the above-mentioned plan and bringing about, in the appropriate ways you know better than I, the resolution of this affair, so that while I am still alive I may know the outcome of my long and hard work.

⟨218⟩ Your Most Illustrious Lordship will receive this from the hand of the above-mentioned messenger, and I shall eagerly wait to hear from Mr. Geri[45] what Your Most Illustrious Lordship has decided about this. Now I reverently kiss your hands and pray for your happiness. Finally, since His Most Serene Highness is kind enough to show his worry about my condition, Your Most Illustrious Lordship can explain to him that I should get by reasonably well if I were not afflicted with mental disturbances.

Bellosguardo,[46] 7 March 1631.[47]
To Your Most Illustrious Lordship.
>   Your Most Devout and Most Obliged Servant,
>   Galileo Galilei.

The Vatican Secretary[48] *to the Tuscan Ambassador*[49]
(25 April 1631)

Most Illustrious, Distinguished, and Honorable Lord and Patron:
Mr. Galilei had the endorsement of my simply signing the imprimatur so that he could show it to His Most Serene Highness; but he had promised me to correct the book and change some details in accordance with our agreement, and then to return and print it in Rome, where with the help of Monsignor Ciampoli[50] all differences would have been resolved.

Undoubtedly Father Stefani has judiciously examined the book. However, he does not know the intentions of His Holiness, and so he cannot give the endorsement which only I can give, if the book is to be printed without the risk of some displeasure either on his part or on mine, depending on enemies who might find in it something contradicting the prescribed instructions. I have no greater concern than to serve the Most Serene Highness, my Lord the Grand Duke, but I should like to do it in such a way that a person protected by such a great Lord is free from any danger of harm to his reputation. This I cannot do by simply giving permission to print, which for your city is not within my jurisdiction, but only by ensuring that it conforms to the rule given him on orders from His Holiness, and by seeing whether he has followed it. If I receive the book's preface and its ending, I can easily see what suffices for my purposes, and I can testify to having approved the work. On the other hand, if even a copy cannot be sent, I shall write a letter to the Inquisitor, mentioning what must be complied with and explaining the orders I have received, so that he may ensure they are complied with and let the book be freely printed. Or else, if one can find another procedure whereby my signature is not used by Mr. Galilei and my goodwill is not abused, then I shall do all I can in response to the least sign by such Patrons.

At any rate, Your Excellency should assure the interested parties that no living creature (of higher, lower, or equal rank) has spoken to me about this matter, except friends common to ⟨255⟩ Mr. Galilei and myself; nor should it be thought that there is any machination by enemies, for in truth there is not. May Your Excellency forgive the tardiness of my reply, and, while I kiss your hands with all due reverence, allow me to write since I have been hindered from being able to reply in person.

At home, 25 April 1631.

To Your Excellency.

> Your Most Devout and Most Obliged Servant,
> Niccolò Riccardi.

Galileo to the Tuscan Secretary of State (3 May 1631)

Most Illustrious Lord and Most Honorable Patron:

I have seen what the Most Reverend Father Master of the Sacred Palace writes in regard to the printing of my Dialogue. From that I learn, to my great disgust, how His Paternity, after having held me up for about a year without ever coming to any conclusion, now plans to do the same with the Most Serene Grand Duke our Lord; that is, he temporizes and proceeds with words devoid of any effect, which I do not think is to be lightly tolerated.

In his 19 April letter the Most Distinguished Lord Ambassador writes that he had agreed with the Father Master that His Paternity would order the book to be printed here, but with certain requests and declarations which he would have sent in a letter. This did not happen until ⟨259⟩ a week later, perhaps on account of his involvement in the Church services for the holy days. On 25 April he sent a letter written by his own hand, the one which the Lord Ambassador sent to Your Most Illustrious Lordship and which you sent to me. According to the agreement made with the Most Distinguished Lord Ambassador, this should contain the order to print the work here and the declarations which His Paternity wanted in it. However, the truth is that in the letter there is neither an order to print, nor any declarations, nor anything else, but rather new delays based upon certain claims and questions of his; but it is months and months since I have satisfied all of these, in a way that I want to show to the Grand Duke and to Your Most Illustrious Lordship, and to anyone who would want to know. Now, seeing that here one is sailing in an ocean without shores or harbors, and being infinitely concerned about the publication of my book as a reward for so much work of mine, I have been thinking of various ways of proceeding, but they all need the authorization of the Most Serene Grand Duke. To be able to arrive at some conclusion, I think it would be very proper that some day, as soon as possible, His Most Serene Highness would agree to do the following: the Most Reverend Father Inquisitor and the very Reverend Father Stefani (who has already reviewed and severely examined my book) should be called to a meeting with himself, Your Most Illustrious Lordship, the Most Illustrious Lord Count Orso,[51] and any other consultant His Most Serene Highness wishes; I should come bringing the work with all the censures and emendations made by the Father Master of the Sacred Palace himself, by his associate Father Visconti, and by Father Stefani; then from seeing them the Father Inquisitor himself could understand how minor are the things that had been noted and that have been emended. Furthermore, he and all those present could see with how much submission and reverence I agree to give the label of dreams, chimeras, misunderstandings, paralogisms, and conceits to all those reasons and arguments which the authorities regard as favoring opinions they hold to be untrue; they would also understand how true is my claim that on this topic I have never had any opinion or intention but that held by the holiest and most venerable Fathers and Doctors of the Holy Church. This is all the more appropriate inasmuch as the Father Master himself writes that, if need be, he will write ⟨260⟩ to the Father Inquisitor, explaining to him what the book must comply with, and that once this has been observed, the latter may permit the work to be printed.

Thus I beg Your Most Illustrious Lordship to do me the favor of finding out from the Most Serene Patron whether he is satisfied with what I am proposing. If he is, I will make every effort to come to Court at the prearranged time, with the hope of showing to His Highness and to all how ill informed about my opinions are those who say they are disturbing; for the opinions which are disturbing are absolutely not mine, and mine are those which are held by St. Augustine, St. Thomas, and all the other sacred authors.

Mr. Niccolò Aggiunti, who came to visit me just now, will bring this letter to Your Most Illustrious Lordship; to importune you less, he will also return to hear what His Most Serene Highness has decided and will inform me of it. In the meantime, I reverently kiss your hands and pray for your complete happiness.

Bellosguardo, 3 May 1631.

To Your Most Illustrious Lordship.

Your Most Devout and Most Obliged Servant,
Galileo Galilei.

The Vatican Secretary to the Florentine Inquisitor
(24 May 1631)

Very Reverend and Most Honorable Father Inquisitor:

Mr. Galilei is thinking of publishing there a work of his, formerly entitled *On the Ebb and Flow of the Sea*,[52] in which he discusses in a probable fashion the Copernican system and motion of the earth, and he attempts to facilitate the understanding of that great natural mystery by means of this supposition, corroborating it in turn because of this usefulness. He came to Rome to show us the work, which I endorsed, with the understanding that certain adjustments would be made to it and shown back to us to receive the final approval for printing. As this cannot be done due to current restrictions on the roads and the risks for the manuscript, and since the author wants to complete the business there, Your Very Reverend Paternity can avail yourself of your authority and dispatch or not dispatch the book without depending in any way on my review. However, I want to remind you that Our Master[53] thinks that the title and subject should not focus on the ebb and flow but absolutely on the mathematical examination of the Copernican position on the earth's motion, with the aim of proving that, if we remove divine revelation and sacred doctrine, the appearances could be saved with this supposition; one would thus be answering all the contrary indications which may be put forth by experience and by Peripatetic philosophy, so that one would never be admitting the absolute truth of this opinion, but only its hypothetical truth without the benefit of Scripture. It must also be shown that this work is written only to show that we do know all the arguments that can be advanced for this side, and that it was not for lack of knowledge that the decree[54] was issued in Rome; this should be the gist of the book's beginning and ending, which I will send from here properly revised. With this provision the book will encounter no obstacle here in Rome, and Your Very Reverend Paternity will be able to please the author and serve the Most Serene Highness, who shows so much concern in this matter. I remind you that I am your servant and I beg you to honor me with your commands.

Rome, 24 May 1631.

To Your Very Reverend Paternity.

Your Most Devout Servant in the Lord,
Fra Niccolò Riccardi,
Master of the Sacred Palace.

The Florentine Inquisitor to the Vatican Secretary
(31 May 1631)

⟨328⟩ My Most Reverend Father and Most Honorable Lord and Patron:

I am in receipt of the letter of Your Most Reverend Paternity dated the 24th of this month, forwarded to me by this Most Serene Highness, in which you were so kind as to explain to me what has to be done to approve Mr. Galilei's work for printing. Your Most Reverend Paternity may rest assured that I will not fail to execute your orders as diligently as possible, and that I will conduct myself in accordance with your requests on the matter. This Highness is very much interested in the publication of this work, and the said Mr. Galilei appears most ready and most willing for any correction. I have given the work for review to Father Stefani, of your Order, a Father of great merit and a Consultant to the Holy Office. For the preface and the ending then we will wait that they be fixed by the great prudence of Your Most Reverend Paternity. I take this occasion to declare myself your affectionate servant, and I beg you to keep me in your favor and at times to let me be worthy of your orders, which I would regard as a special honor. And finally I piously kiss your hands.

Florence, 31 May 1631.

To Your Most Reverend Paternity.

Your Most Devout and Sincere Servant,
Fra Clemente,[55] Inquisitor of Florence.

The Vatican Secretary to the Florentine Inquisitor
(19 July 1631)

Very Reverend Father Master and Most Honorable Inquisitor:

In accordance with the order of Our Master about Mr. Galilei's book, besides what I mentioned to Your Very Reverend Paternity regarding the body of the work, I send you this beginning or preface[56] to be placed on the first page; the author is free to change or embellish its verbal expressions, as long as he keeps the substance of the content. The ending should be on the same theme.[57] And finally I kiss your hands and declare myself a true servant of Your Very Reverend Paternity.

Rome, 19 July 1631.

To Your Very Reverend Paternity.

Your Pious and Most Obliged Servant,
Fra Niccolò Riccardi,
Master of the Sacred Palace.

The Tuscan Ambassador to Galileo (19 July 1631)

My Very Illustrious and Most Honorable Sir:
After an infinity of cares, finally the preface to your distinguished
work has been corrected, as you can see from the enclosed note addressed to the Father Inquisitor, which I send you with the seal broken,
the way it was delivered to me. The Father Master of the Sacred Palace
indeed deserves to be pitied, for exactly during these days when I was
spurring and bothering him, he has suffered embarrassment and very
great displeasure in regard to some other works recently published, as
he must have done at other times too; he barely complied with our
request, and only because of the reverence he feels for the Most Serene
name of His Highness our Master and for his Most Serene House.

⟨285⟩ I rejoice with you for the termination of this business, as well
as for the tranquillity you will be able to enjoy as a result. Finally, while
I remind you of my special regard and my very ardent desire to serve
you, I beg you for a continuation of your requests, and I kiss your
hands.
Rome, 19 July 1631.
To You Very Illustrious Sir.

Your Most Affectionate Servant,
Francesco Niccolini.

Preface to the Dialogue (1632)[58]

To the Discerning Reader:
Some years ago there was published in Rome a salutary edict which,
to prevent the dangerous scandals of the present age, imposed oppor-
tune silence upon the Pythagorean opinion of the earth's motion. There
were some who rashly asserted that that decree was the offspring of
extremely ill-informed passion, and not of judicious examination; one
also heard complaints that consultants who are totally ignorant of astronomical observations should not cut the wings of speculative intellects by means of an immediate prohibition. Upon noticing the audacity
of such complaints, my zeal could not remain silent. Being fully in-

formed about that most prudent decision, I thought it appropriate to appear publicly on the world scene as a sincere witness of the truth. For at that time I had been present in Rome; I had had not only audiences but also endorsements by the most eminent prelates of that Court; nor did the publication of that decree follow without some prior knowledge on my part. Thus it is my intention in the present work to show to foreign nations that we in Italy, and especially in Rome, know as much about this subject as transalpine diligence can have ever imagined. Furthermore, by collecting together all my own speculations on the Copernican system, I intend to make it known that an awareness of them all preceded the Roman censorship, and that from these parts emerge not only dogmas for the salvation of the soul, but also ingenious discoveries for the delight of the mind.

To this end I have in the discussion taken the Copernican side, proceeding in the manner of a pure mathematical hypothesis and striving in every contrived way ⟨30⟩ to present it as superior to the side of the earth being motionless, though not absolutely but relative to how this is defended by some who claim to be Peripatetics; however, they are Peripatetics only in name since they do not walk around but are satisfied with worshiping shadows, and they do not philosophize with their own judgment but only with the memory of a few ill-understood principles.

Three principal points will be treated. First, I shall attempt to show that all experiments feasible on the earth are insufficient to prove its mobility but can be adapted indifferently to a moving as well as to a motionless earth; and I hope that many observations unknown to antiquity will be disclosed here. Second, I shall examine celestial phenomena, strengthening the Copernican hypothesis as if it should emerge absolutely victorious by means of new speculations; these, however, are meant to be of use for astronomical convenience and not for necessitating nature. Third, I shall propose an ingenious fancy. Many years ago I had occasion to say that the unsolved problem of the tides could receive some light if the earth's motion were granted.[59] Flying from mouth to mouth, this assertion of mine has found charitable people who adopt it as a child of their own intellect. Now, so that no foreigner can ever appear who, strengthened by our own weapons, would throw in our face our inadvertence about such an important phenomenon, I decided to disclose those probable arguments which would render it plausible, given that the earth were in motion. I hope these considerations will show the world that if other nations have navigated more, we have not speculated less, and that to assert the earth's rest and take the contrary solely as a mathematical whim does not derive from ignorance of others' thinking but, among other things, from those reasons provided by piety, religion, acknowledgment of divine omnipotence, and awareness of the weakness of the human mind.

Furthermore, I thought it would be very appropriate to explain these ideas in dialogue form; for this is not restricted to the rigorous observation of mathematical laws, and so it also allows digressions which are sometimes no less interesting than the main topic.

Many years ago in the marvelous city of Venice I had several occasions to engage in conversation with Mr. Giovanfrancesco Sagredo, a man of most illustrious family and sharpest mind. From Florence we were visited by ⟨31⟩ Mr. Filippo Salviati, whose least glory was purity of blood and magnificence of riches; his sublime intellect fed on no delight more avidly than on refined speculation. I often found myself discussing these subjects with these two men, and with the participation of a Peripatetic philosopher[60] who seemed to have no greater obstacle to the understanding of the truth than the fame he had acquired in Aristotelian interpretation.

Now, since Venice and Florence have been deprived of those two great lights by their very premature death at the brightest time of their life, I have decided to prolong their existence, as much as my meager abilities allow, by reviving them in these pages of mine and using them as interlocutors in the present controversy. There will also be a place for the good Peripatetic, to whom, because of his excessive fondness of Simplicius's commentaries, it seemed right to give the name of his revered author[61] without mentioning his own. Those two great souls will always be revered in my heart; may they receive with favor this public monument of my undying friendship, and may they assist me, through my memory of their eloquence, to explain to posterity the above-mentioned speculations.

These gentlemen had casually engaged in various sporadic discussions, and, as a result, in their minds their thirst for learning had been aroused rather then quenched. Thus they made the wise decision to spend a few days together during which, having put aside every other business, they would attend to reflecting more systematically about God's wonders in heaven and on earth. They met at the palace of the most illustrious Sagredo, and, after the proper but short greetings, Mr. Salviati began as follows.

Ending of the Dialogue *(1632)*

SALVIATI. Now, since it is time to put an end to our discussions, it remains for me to ask you to please excuse my faults if, when more calmly going over the things I have put forth, you should encounter difficulties and doubts not adequately resolved. For these ideas are novel, my mind is imperfect, the subject is a great one, and finally I do not ask and have not asked from others an assent which I myself do not give to this fancy; I could very easily ⟨488⟩ regard it as a most unreal chimera and a most solemn paradox. As for you, Mr. Sagredo, although in the discussions we have had you have shown many times by means of strong endorsements that you were satisfied with some of my thoughts, I feel that in part this derived more from their novelty than from their certainty, but much more from your courtesy; for by means of your assent you have wished to give me the satisfaction which one naturally feels from the approval and praise of one's own creations. Moreover, just as I am obliged to you for your politeness, so I appreciate the sincerity of Mr. Simplicio; indeed, I have become very fond of him for defending his teacher's doctrine so steadfastly, so forcefully, and so courageously. Finally, just as I express thanks to you, Mr. Sagredo, for your very courteous feelings, so I beg forgiveness to Mr. Simplicio if I have upset him sometimes with my excessively bold and resolute language; there should be no question that I have not done this out of any malicious motive, but only to give him a greater opportunity to advance better thoughts, so that I could learn more.

SIMPLICIO. There is no need for you to give these excuses, which are superfluous, especially to me who am used to being in social discussions and public disputes; indeed, innumerable times I have heard the opponents not only get upset and angry at each other, but also burst out into insulting words, and sometimes come very close to physical violence. As for the discussions we have had, and especially the last one about the explanation of the tides, I really do not understand them completely. However, from the superficial conception I have been able to grasp, I confess that your idea seems to me much more ingenious than any others I have heard, but I do not thereby regard it as true and conclusive. Indeed, I always keep before my mind's eye a very firm doctrine, which I once learned from a man of great knowledge and eminence, and before which one must give pause. From it I know what you would answer if both of you are asked whether God with His infinite power and wisdom could give to the element water the back and forth motion we see in it by some means other than by moving the containing basin; I say you will answer that He would have the power and the knowledge to do this in many ways, some of them even inconceivable by our intellect. Thus, I immediately conclude that in view of this it would be excessively bold if someone should want to limit and compel divine power and wisdom to a particular fancy of his. ⟨489⟩

SALVIATI. An admirable and truly angelic doctrine, to which there corresponds very harmoniously another one that is also divine. This is the doctrine which, while it allows us to argue about the constitution of the world, tells us that we are not about to discover how His hands built it (perhaps in order that the exercise of the human mind would not be stopped or destroyed). Thus let this exercise, granted and commanded to us by God, suffice to acknowledge His greatness; the less we are able to fathom the profound depths of His infinite wisdom, the more we shall admire that greatness.

SAGREDO. This can very well be the final ending of our arguments over the last four days. Hereafter, if Mr. Salviati wants to take some rest, it is proper that our curiosity grant it to him, but on one condition; that is, when he finds it least inconvenient, he should comply with the wish, especially mine, to discuss the problems which we have set aside and which I have recorded, by having one or two other sessions, as we agreed. Above all I shall be looking forward with great eagerness to hear the elements of our Academician's [62] new science of motion (natural and violent). Finally, now we can, as usual, go for an hour to enjoy some fresh air in the gondola which is waiting for us.

Special Commission Report on the Dialogue
(September 1632) [63]

[I] [64]

In accordance with the order of Your Holiness, we have laid out the whole series of events pertaining to the printing of Galilei's book, which printing then took place in Florence. In essence the affair developed this way.

In the year 1630 Galileo took his book manuscript to the Father Master of the Sacred Palace in Rome in order to have it reviewed for printing. The Father Master gave it for review to Father Raffaele Visconti, an associate of his and a professor of mathematics, who after several emendations was ready to give his approval as usual, if the book were to be printed in Rome.

We have written the said Father to send the said certificate, and we are now waiting for it. We have also written to get the original manuscript, in order to see the corrections made.

The Master of the Sacred Palace wanted to review the book himself; but, in order to shorten the time and facilitate negotiations with printers, he stipulated that it be shown him page by page and gave it the imprimatur for Rome.

Then the author went back to Florence and petitioned the Father Master for permission to print it in that city, which permission was denied. But the latter forwarded the case to the Inquisitor of Florence, thus removing himself from the transaction. Moreoover, the Father Master notified the Inquisitor of what was required for publication, leaving to him the task of having it printed or not.

The Master of the Sacred Palace has shown a copy of the letter he wrote the Inquisitor about this business, as well as a copy of the Inquisitor's reply to the said ⟨325⟩ Master of the Sacred Palace. In it the Inquisitor says that he gave the manuscript for correction to Father Stefani, Consultant to the Holy Office.

After this the Master of the Sacred Palace did not hear anything, except that he saw the book printed in Florence and published with the Inquisitor's imprimatur, and that there is also an imprimatur for Rome.

We think that Galileo may have overstepped his instructions by asserting absolutely the earth's motion and the sun's immobility and thus deviating from hypothesis; that he may have wrongly attributed the existing ebb and flow of the sea to the nonexistent immobility of the sun and motion of the earth, which are the main things; and that he may have been deceitfully silent about an injunction given him by the Holy Office in the year 1616, whose tenor is: "that he abandon completely the above-mentioned opinion that the sun is the center of the world and the earth moves, nor henceforth hold, teach, or defend it in any way whatever, orally or in writing; otherwise the Holy Office would start proceedings against him. He acquiesced in this injunction and promised to obey."[65]

One must now consider how to proceed, both against the person and concerning the printed book.

[II]

In point of fact:

1. Galilei did come to Rome in the year 1630, and he brought and showed his original manuscript to be reviewed for printing. Though he had been ordered to discuss the Copernican system only as a pure mathematical hypothesis, one found immediately that the book was not like this, but that it spoke absolutely, presenting the reasons for and against though without deciding. Thus the Master of the Sacred Palace determined that the book be reviewed and be changed to the hypothetical mode: it should have a preface to which the body would conform, which would describe this manner of proceeding, and which would prescribe it to the whole dispute to follow, including the part against the Ptolemaic system, carried on merely *ad hominem* [66] and to show that in reproving the Copernican system the Holy Congregation had heard all the arguments.

2. To follow this through, the book was given for review with these orders to Father Raffaello Visconti, an associate of the Master of the Sacred Palace, since he was a professor of mathematics. He reviewed it and emended it in many places, informing the Master about others disputed with the author, which the Master took out without further discussion. Having approved it for the rest, Father Visconti was ready to give it his endorsement to be placed at the beginning of the book as usual, if the book were to be printed in Rome as it was then presumed.

We have written the Inquisitor to send this endorsement to us and are expecting to receive it momentarily, and we have also sent for the original so that we can see the corrections made.

3. The Master of the Sacred Palace wanted to review the book himself, but since the author complained about the unusual practice of a second revision and about the delay, to facilitate the process it was decided that before sending it to press the Master would see it page by page. In the meantime, to enable the author to negotiate with printers, he ⟨326⟩ was given the imprimatur for Rome, the book's beginning was compiled, and printing was expected to begin soon.

4. The author then went back to Florence, and after a certain period he petitioned to print it in that city. The Master of the Sacred Palace absolutely denied the request, answering by saying that the original should be brought back to him to make the last revision agreed upon and that without this he for his part would have never given permission to print it. The reply was that the original could not be sent because of

the dangers of loss and the plague. Nevertheless, after the intervention of His Highness there, it was decided that the Master of the Sacred Palace would remove himself from the case and refer it to the Inquisitor of Florence: the Master would describe to him what was required for the correction of the book and would leave him the decision to print it or not, so that he would be using his authority without any responsibility on the part of the Master's office. Accordingly, he wrote to the Inquisitor the letter whose copy is appended here, labeled *A*, dated 24 May 1631,[67] received and acknowledged by the Inquisitor with the letter labeled *B*,[68] where he says he entrusted the book to Father Stefani, Consultant to the Holy Office there.

Then a brief composition of the book's preface was sent to the Inquisitor, so that the author would incorporate it with the whole, would embellish it in his own way, and would make the ending of the *Dialogue* conform with it. A copy of the sketch that was sent is enclosed labeled *C*,[69] and a copy of the accompanying letter is enclosed labeled *D*.[70]

5. After this the Master of the Sacred Palace was no longer involved in the matter, except when, the book having been printed and published without his knowledge, he received the first few copies and held them in customs, seeing that the instructions had not been followed. Then, upon orders from Our Master, he had them all seized where it was not too late and diligence made it possible to do so.

6. Moreover, there are in the book the following things to consider, as specific items of indictment:

i.[71] That he used the imprimatur for Rome without permission and without sharing the fact of the book's being published with those who are said to have granted it.

ii. That he had the preface printed with a different type[72] and rendered it useless by its separation from the body of the work; and that he put the "medicine of the end"[73] in the mouth of a fool and in a place where it can only be found with difficulty,[74] and then he had it approved coldly by the other speaker by merely mentioning but not elaborating the positive things he seems to utter against his will.

iii. That many times in the work there is a lack of and deviation from hypothesis, either by asserting absolutely the earth's motion and the sun's immobility, or by characterizing the supporting arguments as demonstrative and necessary, or by treating the negative side as impossible.

iv. He treats the issue as undecided and as if one should await rather than presuppose the resolution. ⟨327⟩

v. The mistreatment of contrary authors and those most used by the Holy Church.

vi. That he wrongly asserts and declares a certain equality between the human and the divine intellect in the understanding of geometrical matters.[75]

vii. That he gives as an argument for the truth the fact that Ptolemaics occasionally become Copernicans, but the reverse never happens.[76]

viii. That he wrongly attributed the existing ebb and flow of the sea to the nonexistent immobility of the sun and motion of the earth.

All these things could be emended if the book were judged to have some utility which would warrant such a favor.

7. In 1616 the author had from the Holy Office the injunction that "he abandon completely the above-mentioned opinion that the sun is the center of the world and the earth moves, nor henceforth hold, teach, or defend it in any way whatever, orally or in writing; otherwise the Holy Office would start proceedings against him. He acquiesced in this injunction and promised to obey."

Cardinal Barberini[77] *to the Florentine Nuncio*[78]
(25 September 1632)

Rome, 25 September 1632.

To the Monsignor Bishop of Ascoli, Nuncio in Florence:

Because some suspicious things were discovered in Galileo's works,[79] out of regard for the Most Serene Grand Duke His Holiness appointed a special commission to examine them and see whether one could avoid bringing them to the attention of the Sacred Congregation of the Holy Office; ⟨398⟩ the commission met five times, considered everything carefully, and decided that one could not avoid bringing the business to the attention of the Congregation. To show his goodwill toward His Highness, His Holiness sent this information to His Highness's Lord Ambassador, who in the name of His Highness has implored His Holiness not to forward the case to the Congregation. To the person who brought him the news, the Ambassador said[80] that the decision made little sense in view of the fact that the book had been seen and approved by the Master of the Sacred Palace; but it was pointed out to him that if the book actually contains errors, these should not for that reason be allowed to circulate. When told all this, His Excellency was bound to secrecy by the Holy Office, but he was given permission to convey the information to the Most Serene Grand Duke, who was bound to secrecy in the same way.

Thus the book was brought before the Congregation of the Holy Office, and everything was considered with the utmost seriousness. It was decided to order the Father Inquisitor of your city to call Galileo and in the name of His Holiness give him an injunction to appear before the Father Commissary of the Holy Office for the whole month of October; the Inquisitor is supposed to make Galileo promise to obey this injunction in the presence of witnesses, so that, if he refuses to accept and obey it, they could be examined.

I am communicating all this to you only for your information, so that if anyone speaks to you, you can answer appropriately; for you must not of your own initiative discuss this matter much or at all. I mean that, although Galileo knows that the Holy Congregation sees some errors in that work, nevertheless he plans to send the said books to various parts of the world as if they were under some dispensation. You should try to learn the truth about this, and if you find that they are being sent, you should inform the Lord Cardinals Apostolic Delegates of Bologna and Ferrara, as well as all other officials or Bishops or Inquisitors of places through which they might pass, so that they will be seized. Make sure you learn any way you can when these books will leave your city, so that you can again notify the above-mentioned Delegates and other officials; but you should not confide this in advance to the Bishops and Inquisitors outside the Papal States, for it will be sufficient to inform them after these books have been shipped; and here I am referring to those bales that will not have to go through Bologna, Ferrara, or other terminals in the Papal States, since for these it will be sufficient to inform the Most Eminent Delegates and Governors.

Galileo to Diodati (15 January 1633)

Very Illustrious Sir and Most Honorable Patron:

I owe answers to two letters, one from you and the other from Mr. Pierre Gassendi, written 1 November of last year but received by me only ten days ago. Because I am extremely preoccupied and burdened, I should like this to serve as an answer to both of you, who are very good friends and whose letters deal with the same subject; that is, your having received my *Dialogue*, sent to both, and your having quickly looked at it with praise and approval. I thank you for that and feel obliged, though I shall be waiting for a more frank and critical judgment after you have reread it more calmly, for I fear you will find in it many things to contest.

I am sorry I did not get Morin's and Froidmont's books[81] until six months after the publication of my *Dialogue*, since I would have had the occasion to say many things in praise[82] of both and also to make some observations on certain ⟨24⟩ details, primarily one in Morin and another in Froidmont. As regards Morin, I am surprised by the truly great respect he shows toward judicial astrology and that he should pretend to establish its certainty by his conjectures (which seem to me very uncertain, not to say most uncertain). It will really be astonishing if he has the cleverness to place astrology in the highest seat of the human sciences, as he promises; I shall be waiting with great curiosity to see such a stunning novelty. As for Froidmont, though he appears to be a man of great intellect, I wish he had not committed what I think is truly a serious error, albeit extremely common; that is, to confute Copernicus's opinion, he first begins with sneering and scornful barbs against those who hold it to be true, then ⟨more inappropriately⟩ he wants to establish it primarily with the authority of Scripture, and finally he goes so far as to label it in that regard little less than heretical.

It seems to me one can prove very clearly that this manner of proceeding is far from laudable. For if I ask Froidmont whose works are the sun, the moon, the earth, the stars, their arrangement, and their motions, I think he will answer they are works of God; and if I ask from whose inspiration Holy Scripture derives, I know he will answer that it comes from the Holy Spirit, namely again God. Thus, the world is the works, and the Scripture is the words, of the same God. Then let me ask him whether the Holy Spirit has ever used, spoken, or pronounced words which, in appearance, are very contrary to the truth, and whether this was done to accommodate the capacity of the people, who are for the most part very uncouth and incompetent. I am very sure he will answer, together with all sacred writers, that such is the habit of the Scripture; in hundreds of passages the latter puts forth (for the said reason) propositions which, taken in the literal meaning of the words, would be not mere heresies, but very serious blasphemies, by making God himself subject to anger, regret, forgetfulness, etc. However, suppose I ask him whether, to accommodate the capacity and belief of the same people, God has ever changed his works; or whether nature is God's inexorable minister, is deaf to human opinions and desires, and has always conserved and continues to conserve her ways regarding the

motions, shapes, and locations ⟨25⟩ of the parts of the universe. I am
certain he will answer that the moon has always been spherical, although for a long time common people thought it was flat; in short, he
will say that nothing is ever changed by nature to accommodate her
works to the wishes and opinions of men. If this is so, why should we,
in order to learn about the parts of the world, begin our investigation
from the words rather than from the works of God? Is it perhaps less
noble and lofty to work than to speak? If Froidmont or someone else
had established that it is heretical to say the earth moves, and that demonstrations, observations, and necessary correspondences show it to
move, in what sort of plot would he have gotten himself and the Holy
Church? On the contrary, were we to give second place to Scripture, if
the works were shown to be necessarily different from the literal meaning of the words, then this would in no way be prejudicial to Scripture;
and if to accommodate popular abilities the latter has many times attributed the most false characteristics to God himself, why should it be
required to limit itself to a very strict law when speaking of the sun and
the earth, thus disregarding popular incapacity and refraining from attributing to these bodies properties contrary to those that exist in reality? If it were true that motion belongs to the earth and rest to the
sun, no harm is done to Scripture, which speaks in accordance with
what appears to the popular masses.

Many years ago, at the beginning of the uproar against Copernicus,
I wrote a very long essay[83] showing, largely by means of the authority
of the Fathers, how great an abuse it is to want to use Holy Scripture so
much when dealing with questions about natural phenomena, and how
it would be most advisable to prohibit the involvement of Scripture in
such disputes; when I am less troubled, I shall send you a copy. I say
less troubled because at the moment I am about to go to Rome, summoned by the Holy Office, which has already suspended my *Dialogue*.
From reliable sources I hear the Jesuit Fathers have managed to convince some very important persons that my book is execrable and more
harmful to the Holy Church than the writings of Luther and Calvin.
Thus I am sure it will be prohibited, despite the fact that to obtain the
license I went personally to Rome and delivered it ⟨26⟩ into the hands
of the Master of the Sacred Palace; he examined it very minutely (changing, adding, and removing as much as he wanted), and after licensing it
he also ordered it to be reviewed again here. This reviewer did not find
anything to modify, and so, as a sign of having read and examined it
most diligently, he resorted to changing some words; for example, in

many places he said *universe* instead of *nature*, *title* instead of *attribute*, *sublime* mind in place of *divine*; and he asked to be excused by saying that he predicted I would be dealing with very bitter enemies and very angry persecutors, as indeed it followed. The publisher says that so far this suspension has made him lose a profit of 2000 scudi, since not only could he have sold the thousand volumes he had already printed, but he could have reprinted twice as many. As for me, to my other troubles is added the following very serious one—namely, to be unable to pursue the completion of my other works (especially the one on motion), so as to publish them before I die.

I read with special pleasure Mr. Pierre Gassendi's Disquisition against Fludd's philosophy, as well as the Appendix on celestial observations.[84] I was unable to observe Mercury or Venus in front of the sun because of rain; but in regard to their smallness, I have been certain of it for a long time, and I am glad that Mr. Gassendi has found this to be a fact. Please share this information with the said gentleman, to whom I send warm greetings, as I also do to the Reverend Father Mersenne. Finally, I kiss your hands with all my heart and pray for your happiness.

Florence, 15 January 1633.

To You Very Illustrious Sir.

Your Most Devout and Most Obliged Servant,
Galileo Galilei.

CHAPTER VIII

Diplomatic Correspondence (1632–1633) [1]

15 August 1632

. . . I have not been able yet to see the Master of the Sacred Palace in regard to the question of Mr. Galilei. However, because I hear that there has been set up a Commission of persons versed in this profession, all unfriendly to Galileo, responsible to the Lord Cardinal Barberini,[2] I have decided to speak about it to His Eminence himself at the earliest opportunity. Furthermore, because they are thinking of calling a mathematician from Pisa, named Mr. Chiaramonti[3] and rather unfriendly to Mr. Galileo's opinions, it will be necessary that His Highness have someone talk to him, to make sure he pursues the cause of truth here, rather than his emotional feelings. . . .

---

22 August 1632

My Most Illustrious and Most Honorable Lord:

I have managed to make a very eloquent plea on behalf of Mr. Galilei, in accordance with the instructions I received, namely that his book be allowed ⟨375⟩ to be published since it is already in print with the required licenses, has been reviewed and examined here and in Florence, and has had the beginning and the ending revised as the authorities wanted. Furthermore, I have petitioned that some neutral persons be appointed to the Commission which has been set up on the matter since the present members are opponents of the same Mr. Galilei. However, to these things and all the others I proposed to the Lord Cardinal Barberini I have received no answer from His Eminence, except that he will present all to the Pope and that one is dealing here with the interests of a friend of His Holiness, who has affection and high regard for him; nor did His Eminence reveal any other details, as if it were a business of great secrecy, though he did show goodwill toward Mr. Galilei. I also hear from friends that they are not thinking of prohibiting it, but rather of amending a few words. Nevertheless, we will have to wait for its resolution. Finally, I express my reverence for Your Most Illustrious Lordship.

Rome, 22 August 1632.
To Your Most Illustrious Lordship.
Lord Balì[4] Cioli.

Your Most Obliged Servant,
Francesco Niccolini.

---

28 August 1632

My Most Illustrious and Most Honorable Lord:

I have explained to the Lord Cardinal Barberini everything which Your Most Illustrious Lordship ordered me to explain on behalf of Mr. Galilei. Although His Eminence paid attention to everything, he did not reply anything definite, except that I should speak of it with the Master of the Sacred Palace; the latter says he can defend himself very well in regard to what is presumed about the revisions and the licenses to print the book, and that he can tell me how. When I explained to His Eminence that the book had been handed over by the author to the supreme authority here, he was a little put off, and then said: This must mean that the Master of the Sacred Palace is the supreme authority. For the rest he did not get involved, either in regard to sending the complaints to your city or about the other details, except that I should deal with the Master of the Sacred Palace; this I will do the day after tomorrow since so far I have not succeeded. Then I will inform Your Most Illustrious Lordship of what I learn. In the meantime I kiss your hands.

Rome, 28 August 1632.
To Your Most Illustrious Lordship.
Lord Balì Cioli.

> Your Most Obliged Servant,
> Francesco Niccolini.

---

5 September 1632

My Most Illustrious and Most Honorable Lord:

Yesterday I did not have the time to report to Your Most Illustrious Lordship what had transpired (in a very emotional atmosphere) between myself and the Pope in regard to Mr. Galilei's work. I appreciated the opportunity because I was able to say certain things to His Holiness himself, though without any profit. As for me, I too am beginning to believe, as Your Most Illustrious Lordship well expresses it, that the sky is about to fall. While we were discussing those delicate subjects of the Holy Office, His Holiness exploded into great anger, and suddenly he told me that even our Galilei had dared entering where he should not have, into the most serious and dangerous subjects which could be stirred up at this time. I replied that Mr. Galilei had not published without the approval of his ministers and that for that purpose I myself had obtained and sent the prefaces to your city. He answered, with the same outburst of rage, that he had been deceived by Galileo and Ciampoli, that in particular Ciampoli had dared tell him that Mr. Galilei was ready to do all His Holiness ordered and that everything was fine, and that this was what he had been told, without having ever seen or read the work; he also complained about the Master of the Sacred Palace, though he said that the latter himself had been deceived by having his written endorsement of the book pulled out of his hands with beautiful words, by the book being then printed in Florence on the basis of other endorsements but without complying with the form given to ⟨384⟩ the Inquisitor, and by having his name printed in the book's list of imprimaturs even though he has no jurisdiction over publications in other cities. Here I interjected that I knew His Holiness had appointed a Commission for this purpose, and that, because it might have members who hate Mr. Galilei (as it does), I humbly begged His Holiness to agree to give him the opportunity to justify himself. Then His Holiness answered that in these matters of the Holy Office the procedure was simply to arrive at a censure and then call the defendant to recant. I replied: Does

it thus not seem to Your Holiness that Galileo should know in advance
the difficulties and the objections or the censures which are being raised
against his work, and what the Holy Office is worried about? He answered violently: I say to Your Lordship that the Holy Office does not
do these things and does not proceed this way, that these things are
never given in advance to anyone, that such is not the custom; besides,
he knows very well where the difficulties lie, if he wants to know them,
since we have discussed them with him and he has heard them from
ourselves. I replied begging him to consider that the book was explicitly
dedicated to our Most Serene Patron and that they were dealing with
one of his present employees, and saying that because of this too I hoped
he would be helpful and would also order his ministers to take it into
consideration. He said that he had prohibited works which had his pontifical name in front and were dedicated to himself, and that in such
matters, involving great harm to religion (indeed the worst ever conceived), His Highness too should contribute to preventing it, being a
Christian prince; furthermore, that because of this, I should clearly
write to the Most Serene Highness to be careful not to get involved, as
he had in the other case of Alidosi, because he would not come out of
it honorably. I retorted that I was sure I would receive orders to trouble
him again, and that I would do it, but that I did not believe His Holiness
would bring about the prohibition of the already approved book without at least hearing Mr. Galilei first. His Holiness answered that this
was the least ill which could be done to him and that he should take
care not to be summoned by the Holy Office; that he has appointed a
Commission of theologians and other persons versed in various sciences, serious and of holy mind, who are weighing every minutia, word
for word, since one is dealing with the most perverse subject one could
ever come across; and again that his complaint was to have been deceived by Galileo and Ciampoli. Finally, he told me to write to our Most
Serene Patron that the doctrine is extremely perverse, that they would
review everything with seriousness, and that His Highness should not
get involved but should go slow; furthermore, not only did he impose
on me the secret about what he had just told me, but he charged me to
report that he also was imposing it on His Highness. He added that he
has used every civility with Mr. Galilei since he explained to the latter
what he knows, since he has not sent the case to the Congregation of
the Holy Inquisition, as is the norm, but rather to a special Commission
newly created, which is something, and since he has used better man-

ners with Galileo than the latter has used with His Holiness, who was deceived. Thus I had an unpleasant meeting, and I feel the Pope could not have a worse disposition toward our poor Mr. Galilei. Your Most Illustrious Lordship can imagine in what condition I returned home yesterday morning.

This past Monday I had gone to meet with the Master of the Sacred Palace. After explaining to him all the points from your letter, and after also calming him in regard to his complaints, I gathered some hopeful signs rather than anything else; ⟨385⟩ in particular, he seemed to believe that one would not go as far as prohibiting the book, but that it would only be corrected and emended in some points which are really bad. He also said that if he could tell me something in advance, without detriment to him and without disobeying orders, he would do it; but that he too had to move cautiously since he had already gone through storms in this regard and had helped himself as well as he knew how. He complains that the form given in his letter to the Inquisitor was not observed, that the declaration to be printed at the beginning is in a different typeface and is not linked to the rest of the work, and that the ending does not correspond to the beginning at all.

As for me, if I have to express my sense to Your Most Illustrious Lordship, I believe it is necessary to take this business without violence and to deal with the ministers and with the Lord Cardinal Barberini rather than with the Pope himself, for when His Holiness gets something into his head, that is the end of the matter, especially if one is opposing, threatening, or defying him, since then he hardens and shows no respect to anyone. The best course is to temporize and try to move him by persistent, skillful, and quiet diplomacy, involving also his ministers, depending on the nature of the business. For example, in the case of Mr. Mariano[5] we should have tried to convince the Nuncio and have him do the writing and the pleading, without us entering into the merits of the case and especially without us writing briefs, which may have given him the opportunity to show that he was very learned and knew more than our experts and to give contrary advice; if we had done that, we would not have exacerbated the feelings of the Pope, to whom one must not give indication of wanting to dispute the administration of justice.

The strong letter which Your Most Illustrious Lordship wrote on the 30th in regard to Mr. Galilei, and which I have just now received, does not seem well balanced to me now that I have heard the Pope, since by making an uproar we will exasperate and spoil the situation. However, I must only obey because my will should depend entirely on the orders of the Patrons. This is really going to be a troublesome affair. I am thinking of going again to talk with the Master of the Sacred Palace to tell him what I gathered from His Holiness, as well as to hear what he thinks of it and what his attitude is now. However, the matter proceeds with extreme secrecy. Finally, I express my reverence to Your Most Illustrious Lordship.

Rome, 5 September 1632.
To Your Most Illustrious Lordship.
Lord Balì Cioli.

Your Most Obliged Servant,
Francesco Niccolini.

---

11 September 1632

My Most Illustrious and Most Honorable Lord:
I have shared with the Father Master of the Sacred Palace the content of your letter dated the 30th of last month, in regard to the business of Mr. Galilei. I decided to do this not so much because of the friendliness and trust which exists between us, but more because of what the Pope told me at the last audience on this matter, as I also reported in previous letters. He answered and advised that I should express the feelings of His Most Serene Highness by means of such complaints if we want to ruin Mr. Galilei and to break with His Holiness, but that if we want to help him I should completely abandon such expressions of complaints; for just as there is no doubt that Mr. Galilei will benefit by temporizing, so we are sure that for now nothing but harm will follow by direct confrontation. In fact, the Pope believes that the Faith is facing many dangers and that we are not dealing with mathematical subjects here but with Holy Scripture, religion, and Faith. Moreover, in printing the book, the manner prescribed and the instructions given were not followed, and the author's opinion is not only mentioned in it but in many places openly declared in an inadmissible manner, so much so that everyone wonders how they let it be printed in your city. He is inclined to believe that ⟨389⟩ if it had been printed here, by reviewing it page by page, as agreed, it would have been published in a form that would be acceptable; and I too believe it was a great error to print it in Florence. He says further that, things being as they are, he feels, indeed he is sure,

that the best help one can give Mr. Galilei is to proceed slowly and without noise. In the meantime, His Most Reverend Paternity is reviewing the work and is trying to fix it in certain places so that it is acceptable; when this is completed he intends to bring it to the Pope and tell him that he is sure it can be allowed to circulate and that His Holiness now has the opportunity of using his customary mercy with Mr. Galileo. After this one could then more appropriately say a few words in the name of His Highness, adding some modest expression of complaint, which might serve to make the Pope yield more easily and agree to let it be published. For the rest, he says that to proceed otherwise seems to him not only a waste of time but harmful to the case, and that to ask for Father Campanella[6] and Father Benedetto[7] as lawyers and attorneys is something impossible to obtain, even if the Holy Office wanted to proceed this way; for the first wrote an almost similar work which was prohibited and as a criminal could not conduct a defense, while the other could not be heard on account of his lack of self-confidence and for other reasons. As for the members of this Commission, he says that he himself in particular is obliged to defend Mr. Galilei, because of his friendship with the latter and with my family,[8] chiefly because of his desire and obligation to serve the Most Serene Patron and also because of his having approved the book; furthermore, the Papal Theologian[9] is truly full of goodwill, and that Jesuit[10] was proposed by himself, is a confidant of his, and has undoubtedly good intentions; thus he does not see why we should complain about them. However, above all he says, with the usual confidentiality and secrecy, that in the files of the Holy Office they have found something which alone is sufficient to ruin Mr. Galilei completely; that is, about twelve years ago, when it became known that he held this opinion and was sowing it in Florence, and when on account of this he was called to Rome, he was prohibited from holding this opinion by the Lord Cardinal Bellarmine, in the name of the Pope and the Holy Office. So he says he is not really surprised that His Highness is acting with so much concern, for he has not been told all the circumstances of this business. In short, he begs His Highness to believe him that no help can be provided for Mr. Galilei, except by proceeding very quietly for the moment, and on this he gives his word and swears on his honor and on his soul; he adds that if it turns out otherwise, he promises to give himself up to His Highness in Florence to be punished, even by being beheaded. In the meantime, he begs that his devotion to His Highness, which makes him speak with such trust and express these things, not be held against him. Finally, he adds that the Pope can say many more things on this subject than he can. And now I kiss your hands.

Rome, 11 September 1632.
To Your Most Illustrious Lordship.
Lord Balì Cioli.

Your Most Obliged Servant,
Francesco Niccolini.

---

18 September 1632

My Most Illustrious and Most Honorable Lord:

Three days ago His Holiness sent to me Mr. Pietro Benessi, one of his secretaries, through whom I was given the following information. As a sign of his regard toward the person of our Most Serene Lord, His Holiness was taking the unusual step of letting me know that he could not avoid forwarding to the Congregation of the Holy Inquisition the book by Mr. Galilei on the Copernican system of the earth's motion; this happened after His Holiness, on account of the concern shown by His Highness, took the unusual step of having it accurately and carefully examined, word for word, by a special Commission of persons who are very learned and very competent in theology and in other disciplines, to see whether one could have avoided forwarding it to the Holy Office; finally, however, after the above-mentioned care, it was judged that it could not in any way be allowed to circulate without a diligent examination by the same Holy Inquisition, which could then judge what to do with it; I was supposed to accept all of this as a sign of the paternal affection His Holiness feels toward His Most Serene Highness, whom he was binding to secrecy and submitting to the orders of the Holy Office in this regard, as he was also doing with me, to ensure we did not speak about it to anyone and did not inform anyone without incurring the usual censures. Your Most Illustrious Lordship can imagine with what troubled heart I received this message. I replied that His Highness would find it strange that a book should be subjected to the will of the Holy Office after being approved by this Holy See and receiving permission to be printed by the Master of the Sacred Palace, and that I should like for His Holiness to agree to allow Mr. Galileo to defend himself somehow, as I had already petitioned. Mr. Benessi answered that he had no other information on the matter, and that he did not really know what to say; however, according to what he had heard from His Holiness when he received his orders, he felt he could tell me that it was not the first time that books already approved by Inquisitors were then rejected and prohibited here, because this had happened many times; furthermore, the Holy Office is not in the habit of hearing defenses, as I was requesting. I replied to the first point that, nevertheless, perhaps the books he was talking about had been approved by Inquisitors of other states, outside Rome, but that here we are dealing ⟨392⟩ with an approval given in Rome, with the participation of the Master of the Sacred Palace and others known even to His Holiness himself. He retorted that in cases where religion might suffer damage, it was less harmful to overreact occasionally than to be remiss as a result of the reasons I mentioned, and thus to endanger Christianity with some sinister opinion; furthermore, he had been told by His Holiness that, since we are dealing with dangerous dogmas, His Highness should put aside all respect and affection toward his Mathematician and be glad to contribute himself to shielding Catholicism from any danger. Finally, he repeated that His Holiness had wanted to proceed by disclosing this information so that His Highness would know about the decision taken by the Congregation, as a sign of good relations and regard toward his Most Serene name. I asked him to humbly kiss for me the feet of His Holiness, and I told him I would inform my Most Serene Patron of His Holiness's commands, though I said I hated to do it because of the displeasure I would bring him.

Then I nevertheless deemed it necessary to speak to His Holiness myself, which I did this morning. After repeating to him what had been told me in his name, I protested that he could have given Mr. Galilei the opportunity to speak and justify himself because, as long as this matter is still in the hands of a special committee, which has nothing to do with the Holy Office and is not the same as its own Congregation, there would be no prejudice to the regulations and procedures of that tribunal, which merely censures, prohibits, and orders defendants to recant; furthermore, I said that the Most Serene Grand Duke, my Lord, reverently implored to be rendered thereby obliged to His Holiness, who should have no doubt about new examples and new proofs of such obligation. However, he answered that it was the same thing—that the unusual committee had been formed only to please the Most Serene Patron as well as Mr. Galilei and to see whether it was possible not to introduce this business to the Holy Office; and that I should be glad for what had been shared with me so far, outside usual procedure. I replied by again humbly begging him to consider that Mr. Galilei is Mathematician to His Highness, currently employed and salaried by him, and also universally known as such. His Holiness answered that this was another reason why he had gone out of the ordinary in this case and that Mr. Galileo was still his friend, but these opinions were condemned about sixteen years ago and Galileo had gotten himself into a fix which he could have avoided; for these subjects are troublesome and dangerous, this work of his is indeed pernicious, and the matter is more serious than His Highness thinks. Then he started to tell me about this matter and these opinions, but with the explicit order not to reveal these things even to His Highness, under penalty of censure; although I begged to be able to report them at least to His Highness only, he answered that I should be glad to have known them from him in confidence and as a friend, and not as a minister. I asked him whether among the members of the Congregation of the Inquisition there are any who understand mathematical subjects; he answered that there were Cardinals Bentivoglio and Verospi and some others, and he mumbled that there might also be some of those from the committee. Then he added, telling me to report it fully to His Most Serene Highness, that one must be careful not to let Mr. Galilei spread troublesome and dangerous opinions under the pretext of running a certain school for young people, because he had heard something (I know not what); furthermore, ⟨393⟩ His Highness should please be careful and have someone be vigilant to ensure some error is not sown throughout the state, which might cause him trouble. I replied I did not believe he could dissent from the true Catholic dogmas in any way, but everyone in this world has those who envy him and wish him ill. Although His Holiness uttered "Enough! Enough!", nevertheless I went on to add the following thought: since Mr. Galilei had once received the instructions he had to observe in printing his book, and he presumably did not comply with them, His Holiness could now have it changed in accordance with those instructions and let it circulate, without having to prohibit the work as a whole. However, in regard to this he answered that the Master of the Sacred Palace was himself at fault. Then in an amicable way he related to me the story of a virtuoso who once sent one of his works to Cardinal Alciato to have it reviewed; since it was beautifully written, in order not to have the sheets of paper messed up, he asked that places deserving correction be marked with some wax; when the Cardinal returned the book to the virtuoso without any mark, the latter then went to thank him and to rejoice with him at the fact that he had not marked anything, for there were none of the marks requested; but the Cardinal answered that he had not used the wax because he would have had to go to a store, ask for one of those vases where liquefied wax is kept, and throw the whole book there in order to censure it; that is how he clarified his response. After laughing a little, I then added again that I hoped nevertheless His Holiness would order that Mr. Galilei's work be harmed as little as possible. I also begged him to allow me to report these things to Your Most Illustrious Lordship because His Highness would need to reply and give me instructions, and he is not in the habit of writing himself, but rather I corresponded with you. The Pope reflected on it a little, and then he answered that, since I said that His Highness does not do his own writing, he was satisfied to let you too know this, but you are bound by the same censures of the Holy Office and by the injunction not to speak or report to others except to His Highness, and he charged me to mention this to you explicitly. Thus, Your Most Illustrious Lordship can explain all of this to the Most Serene Patron and can instruct me how to proceed, while I have the added heavy burden of having to write and copy this troublesome and very long correspondence by my own hand. Finally, I kiss your hands.

Rome, 18 September 1632.

To Your Most Illustrious Lordship.

Your Most Obliged Servant,
Francesco Niccolini.

---

24 October 1632

. . . In regard to the business of Mr. Galileo, I want to mention the copy of his letter to the Lord Cardinal Barberini, which he himself sent me. I do not think it is at all advisable to give it to him; for His Eminence will immediately hand it over to the Congregation, where it will be scrutinized and weighed. In particular they will want to know the identity of that important person mentioned in it whom he does not want to name;[11] they will want to know that ⟨419⟩ from him personally, and he will certainly be placed under house arrest and forced to recant or to write against what he published, without hope that his reasons will be accepted or perhaps even heard. I do not think we can do anything else but ask for the desired delay or change of venue, because the other requests cannot be obtained and have already been denied several times during my negotiations with His Holiness himself, as His Highness may have heard from my letters. As soon as the Lord Cardinal Barberini returns to Rome, I will petition for a delay or change of venue, and then I shall report what His Eminence tells me.

I wanted to see Father Benedetto, but he is still at Castel Gandolfo for the reasons I wrote about yesterday evening to Mr. Galilei himself. However, being at Court, he may have negotiated something in regard to Mr. Galilei's letters, which I sent him at home and which will have been forwarded to him out of town. For the rest I refer to what I wrote to Mr. Galilei. . . .

---

6 November 1632

My Most Illustrious and Most Honorable Lord:

In accordance with the instructions received, I informed the Lord Cardinal Barberini, in the name of the Most Serene Patron, about Mr. Mariano Alidosi coming to Rome. . . .

I also made sure to use the occasion to mention Mr. Galilei's desires in regard to his having to appear here. As His Eminence is very circumspect when speaking about things of the Holy Office, on account of the prohibitions governing it, he did not say anything specific. Nevertheless, it seems that he showed a good disposition toward Mr. Galilei and that one can hope for some delay in his coming here, if not for a decision to hear the case there. He managed to say only that he will bring up the matter and will see what can be done. . . .

---

13 November 1632

My Most Illustrious and Most Honorable Lord:

After giving Mr. Galilei's letter to the Lord Cardinal Barberini, this week I did a number of errands on his behalf without naming His Highness but acting as if I were taking the initiative. I discussed his petitions with the Lord Cardinal Ginetti, who is very close to the Pope and one of the cardinals of the Congregation of the Holy Office, and with Mr. Boccabella, an Assessor of the same Congregation; I pointed out his age of seventy-five years,[12] his ill health, and the danger to his life from the trip and from the quarantine, without proper accommodations and without any amenities. However, because these people listen but do not respond, this morning I discussed it with His Holiness himself. After mentioning that Mr. Galilei is ready to obey and to comply with what he will be ordered to do, I undertook to explain to His Holiness the same things at great length, to move him to pity poor Mr. Galileo, who is now so old and whom I love and adore; I was presuming that His Holiness might have seen the letter Mr. Galilei wrote to the Lord Cardinal his nephew. However, His Holiness told me that he had seen the letter and that, in short, there is no way of avoiding Mr. Galilei's coming to Rome. I replied that his age was such that His Holiness was taking the risk of not holding the trial either there or here, and I tried to argue that with the discomfort combined with the grief he might pass away during the journey. His Holiness answered that he could come slowly in a litter with all comforts, for indeed it was necessary to examine him personally, and that God would hopefully forgive his error of having gotten involved in an intrigue like this after His Holiness himself (when he was cardinal) had delivered him from it. I said that the approval given here to the book had caused all this since, in light of the written endorsement and the orders given to the Florentine Inquisitor, one had felt safe and had had no suspicions in this regard. However, I was interrupted by being told ⟨429⟩ that Ciampoli [13] and the Master of the Sacred Palace had behaved badly and that subordinates who do not do what their masters want are the worst possible servants; for, when asking Ciampoli many times what was happening with Galilei, His Holiness had never been told anything but good and had never been given the news that the book was being printed, even when he was beginning to smell something. Finally, he reiterated that one is dealing with a very bad doctrine.

I then reported all this to the Lord Cardinal Barberini, and I tried to move him to pity too by mentioning the same things; but I did not get anything, except that His Eminence asked me what the Pope had said and promised that they would facilitate the quarantine. Furthermore, neither His Holiness nor the Lord Cardinal had told me anything in regard to a delay in the trial, perhaps because they had not yet considered the matter; so today I sent my secretary to Mr. Boccabella to know what I should report in this regard. He said he would forcefully raise the issue at his next audience, despite the fact that tonight, pursuant to the orders of the Holy Congregation, they would be formally writing to Florence that Mr. Galilei must come to Rome. Next week I will try to learn what has been granted, and then I shall inform Your Most Illustrious Lordship; in the meantime tonight I am writing about the same things to Mr. Galilei. Finally, I kiss your hands.

Rome, 13 November 1632.
To Your Most Illustrious Lordship.
Lord Balì Cioli.

Your Most Obliged Servant,
Francesco Niccolini.

---

11 December 1632

... This morning I undertook some new efforts on behalf of Mr. Galilei by mentioning to a number of people the things which Your Most Illustrious Lordship wrote me about and which he himself also mentioned in a letter to me, in order to see whether one could obtain a delay. Finally, however, not only do I believe this to be impossible, but I feel it is necessary for him to decide to come however he can; he could go somewhere in the state of Siena to stay there at least twenty days, thus beginning the quarantine, because this promptness will also benefit him a great deal. In regard to the point about where he should come to reside, it is impossible to learn anything about that, and it is enough to say we are dealing with the Congregation of the Holy Office, whose goings-on are so secret and none of whose members opens his mouth because of the censures that are in force. He could come directly to my house here; what may happen thereafter, I have no way of knowing. Furthermore, Monsignor Boccabella gives the friendly advice that for his own sake he should come as soon as possible, rather than continuing ⟨439⟩ with further delays; for they would take into consideration the possibility that it might suffice as a penalty, his having to make the journey from there, at a time like this, and at his age of seventy-five years. However, Your Most Illustrious Lordship must report these things to him orally in order to save Monsignor Boccabella from any breach of secret; moreover, when here he should never name this person. . . .

---

26 December 1632

My Most Illustrious and Most Honorable Lord:
This affair of Mr. Galileo is turning also against the Master of the Sacred Palace, which pains me very much. For it is true that he gave the book his written endorsement, which ⟨444⟩ he should have never done, as the general of the Dominicans and everyone else says; and he did send there that corrected preface and those orders to the Inquisitor,

though unwillingly, and only out of the regard he feels for the Most Serene Patron and out of the intimate friendship he has with my family.

As for Mr. Galilei himself, I showed his last letters again to the Monsignor Assessor of the Holy Office. Although he knows that their allegations deserve compassion, nevertheless he feels embarrassed to mention them to the Pope because His Holiness has a very bad disposition about the matter. He wanted at least to have in hand those medical certificates,[14] so as to have a pretext to begin mentioning it to His Holiness, for otherwise he does not know how to bring it up. He also wishes that Mr. Galilei at least had left Florence to show obedience, and then if he had become ill he could have received greater favor. I do not know what more I can do in this matter, besides what I have been able to obtain so far on behalf of Mr. Galileo. In the meantime, His Highness should think of how to respond if the Nuncio receives some strange order, as some here think. Finally, I express my reverence to you.

Rome, 26 December 1632.

To Your Most Illustrious Lordship.

Lord Balì Cioli.

> Your Most Obliged Servant,
> Francesco Niccolini.

---

15 January 1633

My Most Illustrious and Most Honorable Lord:

The Congregation of the Holy Office received the certificate of Mr. Galilei's ill health, and I tried to learn from the Monsignor Assessor whether it was being approved, as one would hope, and whether they would grant him the favor of a delay in his coming here. He said in confidence that they are paying little attention to this certificate, indicating by the movements of his head and by the tone of his voice that they did not like it and believe it was written just to please him. He did not know what else to say, except that he would deem it very appropriate and in Mr. Galilei's interest for him to decide to make whatever arrangements he can for his comfort and to come; for otherwise they may very well take some extraordinary step against him. I do not think it would be advisable to say nothing to him, though on the other hand I would not want to trouble unnecessarily this poor old man. I thought it best to report this to Your Most Illustrious Lordship so that you can let him know whatever you deem most effective, in the manner you judge most appropriate. Finally, I kiss your hands.

Rome, 15 January 1633.
To Your Most Illustrious Lordship.
Lord Balì Cioli.

Your Most Obliged Servant,
Francesco Niccolini.

---

14 February 1633

My Most Illustrious and Most Honorable Lord:

Last night Mr. Galilei arrived at this house in good health. Today he went to see Monsignor Boccabella, not in the latter's capacity as a minister of the Holy Office (since it is already two weeks that he left the office of Assessor), but in his role as a friend who has always shown great compassion and liking toward Mr. Galilei. With the pretext of expressing his gratitude for such a good disposition, he asked for advice on how he should behave and began receiving some suggestions. He also presented himself, ⟨41⟩ of his own accord, to the new Assessor, and he tried to do the same with the Father Commissary, though the latter was not there. Furthermore, a friend of the same Father, Mr. Girolamo Matti, had already done some services on behalf of Mr. Galilei and had offered to continue, not so much out of love for his singular qualities but in order to please His Highness; thus, I thought it would be good for him to see and speak to this man for the same purpose, as indeed it happened. Today there has been no time to do anything else. Tomorrow, I myself will try to see the Lord Cardinal Barberini to recommend Mr. Galilei to him and have His Eminence intercede with His Holiness (if he wishes) to allow Mr. Galilei to remain in this house (if possible) without being taken to the Holy Office; this would be out of regard for his age, his reputation, and his promptness in obeying. I shall inform Your Most Illustrious Lordship of what happens, and I kiss your hands.

Rome, 14 February 1633.
To Your Most Illustrious Lordship.

Your Most Obliged Servant,
Francesco Niccolini.

---

16 February 1633

My Most Illustrious and Most Honorable Lord:

I continue to assist Mr. Galilei by every possible means. The Lord

Cardinal Barberini has warned him not to socialize and not to bother talking with everyone who comes to visit him since for various reasons this could cause harm and prejudice; thus he stays home in seclusion, waiting to be told something. In the meantime, the Commissary of the Holy Office has promised to report to His Holiness and to these other Lords that he is ready to obey, which is regarded as a very important point. Although the affairs of this Tribunal can never be discussed with certainty and with clarity, nevertheless, from the little light which has emerged, it seems that no great disaster is about to happen. The Lord Cardinal Barberini does not usually attend meetings of the Congregation of the Holy Office, especially those on Wednesday, which are held at the Minerva; but this morning he went, and perhaps they discussed the manner of proceeding in this trial. This is a guess, however, for it may also be that His Eminence went instead in regard to the business of the dispensation of Mantua, though Father Bombino does not know whether that case has been presented yet. . . .

---

19 February 1633

. . . I have already informed Your Most Illustrious Lordship of Mr. Galilei's arrival and of what we had begun to do on his behalf. Now I can add that I went to see Cardinals Scaglia and Bentivoglio to recommend him to them, and I found them very well disposed. The Commissary has told him the same thing which the Lord Cardinal Barberini had communicated to him, namely that he should be satisfied to keep himself in seclusion, without going out, and even without socializing at the house if possible. The Commissary stated that this was not meant to be an order from himself or from the Holy Congregation, but rather a friendly warning motivated by the prejudice and harm which might otherwise result. Since he is going along with this, I am trying to help him through friends in every way I deem appropriate to the case. From the way ministers seem to be acting, I am hoping this trial will proceed with restraint, despite the fact that His Holiness feels so negative about this business, as I have reported several times.

I hear that they discussed this matter on Wednesday morning, when the Lord Cardinal Barberini, deviating from habit, attended the meeting of the Congregation of the Holy Office. Thereafter, Mr. Galilei was not told anything; nor did he hear from anyone in that Tribunal except from Monsignor Serristori, one of its Consultants. The latter has come twice, claiming to be acting on his own and to want to visit; but he has always mentioned the trial and discussed various details, and so I believe one may be certain he has been sent to hear what Mr. Galilei says, what his attitude is, and how he defends himself, so that they can then decide what to do and how to proceed. These visits seem to have comforted this good old man by encouraging him and giving him the impression that they are interested in his case and in what decisions are being taken. Nevertheless, sometimes this persecution seems very strange to him. I have advised him to show himself always ready to obey and to submit to whatever he is ordered to do, because this is the way to mitigate the fervor of those who are fiercely excited and treat this trial as a personal thing. . . .

---

27 February 1633 (I)

My Most Illustrious and Most Honorable Lord:

Mr. Galilei is still staying in this house, without having been told any more than what I reported to Your Most Illustrious Lordship in my previous letters. In the meantime, I have not failed to recommend him in the ⟨55⟩ manner allowed by the nature of the Tribunal of the Holy Office, mentioning his readiness to obey and willingness to give satisfaction in every way, and also the respect which his age and his ills deserve. Although I cannot say precisely what stage the trial is at, nor what will happen, nevertheless, from what I gather, the greatest difficulty seems to lie in the claim by these Lords that in the year 1616 Mr. Galilei received an injunction not to dispute about or discuss this opinion. However, he says that the order does not have this form, but rather that he should not hold or defend it; he thinks he can justify himself since his book does not show he holds or defends it, and does not decide anything, but only presents the reasons for one side and for the other. The other things seem to be less significant and also easier to dispose of. Nevertheless, since in this city things frequently turn out very differently than expected, one can be sure only after the event. There are some who feel that it will be difficult for him to avoid being detained by the Holy Office, despite the fact that for now they are proceeding with great affection and calm. I will inform Your Most Illustrious Lordship of what develops, and for now I kiss your hands.

Rome, 27 February 1633.

To Your Most Illustrious Lordship.

Lord Balì Cioli.

> Your Most Obliged Servant,
> Francesco Niccolini.

27 February 1633 (II)

My Most Illustrious and Most Honorable Lord:

Yesterday morning I explained to His Holiness what Your Most Illustrious Lordship had instructed me to say in regard to the alliance which could be made at this time against the Turks. . . .

I informed him of Mr. Galileo's arrival, adding that I hoped His Holiness was convinced of his reverent and most devout observance in ecclesiastical matters, and especially in the matter at hand; for he had come in very high spirits and resolved to submit entirely to his wise judgment and to the most prudent views of the Congregation, and he had even uplifted and consoled me. His Holiness answered that he had done Mr. Galilei a singular favor, not done to others, by allowing him to stay in this house rather than at the Holy Office, and that this kind procedure had been used only because he is a dear employee of the Most Serene Patron and because of the regard due to His Highness; for a Knight of the House of Gonzaga, son of Ferdinando, had been not only placed in a litter and escorted to Rome under guard but was taken to the Castle and kept there for a long time till the end of the trial. I showed myself to be aware of the nature of the favor, and I humbly thanked His Holiness; then I begged him to give orders for expediting the trial, so that ⟨56⟩ the sick old man could return home as soon as possible. He replied that the activities of the Holy Office ordinarily proceeded slowly, and he did not really know whether one could hope for such a quick conclusion, since they were in the process of preparing for the formal proceedings and had not yet finished with that. Then he went on to say that, in short, Mr. Galilei had been ill-advised to publish these opinions of his, and it was the sort of thing for which Ciampoli was responsible; for although he claims to want to discuss the earth's motion hypothetically, nevertheless when he presents the arguments for it he mentions and discusses it assertively and conclusively; furthermore, he had also violated the order given to him in 1616 by the Lord Cardinal Bellarmine in the name of the Congregation of the Index. In his defense I replied saying everything I remembered him expressing and explaining to me on this and related matters; but, as the subject is delicate and troublesome, and as His Holiness gives the impression that Mr. Galileo's doctrine is bad and that he even believes it, the task is not easy. Furthermore, even if they should be satisfied with his answers, they will not want to give the appearance of having made a blunder, after everybody knows they summoned him to Rome.

I strongly recommended him to the protection of the Lord Cardinal Barberini, and I did this all the more gladly inasmuch as I felt I found His Holiness less irritated than usual. His Eminence replied that he felt warmly toward Mr. Galilei and regarded him as an exceptional man, but this subject is very delicate for it involves the possibility of introducing some imaginary dogma into the world, particularly into Florence where (as I know) the intellects are very subtle and curious, and especially by his reporting much more validly what favors the side of the earth's motion than what can be adduced for the other side. I said that perhaps the nature of the situation indicated this, and therefore he was not to blame; but His Eminence answered that I was aware that he knew how to express exquisitely and how to justify wonderfully whatever he wanted. Finally, I kiss your hands.

Rome, 27 February 1633.
To Your Most Illustrious Lordship.
Lord Balì Cioli.

Your Most Obliged Servant,
Francesco Niccolini.

---

6 March 1633

My Most Illustrious and Most Honorable Lord:
About Mr. Galilei I cannot report to Your Most Illustrious Lordship anything more than what I wrote in my past letters, except that I am trying to arrange, if possible, that he be allowed occasionally to go into the garden of the Trinità [15] in order to be able to exercise a little; for it is very harmful to remain always inside the house. However, so far I have not received any answer, nor do I know what we can hope on the matter. . . .

---

13 March 1633

My Most Illustrious and Most Honorable Lord:
This morning I began my discussion with His Holiness by expressing thanks, as Your Most Illustrious Lordship had instructed me to do, for having granted Mr. Galilei the privilege of residing at this house rather than at the Holy Office; further, I begged him with the most appropriate words of which I was capable to expedite the trial. However, His Holiness answered that he had gladly shown this consideration in honor of His Highness, but he did not think there is any way of avoiding summoning Mr. Galilei to the Holy Office when he is to be ⟨68⟩ examined,

because this is the usual procedure and this is the least that can be done. I replied that I hoped His Holiness would double the obligation imposed on His Highness by exempting him from this too; but I was told that he thinks it cannot be avoided. I reiterated that his old age, ill health, and readiness to submit to any censure might render him worthy of any favor; but he again said he does not think there is any way out, and may God forgive Mr. Galilei for having meddled with these subjects. He added that one is dealing with new doctrines and Holy Scripture, that the best course is to follow the common opinion, and that may God also help Ciampoli with these new opinions since he too is attracted to them and is a friend of the new philosophy; further, Mr. Galileo had been his friend, they had conversed and dined several times together familiarly, and he was sorry to have to displease him, but one was dealing with the interests of the faith and religion. I think I went on to add that if he is heard, he will easily give every satisfaction, though with the proper reverence which is due the Holy Office. He replied that Mr. Galilei will be examined in due course, but there is an argument which no one has ever been able to answer: that is, God is omnipotent and can do anything; but if He is omnipotent, why do we want to bind him?[16] I said that I was not competent to discuss these subjects, but I had heard Mr. Galilei himself say that first he did not hold the opinion of the earth's motion as true and then that since God could make the world in innumerable ways, one could not deny that He might have made it this way. However, he got upset and told me that one must not impose necessity on the blessed God; seeing that he was losing his temper, I did not want to continue discussing what I did not understand, and thus displease him, to the detriment of Mr. Galilei. So I said that, in short, he was here to obey and to retract everything for which he could be blamed in regard to religion and that I was not knowledgeable about this science and did not want to utter some heresy by continuing to talk about it; then, in order not to arouse suspicion that I too might offend the Holy Office, I changed the subject. I begged His Holiness to have compassion and to make him a recipient of his mercy, especially by being willing to consider granting him the privilege of not having to leave this house; but he repeated that he would arrange for Mr. Galilei to have certain specific rooms, which are the best and most comfortable in this place. I stated that I would report about this to His Highness and would return to plead if he should order me to do so.

When I returned home I told Mr. Galilei part of what I had discussed with the Pope, but I have not yet told him that they are thinking of summoning him to the Holy Office. I was sure this would worry him a great deal and would make him restless until that time, especially since it is not known yet when they will want him; for in regard to expediting the trial, the Pope told me that he does not know yet what one can expect and they will do whatever they can. Yet a few days ago the Commissary of the Holy Office told my secretary that they were going to start as soon as possible. On the other hand, I do not like His Holiness's attitude, which is not at all mollified. Finally, I kiss your hands.

Rome, 13 March 1633.
To Your Most Illustrious Lordship.
Lord Balì Cioli.

> Your Most Obliged Servant,
> Francesco Niccolini.

---

19 March 1633

My Most Illustrious and Most Honorable Lord:
About Mr. Galilei I cannot add to what I wrote in my past letters, except that, since the Most Serene Patron has written on his behalf ⟨74⟩ to Cardinals Bentivoglio and Scaglia, I think it would be appropriate for him to write also to the other Cardinals of the Congregation. They are the following:[17] Sant'Onofrio, Barberini, Borgia, Gessi, Ginetti, San Sisto, and Verospi. This might encourage them to favor him, and further, if they should learn that His Most Serene Highness had written to the others, they would not then get the idea that they had been esteemed and trusted less than the others. Nevertheless, I defer to your best judgment. In the meantime, one may believe that nothing will be done until after the holidays. For now I understand that the Lord Cardinals Scaglia and Bentivoglio are very much united to protect and favor him. Finally, I express my reverence to Your Most Illustrious Lordship.

Rome, 19 March 1633.
To Your Most Illustrious Lordship.
Lord Balì Cioli.

> Your Most Obliged Servant,
> Francesco Niccolini.

---

9 April 1633

My Most Illustrious and Most Honorable Lord:
Last Wednesday the Lord Cardinal Barberini told my secretary that I should go to see him, and so Thursday afternoon I went to him to

receive his orders. He explained to me he had been ordered by His Holiness and by the Congregation of the Holy Office to inform me that in order to proceed with Mr. Galileo's case they had to summon him to appear before the Holy Office; further, since His Eminence did not know whether they would be finished in a few hours, it might be necessary to detain him there for the convenience of the proceedings. They had wanted me to know this out of regard for the house where he is staying and for me as a minister of His Most Serene Highness, as well as in recognition of the good relations existing between His Most Serene Highness and the Holy See (especially in matters of the Holy Inquisition), ⟨85⟩ and finally as an expression of acknowledgment to a Prince so full of zeal in religious matters. I thanked His Eminence very much for the regard His Holiness and the Holy Congregation were showing toward this Most Serene House and also toward me as its representative. However, I could hide neither the ill health of this good old man, who for two whole nights had constantly moaned and screamed on account of his arthritic pains, nor his advanced age, nor the hardship he would suffer as a result; taking these things into consideration, I thought I could implore His Holiness to reflect whether he could agree to allow him to come back to my house to sleep, and, in order to prevent disclosure of his interrogations, bind him to silence under penalty of censure. The Lord Cardinal did not think one could hope for any favor in this regard, though in the course of the discussion I begged him to reflect on it; on the other hand, he offered all the conveniences one might desire, and he said Mr. Galilei would not be held in a cell or as if in prison, which is the norm in other cases, but would be provided with some good rooms, perhaps even left unlocked. This morning I spoke to His Holiness about it, and, after I expressed appropriate thanks for the advance notice he was so kind to give me, His Holiness said he was sorry that Mr. Galilei had gotten involved in this subject, which he considers to be very serious and of great consequence for religion. Nevertheless, Mr. Galilei tries to defend his opinions very strongly; but I exhorted him, in the interest of a quick resolution, not to bother maintaining them and to submit to what he sees they want him to hold or believe about that detail of the earth's motion. He was extremely distressed by this, and, as far as I am concerned, since yesterday he looks so depressed that I fear greatly for his life. I will try to obtain permission for him to keep a servant and to have other conveniences; nor does any one of us fail to try to console him and recommend him to friends and to those who are participating in these deliberations, for he really deserves the greatest good, and this whole house is extremely fond of him and feels unspeakable sorrow about it.

I will present the letters you sent me to the Lord Cardinals of the Holy Congregation, and, just as I implored His Holiness and the Lord Cardinal for a quick and favorable resolution, so will I be making the same plea with them. Finally, I kiss Your Most Illustrious Lordship's hands.

Rome, 9 April 1633.
To Your Most Illustrious Lordship.
Lord Balì Cioli.

> Your Most Obliged Servant,
> Francesco Niccolini.

---

16 April 1633

My Most Illustrious and Most Honorable Lord:

After reporting to you what the Lord Cardinal Barberini told me in regard to Mr. Galileo, I can add that on Tuesday morning he presented himself to the Father Commissary of the Holy Office, who received him in a friendly manner and had him lodged in the chambers of the prosecutor of that Tribunal, rather than in the cells usually given to criminals; thus, not only does he reside among the officials, but he is free to go out into the courtyard of that house. Nevertheless, he had thought he could return home the evening of the same day because he was examined immediately after his arrival; but the Commissary himself told my secretary, who had accompanied Mr. Galileo, that he can only do what he is ordered to do after he reports to His Holiness about the latter's appearance and about what has been learned from this first examination. It is felt, however, that he will be sent home soon; for in this trial they have proceeded in extraordinary and agreeable ways, out of regard for the promptness His Highness shows in matters ⟨95⟩ of the Holy Inquisition, as I have been told by His Holiness himself and by the Lord Cardinals Barberini and Bentivoglio, and so one can hope that the conclusion will be quick and favorable. Indeed, there is no precedent of anyone ever having been interrogated during a trial without being detained in a prison cell, and in this regard he has profited from being employed by His Highness and from being lodged at this house; nor is there knowledge of anyone else (whether bishop, prelate, or nobleman) who, immediately upon his arrival in Rome, has not been kept at the Castle or at the same palace of the Inquisition, subject to all rigor and strictness.

Furthermore, they even allow his servant to wait on him, to sleep there, and, what is more, to come and go as he pleases, and they allow my own servants to bring him food to his room from here and to return to my house morning and evening. Just as these conveniences are permitted on account of the authority enjoyed by and regard due to this Most Serene House, so it would seem appropriate to convey special thanks to His Holiness, once the present difficulties have been resolved. In the meantime, I myself will be pleading with His Holiness and with the Lord Cardinal; according to the Commissary, the latter is helping Mr. Galilei and has helped him, in front of the Pope, by mitigating the feelings of His Holiness in a significant way. Nevertheless, Mr. Galilei is distressed for being at the Holy Office and he considers the treatment harsh. I will not stop helping him toward a favorable conclusion, as I did by means of the letters from the Most Serene Highness after he left this house; but since in that Tribunal one has to deal with men who do not talk and do not answer, either orally or in writing, so it is more difficult to negotiate with them or fathom their meanings. Indeed, some of the Cardinals to whom I gave His Highness's letters asked to be excused if they will not answer, on account of the prohibition on the matter; some were hesitant even to accept them, thinking they might be subject to censure, but I encouraged them with the example of the Lord Cardinal Barberini and the others who had accepted them. Mr. Galilei must have been enjoined from discussing or disclosing the contents of the cross-examination since he did not want to say anything to Tolomei, my personal secretary, not even whether he can or cannot speak. Finally, I kiss your Most Illustrious Lordship's hands.

Rome, 16 April 1633.
To Your Most Illustrious Lordship.
Lord Balì Cioli.

Your Most Obliged Servant,
Francesco Niccolini.

---

23 April 1633

My Most Illustrious and Most Honorable Lord:
There is no lack of business with the Holy Office. Mr. Mariano Alidosi arrived here and surrendered himself at its prison, according to what I have been told by Monsignor Baffadi, who came to this house Tuesday with a letter by Your Most Illustrious Lordship for me dated the 17th of the past. . . .

As for Mr. Galilei, he is still at the same place, with the same conveniences. He writes to me daily, and I answer telling him my views with frankness without worrying about it; I am beginning to think that this party will end with someone else being blamed. He has been examined only once,[18] and I believe they will free him immediately after His Holiness returns from Castel Gandolfo, which will be for the feast of the Ascension. So far they are not talking about the book's content; they are concerned only with learning why the Father Master of the Sacred Palace gave his imprimatur, since His Holiness says he was never told anything about it, let alone that he ever ⟨104⟩ ordered the license be granted. On the evening before the Pope's departure I decided to recommend him to the Lord Cardinal Antonio.[19] Since I now hear from Mr. Galilei himself what he writes to Mr. Bocchineri,[20] I am becoming convinced that this contact with Antonio has benefited him more than anything else; this Cardinal takes things seriously when one goes to him for help, since he likes to be esteemed. Finally, I kiss Your Most Illustrious Lordship's hands.

Rome, 23 April 1633.
To Your Most Illustrious Lordship.
Lord Balì Cioli.

> Your Most Obliged Servant,
> Francesco Niccolini.

---

1 May 1633

My Most Illustrious and Most Honorable Lord:
Yesterday, when I least expected him, Mr. Galileo was sent back to my house, although the proceedings have not yet ended. This is due to the intervention[21] of the Father Commissary with the Lord Cardinal Barberini, who, acting on his own without the Congregation, had him freed so that he can recover from the discomforts and from his usual indispositions, which kept him in constant torment. The same Father Commissary also lets it be known that he is working ⟨110⟩ to have the trial expedited, so that one can forget about it; if he succeeds, the whole thing will be shortened, and many people will be spared trouble and dangers.

About Mr. Mariano Alidosi I can only say that, after he was imprisoned, Monsignor Baffadi has not told me anything else. However, he does not enjoy the conveniences and advantages granted to Mr. Galileo, and he is subject to every austerity. Finally, I kiss Your Most Illustrious Lordship's hands.

Rome, 1 May 1633.
To Your Most Illustrious Lordship.

Your Most Obliged Servant,
Francesco Niccolini.

3 May 1633

... As I mentioned in previous letters, Mr. Galilei was allowed to return to this house, where he seems to have regained his good health. Because he desires ⟨111⟩ to have his trial brought to a final conclusion, the Father Commissary of the Holy Office has told him he intends to come to see him for this purpose. The latter continues to do us all possible favors in this business and to show himself very well disposed toward that Most Serene House, so I leave nothing undone to preserve and increase this good disposition of his. . . .

15 May 1633

... Mr. Galilei is in very good health, but his trial is still not being expedited. Furthermore, he is sequestered in this house, somewhat displeased for being unable to engage in exercise. In regard to what Your Most Illustrious Lordship tells me, namely that His Highness does not intend to pay for his expenses here beyond the first month,[22] I can reply that I am not about to discuss this matter with him while he is my guest; I would rather assume the burden myself. The expenses will not exceed 14 or 15 scudi a month, including everything; thus, if he were to stay here six months, they would add up to 90 or 100 scudi between him and a servant. . . .

22 May 1633

My Most Illustrious and Most Honorable Lord:
From what Your Most Illustrious Lordship will read here, you will be able to see that no time has been lost at all since yesterday morning, after I returned from the audience. Please forgive me if I do not answer some details because in the next mail I will make up for whatever I miss.

I spoke with His Holiness about the business of Mr. Galileo and was told by him and the Lord Cardinal Barberini that the trial can be easily concluded at the second Inquisition meeting, which will be Thursday of next week. I can very well imagine the prohibition of the book, unless they decide instead to have it include an apology written by Mr. Galilei himself, as I proposed to His Holiness. He will be subject to some salutary penance since they claim he disobeyed orders given to him in 1616 by the Lord Cardinal Bellarmine in regard to the same subject of the earth's motion. I have not yet told him everything because I plan to prepare him slowly in order not to distress him; hence, it will be good not to disclose these views there, so as to prevent his relatives from telling him, especially since there could also be changes. . . .

Rome, 22 May 1633.

To Your Most Illustrious Lordship.

P.S. I forgot to mention that when I spoke with His Holiness about Mr. Galileo, he told me they would also try to expedite the case of Mr. Mariano Alidosi, so His Highness could see that his concerns were taken seriously by His Holiness; furthermore, I believe that about the estate there is nothing to worry about. Then I kissed His Holiness's feet and expressed to him appropriate thanks for that.

Lord Balì Cioli.

Your Most Obliged Servant,
Francesco Niccolini.

---

29 May 1633

My Most Illustrious and Most Honorable Lord:

In the last few days I explained to the Father Commissary of the Holy Office that Mr. Galilei was in need of occasionally going out of the house to have some fresh air and to walk because he was used to exercising and, now being prevented from it, he was not in good health. I pleaded with him that, while one was waiting for the conclusion of the trial, he obtain this favor from the Lord Cardinals of the Holy Office and especially from the Lord Cardinal Barberini. This indeed followed, for the same Father informed me that these Lords had agreed. Thus, now Mr. Galilei goes to these gardens, though in a half-closed carriage. . . .

---

19 June 1633

My Most Illustrious and Most Honorable Lord:

This morning His Holiness displayed very friendly feelings in innumerable ways. . . .

Again I pleaded that Mr. Galilei's trial be brought to an end. His Holiness told me that it has already been concluded, and that one morning next week he will be called to the Holy Office to hear its decision or sentence. Upon hearing this, I implored His Holiness to be willing, as a favor to our master His Most Serene Highness, to mitigate the harshness that he and the Holy Congregation might have thought of having to use in this business; for the Most Serene Highness had, during this trial, incurred many obligations as a result of so many other singular favors, and he was planning to give due thanks personally, once this affair was completely finished. He replied that it was not necessary for His Highness to take this trouble, for he had willingly done every favor to Mr. Galileo out of the warmth he feels toward the Most Serene Patron. However, he said that in regard to the issue, there is no way of avoiding prohibiting that opinion, since it is erroneous and contrary to the Holy Scripture dictated by the mouth of God; and in regard to the person, as ordinarily and usually done, he would have to remain imprisoned here for some time because he disobeyed the orders he received in the year 1616; but as soon as the sentence is published, His Holiness will see me again and will discuss with me what can be done to cause the least pain and the least affliction to him, for there is no way of avoiding some personal punishment. I then humbly implored him again to show his usual compassion toward the advanced age of seventy years of this good old man, and also toward his sincerity. However, he indicated believing that there is no way of avoiding relegating him at least to some convent like Santa Croce for some time; but he did not yet know very well what the Congregation would decide, although all of it united and with no one dissenting[23] was inclined in this direction about penalizing him. Nevertheless, His Holiness wants it declared, in order to discourage others, that every punishment has been mitigated out of regard for our master the Most Serene Grand Duke, and this is the real and only reason why all possible accommodations have been and will be made.

So far to Mr. Galileo I have only mentioned the imminent conclusion of the trial and the prohibition of the book. However, I told him nothing about the personal punishment, in order not to afflict him by telling him everything at once; furthermore, His Holiness ordered me not to tell him in order not to torment him yet and because things will perhaps change through deliberations. Thus I also think it proper that no one there inform him of anything. . . .

CHAPTER IX

The Later Inquisition
Proceedings (1633)

Galileo's First Deposition (12 April 1633)

Summoned, there appeared personally in Rome at the palace of the
Holy Office, in the usual quarters of the Reverend Father Commissary,
fully in the presence of the Reverend Father Fra Vincenzo Maculano of
Firenzuola, ⟨337⟩ Commissary General, and of his assistant Reverend
Father Carlo Sinceri, Prosecutor of the Holy Office, etc.

Galileo, son of the late Vincenzio Galilei, Florentine, seventy years
old, who, having taken a formal oath to tell the truth, was asked by the
Fathers the following:

Q:[1] By what means and how long ago did he come to Rome.

A: I arrived in Rome the first Sunday of Lent, and I came in a litter.

Q: Whether he came of his own accord, or was called, or was ordered by someone to come to Rome, and by whom.

A: In Florence the Father Inquisitor ordered me to come to Rome
and present myself to the Holy Office, this being an injunction by the
officials of the Holy Office.

Q: Whether he knows or can guess the reason why he was ordered
to come to Rome.

A: I imagine that the reason why I have been ordered to present
myself to the Holy Office in Rome is to account for my recently printed
book. I imagine this because of the injunction to the printer and to
myself, a few days before I was ordered to come to Rome, not to issue
any more of these books, and similarly because the printer was ordered

by the Father Inquisitor to send the original manuscript of my book to the Holy Office in Rome.

Q: That he explain the character of the book on account of which he thinks he was ordered to come to Rome.

A: It is a book written in dialogue form, and it treats of the constitution of the world, that is, of the two chief systems, and the arrangement of the heavens and the elements.

Q: Whether, if he were shown the said book, he is prepared to identify it as his.

A: I hope so; I hope that if the book is shown me I shall recognize it.

And having been shown one of the books printed in Florence in 1632, whose title is *Dialogue of Galileo Galilei Lincean* etc.,[2] which examines the two systems of the world, and having looked at it and inspected it carefully, he said: I know this book very well; it is one of those printed in Florence; and I acknowledge it as mine and written by me.

Q: Whether he likewise acknowledges each and every thing contained in the said book as his.

A: I know this book shown to me, for it is one of those printed in Florence; and I acknowledge all it contains as having been written by me.

Q: When and where he composed the said book, and how long it took him. ⟨338⟩

A: In regard to the place, I composed it in Florence, beginning ten or twelve years ago; and it must have taken me seven or eight years, but not continuously.

Q: Whether he was in Rome other times, especially in the year 1616, and for what occasion.

A: I was in Rome in the year 1616; then I was here in the second year of His Holiness Urban VIII's pontificate; and lastly I was here three years ago, the occasion being that I wanted to have my book printed. The occasion for my being in Rome in the year 1616 was that, having heard objections to Nicolaus Copernicus's opinion on the earth's motion, the sun's stability, and the arrangement of the heavenly spheres, in order to be sure of holding only holy and Catholic opinions, I came to hear what was proper to hold in regard to this topic.

Q: Whether he came of his own accord or was summoned, what the reason was why he was summoned, and with which person or persons he discussed the above-mentioned topics.

A: In 1616 I came to Rome of my own accord, without being summoned, for the reason I mentioned. In Rome I discussed this matter with some cardinals who oversaw the Holy Office at that time, especially with Cardinals Bellarmine, Aracoeli, San Eusebio, Bonsi, and d'Ascoli.[3]

Q: What specifically he discussed with the above-mentioned cardinals.

A: The occasion for discussing with the said cardinals was that they wanted to be informed about Copernicus's doctrine, his book being very difficult to understand for those who are not professional mathematicians and astronomers. In particular they wanted to understand the arrangement of the heavenly spheres according to Copernicus's hypothesis, how he places the sun at the center of the planets' orbits, how around the sun he places next the orbit of Mercury, around the latter that of Venus, then the moon around the earth, and around this Mars, Jupiter, and Saturn; and in regard to motion, he makes the sun stationary at the center and the earth turn on itself and around the sun, that is, on itself with the diurnal motion and around the sun with the annual motion.

Q: Since, as he says, he came to Rome to be able to have the resolution and the truth regarding the above, what then was decided about this matter.

A: Regarding the controversy which centered on the above-mentioned opinion of the sun's stability and earth's motion, it was decided by the Holy Congregation of the Index that this opinion, taken absolutely, is repugnant to Holy Scripture and is to be admitted only suppositionally,[4] in the way that Copernicus takes it.

Q: Whether he was then notified of the said decision, and by whom.

A: I was indeed notified of the said decision of the Congregation of the Index, and I was notified by Lord Cardinal Bellarmine.

Q: What the Most Eminent Bellarmine told him about the said decision, whether he said anything else about the matter, and if so what. ⟨339⟩

A: Lord Cardinal Bellarmine told me that Copernicus's opinion could be held suppositionally, as Copernicus himself had held it. His Eminence knew that I held it suppositionally, namely in the way that Copernicus held it, as you can see from an answer by the same Lord Cardinal to a letter of Father Master Paolo Antonio Foscarini, Provincial of the Carmelites; I have a copy of this, and in it one finds these words: "I say that it seems to me that Your Paternity and Mr. Galileo are proceeding prudently by limiting yourselves to speaking suppositionally and not absolutely."[5] This letter by the said Lord Cardinal is dated 12 April 1615. Moreover, he told me that otherwise, namely taken absolutely, the opinion could be neither held nor defended.

Q: What was decided and then made known to him precisely in the month of February 1616.

A: In the month of February 1616, Lord Cardinal Bellarmine told me that since Copernicus's opinion, taken absolutely, was contrary to Holy Scripture, it could be neither held nor defended, but it could be taken and used suppositionally. In conformity with this I keep a certificate by Lord Cardinal Bellarmine himself, dated 26 May 1616, in which he says that Copernicus's opinion cannot be held or defended, being against Holy Scripture. I present a copy of this certificate, and here it is.[6]

And he showed a sheet of paper with twelve lines of writing on one side only, beginning "We, Robert Cardinal Bellarmine, have" and ending "on this 26th day of May 1616," signed "The same mentioned above, Robert Cardinal Bellarmine." This evidence was accepted and marked with the letter *B.*

Then he added: I have the original of this certificate with me in Rome, and it is written all in the hand of the above-mentioned Lord Cardinal Bellarmine.

Q: Whether, when he was notified of the above-mentioned matters, there were others present, and who they were.

A: When Lord Cardinal Bellarmine notified me of what I mentioned regarding Copernicus's opinion, there were some Dominican Fathers present, but I did not know them nor have I seen them since.

Q: Whether at that time, in the presence of those Fathers, he was given any injunction either by them or by someone else concerning the same matter, and if so what.

A: As I remember it, the affair took place in the following manner. One morning Lord Cardinal Bellarmine sent for me, and he told me a certain detail that I should like to speak to the ear of His Holiness before telling others; but then at the end he told me that Copernicus's opinion could not be held ⟨340⟩ or defended, being contrary to Holy Scripture. I do not recall whether those Dominican Fathers were there at first or came afterward; nor do I recall whether they were present when the Lord Cardinal told me that the said opinion could not be held. Finally, it may be that I was given an injunction not to hold or defend the said opinion, but I do not recall it since this is something of many years ago.

Q: Whether, if one were to read to him what he was then told and ordered with injunction, he would remember that.

A: I do not recall that I was told anything else, nor can I know whether I shall remember what was then told me, even if it is read to me. I am saying freely what I recall because I do not claim not to have in any way violated that injunction, that is, not to have held or defended at all the said opinion of the earth's motion and sun's stability.

And having been told that the said injunction,[7] given to him then in the presence of witnesses, states that he cannot in any way whatever hold, defend, or teach the said opinion, he was asked whether he remembers how and by whom he was so ordered.

A: I do not recall that this injunction was given me any other way than orally by Lord Cardinal Bellarmine. I do remember that the injunction was that I could not hold or defend, and maybe even that I could *not teach*. I do not recall, further, that there was the phrase *in any way whatever*, but maybe there was; in fact, I did not think about it or keep it in mind, having received a few months thereafter Lord Cardinal Bellarmine's certificate dated 26 May which I have presented and in which is explained the order given to me not to hold or defend the said opinion. Regarding the other two phrases in the said injunction now mentioned, namely *not to teach* and *in any way whatever*, I did not retain them in my memory, I think because they are not contained in the said certificate, which I relied upon and kept as a reminder.

Q: Whether, after the issuing of the said injunction, he obtained any permission to write the book identified by himself, which he later sent to the printer.

A: After the above-mentioned injunction I did not seek permission to write the above-mentioned book which I have identified, because I do not think that by writing this book I was contradicting at all the injunction given me not to hold, defend, or teach the said opinion, but rather that I was refuting it.

Q: Whether he obtained permission for printing the same book, by whom, and whether for himself or for someone else.

A: To obtain permission to print the above-mentioned book, although I was receiving profitable offers from France, Germany, and Venice, I refused them and spontaneously came to Rome three years ago to place it into the hands of the chief censor, namely the Master of the Sacred Palace, ⟨341⟩ giving him absolute authority to add, delete, and change as he saw fit. After having it examined very diligently by his associate Father Visconti, the said Master of the Sacred Palace reviewed it again himself and licensed it; that is, having approved the book, he gave me permission but ordered to have the book printed in Rome. Since, in view of the approaching summer, I wanted to go back home to avoid the danger of getting sick, having been away all of May and June, we agreed that I was to return here the autumn immediately following. While I was in Florence, the plague broke out and commerce was stopped; so, seeing that I could not come to Rome, by correspondence I requested of the same Master of the Sacred Palace permission for the book to be printed in Florence. He communicated to me that he would want to review my original manuscript, and that therefore I should send it to him. Despite having used every possible care and having contacted even the highest secretaries of the Grand Duke and the directors of the postal service, to try to send the said original safely, I received no assurance that this could be done, and it certainly would have been damaged, washed out, or burned, such was the strictness at the borders. I related to the same Father Master this difficulty concerning the shipping of the book, and he ordered me to have the book again very scrupulously reviewed by a person acceptable to him; the person he was pleased to designate was Father Master Giacinto Stefani, a Dominican, professor of Sacred Scripture at the University of Florence, preacher for the Most Serene Highnesses, and consultant to the Holy Office. The book was handed over by me to the Father Inquisitor of Florence and by the Father Inquisitor to the above-mentioned Father Giacinto Stefani; the latter returned it to the Father Inquisitor, who sent it to Mr. Niccolò dell'Antella, reviewer of books to be printed for the Most Serene Highness of Florence; the printer, named Landini, received it from this Mr. Niccolò and, having negotiated with the Father Inquisitor, printed it, observing strictly every order given by the Father Master of the Sacred Palace.

Q: Whether, when he asked the above-mentioned Master of the Sacred Palace for permission to print the above-mentioned book, he revealed to the same Most Reverend Father Master the injunction previously given to him concerning the directive of the Holy Congregation, mentioned above.

A: When I asked him for permission to print the book, I did not say anything to the Father Master of the Sacred Palace about the above-mentioned injunction because I did not judge it necessary to tell it to him, having no scruples since with the said book I had neither held nor defended the opinion of the earth's motion and sun's stability; on the contrary, in the said book I show the contrary of Copernicus's opinion and show that Copernicus's reasons are invalid and inconclusive.

With this the deposition ended, and he was assigned a certain room in the dormitory of the officials, located in the Palace of the Holy Office, in lieu of prison, with ⟨342⟩ the injunction not to leave it without special permission, under penalty to be decided by the Holy Congregation; and he was ordered to sign below and was sworn to silence.

I, Galileo Galilei, have testified as above.

Oreggi's[8] Report on the Dialogue *(17 April 1633)*

The 17th day of the month of April of the year of our Lord 1633.

In the work entitled *Dialogue of Galileo Galilei etc. on the Two Chief World Systems, Ptolemaic and Copernican*, the opinion is held and defended which teaches that the earth moves and the sun stands still, as one gathers from the whole thrust of the work, and especially from the comments in the report[9] which, by order of His Holiness, the very reverend Father Niccolò Riccardi, Master of the Sacred Apostolic Palace, and Agostino Oreggi, Theologian to His Holiness, both Consultants to the Holy Office, presented to the Most Eminent and Most Reverend Cardinals Inquisitors-General against heretical depravity.

So it seems to me, Agostino Oreggi, Theologian to His Holiness and Consultant to the Holy Roman Inquisition.

Inchofer's Report on the Dialogue *(17 April 1633)*[10]

[I][11]

⟨349⟩ I am of the opinion that Galileo not only teaches and defends the immobility or rest of the sun or center of the universe, around which both the planets and the earth revolve with their own motions, but also that he is vehemently suspected of firmly adhering to this opinion, and indeed that he holds it.

Melchior Inchofer.

[II]

The reasons for the point about the stopping, rest, or immobility of the sun, and about its being the center of the universe around which the planets and the earth move, are in the various chapters the same as those given for the other point about the turning of the earth. Indeed these two ideas, that the earth moves and that the sun stands still and is the center, are correlative of each other in the Copernican system.

Therefore all those reasons on which Galileo—in a categorical, absolute, and nonhypothetical manner—grounds the earth's motion necessarily also prove or assume the immobility and central position of the sun.

In actual detail and in absolute terms, he says on page 25,[12] "Aristotle will never prove that the earth is at the center,"[13] despite what he adds in the margin, "The sun is more probably at the center than the earth is," as though the assertion were not absolute. The probability is so much greater as to be overwhelming, and so on page 316[14] he shows absolutely and demonstratively that the sun is the center and the earth, like the other planets, moves around it. He concludes this "from the most conclusive observations," as he himself puts it.

In truth he shows his intent first, affirmatively, on pages 318, 319, 321, 323, 324, and 325,[15] and later, while rejecting the diurnal motion of the celestial sphere and destroying the Ptolemaic system with as much power as he can. From these things he finally infers that the sun is the center around which the bodies of the world and the earth turn. See pages 332, 333, and 334.[16]

On the question of Galileo's mental attitude, it is certain on the basis of the reasons given under each heading that he teaches, defends, and holds the opinion of the earth's motion and the sun's immobility and central position; in addition, however, all these things are shown very powerfully by that somewhat lengthy treatise of the same Galileo which, before the publication of the present *Dialogue*, he presented to the Grand Duke of Florence in support of his cause and in which he not only proved Copernicus's opinion but established it by solving scriptural difficulties as well as he could.[17]

⟨350⟩ Moreover, in discussing the passages from Scripture, especially those about the sun's motion, he did his best to show that Scripture speaks with a meaning adapted to popular opinion and that in reality the sun does not move. Then he ridiculed those who are strongly committed to the common scriptural interpretation of the sun's motion as if they were small-minded, unable to penetrate the depth of the issue, half-witted, and almost idiotic.

I have read this composition, and, if I am not deceived, here in Rome it passed through the hands of quite a few. These things are said in confirmation of previous claims.

Melchior Inchofer.

[III]

I am of the opinion not only that Galileo teaches and defends the view of Pythagoras and Copernicus but also, if we consider the manner of proceeding, of reasoning, and then of expression, that he is vehemently suspected of firmly adhering to it, and indeed that he holds it.

Melchior Inchofer.

[IV]

Reasons why it appears that Galileo teaches, defends, and holds the opinion of the earth's motion.

1. It is beyond question that Galileo teaches the earth's motion in writing. Indeed his whole book speaks for itself. Nor can one teach in any other way those of future generations and those who are absent than through writing or by tradition.

2. A function of a teacher is to transmit those concepts of a discipline which he regards easier and more effective for acquiring willing and well-disposed disciples, especially in the case of an allegedly new discipline, which entices curious minds with wonder. How expert and adroit Galileo shows himself to be in this sort of activity is something that the whole book makes clear to the careful reader.

3. Furthermore, one who teaches those who oppose his doctrine tries, as much as he can, to answer their objections and to expose their difficulties and even falsehoods. In this whole work Galileo argues for nothing more than to establish the doctrine of the earth's turning and to thoroughly defeat the contrary doctrine.

4. Next, something extraordinary happens, namely that Galileo attributes to the earth's turning all kinds of effects visible in nature, whose true causes are not unknown and have been found by others, as if it were the only genuine and proper cause; and so he teaches ad nauseam that the phenomena of sunspots, tides, and terrestrial magnetism are of this sort. This, most assuredly, is a sign not only of wishing to teach but of actually teaching, even explaining many topics neither Copernicus nor his followers had thought about, as the author himself wishes to claim. ⟨351⟩

5. Galileo additionally deplores the fact that this opinion is understood by so few and that so many are committed to the long-established opinion; thus he tries to have Simplicio unlearn it and, in his person, draw to his own opinion, if he can, all Peripatetics. Without a doubt he acts with the solicitude of a diligent teacher who desires to have and to

make disciples. As St. Augustine says in his commentary on Psalm 118 (section 17),[18] to teach is nothing more than to impart knowledge, which is so connected to the teaching that one cannot exist without the other; if so, then it is clear that Galileo really and truly teaches this opinion, and indeed that under the name of "the Academician"[19] he acts as a teacher of those to whom he introduces the speakers in his *Dialogue*. Nor is the process of teaching or learning ever easier than when doctrines are expounded by means of a dialogue, as is well known from countless examples of great men.

Enough about the first point concerning the teaching of the doctrine in writing. Actually, that the same thing is not new for Galileo appears evident from the little book published a long time before, in which he is praised and defended for the same doctrine.

As regards the second point, that is, whether he defends it, though it can be easily deduced from what has been said, nevertheless undoubtedly the answer appears affirmative in this way:

1. Because if someone is said to defend an opinion he merely supports without refutation or destruction of the contrary view, how much more can this be said of one who defends it so as to wish the contrary completely destroyed? Hence, in law, to defend is sometimes equated with to attack.[20]

2. Because Copernicus, satisfied with a simple system, merely explained celestial phenomena in terms of this hypothesis by what, so he thought, was an easier method; whereas Galileo, having searched for many additional arguments, confirms Copernicus's discoveries and introduces new ones, and this is a twofold defense.

3. Because this time Galileo's principal aim was to attack Father Christopher Scheiner,[21] who more recently than anyone else had written against the Copernicans. But this is precisely to defend and to wish to strengthen the opinion of the earth's motion, lest perhaps it be ruined when attacked by others.

4. Because there is no better means of defense, indeed the strongest, than the one used by Galileo, namely, to present contrary arguments and in the process to refute and weaken them so that they seem without force, without reason, and accordingly without any ability and judgment on the part of one's opponents.

5. Because if he had undertaken this discussion only with the purpose of engaging in a disputation and exercising the mind, he would not have waged such arrogant war against Ptolemaics and Aristotelians, nor would he have ridiculed Aristotle and his followers so insolently; rather he should have modestly proposed arguments for investigating and es-

tablishing the truth, and not impugning it, which granted he did not understand.

Enough about the second point concerning the defense published in writing. From these things one can also infer a conclusion concerning the defense carried on orally.

⟨352⟩ As regards the third point, that is, whether Galileo holds the opinion of the earth's physical motion so that he might be proved guilty of regarding it as true, an affirmative answer may be seen in two ways: first, by means of necessary consequences; second, by Galileo's own words, which are absolute and categorical, or certainly equivalent. Moreover, I submit that the mind of a speaker is tied to his words, and there is no value in any nice-sounding declarations he may be wont to make lest he be viewed as sinning against the Decree.[22] Indeed my judgment will be based on facts indicating the contrary. But let us come to the evidence.

1. Because the reason which he pretends moved him to writing is empty and frivolous and insufficient to induce a wise man to undertake such a work; that is, that foreigners had murmured against the Decree and denounced the consultants to the Holy Congregation to be ignorant of astronomy. On this matter I did not see any publications by a single foreigner in which there is any mention of the Decree, or of the consultants, of whom even the designation is unknown to them. It is certain that among Catholics no one would have dared. And then, if this reason moved Galileo, why did he not undertake to defend the Decree and the Holy Congregation together with its consultants? Indeed, this would have been arguably preferable for him in order to respond to the cause for which he was writing. But this is so far from Galileo's preference that he tries to strengthen the Copernican opinion with new arguments of which foreigners would never think in this connection; and he writes in Italian, certainly not to extend the hand to foreigners or other learned men, but rather to entice to that view common people in whom errors very easily take root.

2. Anyone who discusses something for the sake of the argument, and not because he really is convinced of it, proceeds problematically either by laying down neither part as more certain than the other or, having rejected the one, by adhering to the other which he regards as more certain. Galileo everywhere makes pronouncements by means of theorems and solid demonstrations (in his view), as if he wanted the opinion of the earth's immobility to be completely discarded.

3. Galileo promises to proceed in the manner of a mathematical hypothesis, but a mathematical hypothesis is not established by physical

and necessary conclusions. For example, a mathematician assumes the existence of an infinite line, and from this he concludes that a triangle made of infinite lines is of infinite capacity; nevertheless, he neither proves nor believes that an infinite line exists, taking infinity in a strict sense. So Galileo should have posited the earth's motion as something he intended to analyze deductively, not as something to be proved true by destroying the opposite view, as indeed he does in the entire work.

4. Theologians inquire whether God exists, not because a Christian theologian doubts that he exists, but in order to show that, even independently of faith, God's existence can be proved with many arguments intelligible to us (as they say) and by refuting arguments to the contrary. If Galileo wanted to proceed hypothetically, he should have produced the arguments that seem to indicate the earth's motion; then, after criticizing them, ⟨353⟩ he should have either supposed or proved the contrary, certainly not confuted it. And, indeed, I say this for the case where one is not proceeding purely mathematically, but rather, as Galileo does, with physical disputations intermingled; otherwise a mere supposition suffices for the mathematician, without assuming and requiring any demonstration of it.

5. Philosophers also inquire whether the world could have existed from eternity; yet no Christian says that it has existed from eternity, but merely that, assuming it existed from eternity, such and such consequences would follow necessarily or probably. So, in order to restrict himself to a pure mathematical hypothesis, Galileo did not have to prove absolutely that the earth moves, but only to conceive its motion in the imagination without assuming it physically, and thereby explain celestial phenomena and derive the numerical details of the various motions.

6. If Galileo did not firmly adhere to the view of the earth's motion as if he believed it true, he would not have fought so vigorously in its favor; nor would he have regarded so despicably those who believe the opposite, as if he thought them not to be numbered among human beings (page 269).[23] What Catholic ever conducted such a bitter dispute against heretics, even regarding a truth of faith, as Galileo does against those who maintain the earth's immobility, especially when not attacked by anybody? Surely, if this is not to defend an opinion to which someone adheres firmly, then, disregarding matters of faith, I do not know how you would discern any sign that a person is of this or that opinion, except that he defends it with every effort.

7. If Galileo had attacked someone in particular who may not have argued very strongly for the earth's immobility or may not have com-

pletely convinced the Copernicans, many things could be interpreted in a favorable light on his behalf. But since he declares war on everybody and regards as dwarfs all who are not Pythagorean or Copernican, it is clear enough what he has in mind; this is especially so since he praises greatly and prefers to others William Gilbert, a perverse heretic and a quarrelsome and sophistical defender of this opinion.

And so these are all the particular reasons which for me render Galileo vehemently suspected of being of the opinion that the earth physically moves. Without doubt nowhere in this whole work does it emerge that he feels otherwise. Though he sometimes says that he does not want to make a decision at all, he does this because, after the wounds inflicted upon his targets, he wants to ensure that he not be viewed as having wounded them deliberately.

Let us now come to the other part of the claim made a little while ago, in order to show that Galileo asserts this opinion in absolute or at least equivalent terms.

1. On page 108 we have this: "I cannot bring myself to believe that you can find anybody who would think it *more reasonable and credible* that the celestial sphere is the one which does the turning and the terrestrial globe the one that stands still."[24]

2. Page 113 in the seventh confirmation: "If the diurnal rotation is attributed to the heavens . . . it seems to me that this involves a great difficulty, and I would be unable to understand how the earth, being a body suspended and balanced on its center, ⟨354⟩ indifferent to motion and to rest . . . would not give in and be made to rotate."[25]

3. Page 110: He proves the earth's motion by that physical principle according to which nature does not accomplish with many instruments what it can bring about with a few, and it is in vain to do by means of many things what can be done by means of fewer.[26]

4. Page 122: "Considering those things, I began to believe that if someone abandons an opinion imbibed with mother's milk and followed by countless people, in order to arrive at another one accepted by very few and denied by all schools and which really looks like a very great paradox, then he must necessarily have been moved, not to say forced, by more powerful reasons."[27]

5.[28] Page 370: He does not believe that anyone has taken seriously the earth's motion; it is only because people find it written that the earth does not move that they follow this opinion.[29]

6. Page 366: He calls the proposition that the heavens move a deep-rooted impression, as if it were not after all a true opinion.[30]

7. Page 399: "I confess that I have never heard anything more wonderful than this, nor can I believe that the human intellect has ever engaged in subtler speculation."[31]

8. Pages 48 and 49: Where Simplicio opposes the overthrowing of Aristotelian philosophy as a consequence of the earth's motion, he answers that this is impossible, that it is the human brain that needs to be reconstructed so as to enable it to distinguish truth from falsehood.[32]

9. Page 317: He says that Aristotle placed the terrestrial globe at the center, but that if he were forced by very clear experiments to modify in part this structure and arrangement of the universe, and to confess to have been deceived, etc.[33]

10. Page 317, paragraph "I do not ask": He says that Peripatetics are slaves to Aristotle and would rather say the world is as Aristotle wrote about it than as nature wishes it to be.[34]

11. Same page, paragraph "Do not use": He dislikes calling inconvenient the idea of not placing the earth at the center with the heavens moving around it; rather he says that it might be necessarily true.[35]

12. Page 318, paragraph "Now if": He supposes as true that the earth moves around the center.[36]

13. Same page, paragraph "We conclude": "That the sun is at the center . . . is a conclusion arrived at from very clear and hence very conclusive observations."[37]

14. Page 319: He proves that the effect of diurnal motion on heavenly bodies is, and could be, only to make the universe appear to us to be turning rapidly in the opposite direction.[38]

15. Page 324: He does not regard as human those who hold the earth's immobility.[39] ⟨355⟩

16. Page 325: He shows that those who have embraced Copernicus's opinion were of superior talent, having followed the intellect against sensory experience, and that here reason did violence to the senses.[40]

17. Same page, paragraph "We are going to": Salviati says that he would still be believing the Peripatetics if a superior sense, more excellent than the common and natural ones, had not gotten together with reason.[41]

18. Page 331: He utters an exclamation to Copernicus about how much he would have enjoyed the telescope as a partial confirmation of his system if it had been discovered in his time; and he praises him because through his reasoning he contravened experience.[42]

19. Page 332, paragraph "Finally such": Concerning the earth, he

concludes that it moves around the sun as a very probable and perhaps necessary inference.[43]

20. Page 333: He says that Copernicus, after restoring astronomy based upon Ptolemy's suppositions, judged that if one could save the heavenly appearances by means of assumptions that were false in nature, one could do it much better by means of true suppositions.[44]

21. Page 334, paragraph "You, Mr. Sagredo": He judges that the removal of planetary stations and retrograde motions is sufficient for someone who is not too proud and incorrigible to give assent to Copernicus's doctrine.[45]

22. Page 336, toward the end: He says that Mercury and Venus appear as they do owing to the earth's annual motion, as Copernicus "sharply demonstrates."[46] (Note that on page 27 he says that demonstrations are given only by mathematicians.)[47]

23. Page 337: He says that sunspots force the human intellect to admit the earth's annual motion.[48]

24. Page 344, paragraph "Mr. Simplicio": He speaks of Copernicus's strong arguments and conjectures and very solid observations, given that what Sagredo says is true; nor is it proper to doubt these words, he says.[49]

25. Page 348: Speaking of sunspots, given the proofs, "I myself think it necessarily must be that those who remain stubborn against this doctrine either have not heard or have not understood these so clearly convincing reasons."[50]

26. Pages 348–349, paragraph "I am not going to attribute them": "Granted that one of the two systems is necessarily true and the other necessarily false, it is impossible (remaining within the confines of human doctrines) for the reasons advanced for the true side not to manifest themselves to be just as conclusive as the contrary ones are vain and ineffective." (This passage says more than it appears at first sight.)[51]

27. Page 396: He does not doubt that the science which teaches the earth to be a magnet is to be perfected by means of true and necessary demonstrations.[52] He must say the same ⟨356⟩ of the earth's motion since on page 404 he proves the earth's various motions from the motions of the lodestone.[53]

These are the reasons which led me, for love of truth, to bring a censure of the same, which nevertheless I entrust and submit freely to the better judgment of others.

Melchior Inchofer.

Pasqualigo's[54] *Report on the* Dialogue
(17 April 1633)[55]

[I][56]

I, Zaccaria Pasqualigo, Clerk Regular, Professor of Sacred Theology, in the presence of the Most Eminent and Most Reverend Cardinal Ginetti, Vicar of His Holiness Pope Urban VIII, have been asked whether Galileo Galilei, by the publication of his *Dialogue* where he deals with the Copernican system, has transgressed the injunction by which the Holy Office prohibits him to hold, teach, or defend in any way whatever, orally or in writing, this opinion of the earth's motion and sun's immobility at the center of the world. Having diligently inspected his book, I am of the opinion that he transgressed it as regards the words "teach or defend," since indeed he tries as best he can to support the earth's motion and the sun's immobility, and also that he is strongly suspected of holding such an opinion. And so for the formal declaration of these things I sign with my own hand.

Zaccaria Pasqualigo, Clerk Regular,
Professor of Sacred Theology.

[II]

I, Zaccaria Pasqualigo, Clerk Regular, Professor of Sacred Theology, in the presence of the Most Eminent and Most Reverend Cardinal Ginetti, Vicar of His Holiness Pope Urban VIII, have been asked whether Galileo Galilei, by the publication of his *Dialogue* where he explains the Copernican system, has transgressed the injunction by which the Holy Office prohibits him to hold, teach, or defend in any way whatever, orally or in writing, the opinion of the earth's motion. I am of the opinion that he transgressed it as regards the words "teach or defend," and that such a *Dialogue* renders him strongly suspected of holding this opinion; and I declare this after having diligently reflected upon his book. And so I sign with my own hand.

Zaccaria Pasqualigo, Clerk Regular,
Professor of Sacred Theology.

[III]

Although at the beginning of his book Galileo claims to want to deal with the earth's motion as a hypothesis,[57] in the course of his *Dialogue* he puts the hypothesis aside and proves its motion absolutely, using unconditional arguments. Thus from absolute premises he draws an absolute conclusion, and sometimes he feels that his reasons are convincing.

⟨357⟩ And so he advances his reasons. In the first of them he assumes that the same appearance results whether the earth moves with diurnal motion or whether the stars move; then he argues that since nature does not do by many things what she can do by few, and since all appearances can be saved by merely attributing the diurnal motion to the earth, therefore we must say that nature has not used many diurnal motions for the stars and planets, but only one for the earth (page 109).[58] To confirm this, he adds that if the diurnal motion belongs to the heavens, the planetary spheres must have their swift motion from east to west, contrary to the natural turning proper to them (page 110).[59] He also says that the bigger the sphere the slower it is in its motion, which is why Saturn takes thirty years for its revolution, and therefore the Prime Mobile, being greater than all the others, cannot take twenty-four hours to go through its natural revolution (page 111);[60] and that, carrying along with itself the planetary spheres, the Prime Mobile would also carry the earth, insofar as it is a body suspended (page 113).[61]

Second reason (page 318).[62] With Aristotle he assumes that the center of the world is that around which the heavenly revolutions are made, and from this he infers that the sun is such a center and hence stands still therein. Then, that the heavenly revolutions are made around the sun is derived, he says, from very clear and necessarily conclusive observations, such as the following: that the planets are sometimes nearer the earth and sometimes farther from it, the difference being so great that when Venus is farthest it is six times farther from us than when it is closest, that Mars is then eight times farther and appears sixty times bigger when closest, and that Saturn and Jupiter are very far when in conjunction with the sun and very close at opposition.

Third reason (page 334).[63] Assuming the earth's annual motion, all stations and retrograde motions of the five planets are eliminated, each of the planets always moves in the same direction, and their stations and retrograde motions become merely apparent. To confirm this he gives a demonstration by means of a diagram, which however is subject to difficulties.

Fourth reason, taken from sunspots (pages 339, 346, 347).[64] He says that before making detailed observations of sunspots, he formulated the following judgment: that if the earth were moving with the annual motion along the ecliptic and around the sun, and if the sun at the center were rotating with its axis inclined to, rather than being identical with, the axis of the ecliptic, it would follow that the paths of sunspots would be in a straight line twice a year at six-month intervals and along curved arcs at other times; that for half a year the curvature of these arcs would be in a direction opposite to that for the other half, since the curvature would be toward the upper part of the solar disk for six months and toward the lower part of the disk for the other six; ⟨358⟩ and that the eastern and western endpoints of the spots (by which he means those parts of the solar disk near which the spots appear and disappear) would be at the same level only two days a year, whereas at other times the eastern endpoints would be higher than the western ones for six months, and the western higher than the eastern during the other six. Then he adds that, after diligently observing the motion of these spots, he found that it corresponded in every way to the manner described; and, therefore, from the motion of these spots he infers the earth's motion. He also tries to show that, given the earth's immobility and the sun's motion along the ecliptic, the apparent motion of these spots cannot be saved. This argument is based on a premise about what de facto exists and infers a conclusion about what de facto may exist.

Fifth reason (page 410).[65] Assuming the earth to be motionless, the tides would not occur naturally; and assuming the earth's annual and diurnal motions, the tides would be caused necessarily. Thus he says that from the mixture of the annual and diurnal motions there results an acceleration of motion in some parts of the earth and during the same period a retardation in the others, and on page 420 he shows this in a diagram.[66] For at some parts of the earth there is an adding up of the annual and diurnal motions, where they carry the earth in the same direction, while at the opposite parts, the earth being carried in one direction by the annual motion and in the opposite direction by the diurnal motion, subtracting one motion from the other, there results a great retardation of the absolute motion; this acceleration and retardation necessarily cause the rising and falling of the water since, not being firmly attached to the earth, the water does not necessarily follow its motion, as one can observe in a boat full of water which is moving on a lake and whose motion happens to vary in speed and slowness.

However, he does not untangle the difficulty that, given this doctrine, since the change between greatest acceleration and maximum retardation of the earth's motion occurs at twelve-hour intervals, then high and low tides should also occur at twelve-hour intervals. But experience teaches that they occur every six hours.

Then he accordingly attributes, as to a cause, the monthly period of the tides to the monthly variation in the earth's annual motion caused by the moon's motion. Since the latter moves together with the earth in its orbit and also around the earth, when at the time of conjunction the moon is between the earth and the sun, its motion (and by participation also that of the earth) is faster than when it is farther from the sun, that is, on the other side of the earth and in opposition to the sun. ⟨359⟩ The monthly variation in the tides is caused by this greater and lesser speed.

As for the variation in the tides occurring at the equinoxes and solstices, he also reduces it to a change in the earth's motion: the combination of the annual and diurnal motions produces accelerations in the absolute motion because of the various lines along which the terrestrial globe is carried, as he shows in a diagram. However, all these pictorial demonstrations of his are subject to various difficulties.

In past years Galileo received an injunction from the Holy Office regarding the Copernican opinion of the earth's motion and sun's immobility at the center of the world, to the effect that "he should neither hold, nor teach, nor defend it in any way whatever, orally or in writing." Since he published his *Dialogue* on this topic, the question is whether he transgressed the above-mentioned injunction.

The answer is that he contravened the injunction insofar as it prohibits him from "teaching in any way whatever." First, because the purpose of one who writes and publishes is to teach the doctrine contained in the book; thus St. Thomas says, "Writing is composed for the purpose of impressing a doctrine on the hearts of the readers" (part 3, question 42, article 4).[67] Second, because to teach is nothing but to communicate some doctrine, as St. Augustine says in section 17 of his commentary on Psalm 118 ("What else is teaching but to give knowledge?"); then he adds that, from the viewpoint of the teacher, teaching involves nothing but saying what is necessary in order that a doctrine be understood; and so he claims that if the student does not understand, the teacher can say "I told him what had to be said; if he did not learn, it is because he did not comprehend"; that is, the teacher can say that he did what was necessary to teach; it follows that, by saying what can be said from the viewpoint of reason to instill the Copernican opinion in those who are capable, Galileo is teaching this opinion. Third, because he explains his doctrine in such a way that many find it persuasive, even people knowledgeable in the mathematical sciences. Fourth, because he says (page 213)[68] he deems to have spent his time and words well if he has shown that the opinion of the earth's motion at least is not foolish; and this is nothing other than to argue that it is probable.

He has also transgressed the other clause, that "he should not defend it in any way whatever." For defending an opinion consists precisely in basing it on some reason and unraveling the arguments against it, which he does with every effort throughout the course of the *Dialogue*. Though he claims to be speaking hypothetically, nevertheless, in supporting his opinion he excludes hypotheses, because from premises which are absolute and (at least for him) de facto true, he draws an absolute conclusion. This is apparent from all the reasons he puts forth, and especially from the following: that, since nature abhors superfluity, one should not multiply motions to the point of having as many of them as there are stars (page 109);[69] that very evident and necessarily conclusive ⟨360⟩ observations regarding planetary motions show the sun to be the center of the world (page 318);[70] that, assuming the earth's motion, he judged that certain specific properties should be detected in the motion of sunspots and that then, having made the observations, he found the occurrences to be such that they must correspond to the earth's motion (page 339);[71] and that without the earth's motion the tides cannot be naturally produced (page 410).[72]

Regarding the other point, which prohibits him from "holding" the view, he arouses strong suspicion of having transgressed it. First, in the whole course of the book he appears to adhere closely to this opinion, exerting himself skillfully to transmit it as true and to destroy the opposite; for he tears down all the reasons by which the latter is defended, and he seems to feel that the ones in favor of the earth's motion are efficacious. Second, because he accepts some data from which he thinks he can derive the earth's motion as a true consequence—such as, that the observations pertaining to planetary motions show the revolutions of the planets to be around the sun as center, and he calls these observations very evident and necessarily conclusive for showing that these revolutions are around the sun (page 318);[73] and he says (page 339)[74] to have judged (which is nothing but to have agreed) that, if the earth were moving, then by virtue of this motion one would have to detect some particular properties of sunspot motions; and he claims to have later found by observation that these particulars indeed took place, and then again from these phenonema (for him already proved by experience) he argues for the earth's motion.

I, Zaccaria Pasqualigo, Clerk Regular, Professor of Sacred Theology, in the presence of the Most Eminent and Most Reverend Lord Cardinal Ginetti, Vicar to His Holiness Pope Urban VIII, set forth the above opinion, and I so judge.

Commissary General to Cardinal Barberini
(28 April 1633)[75]

Most Eminent, Most Reverend, and Most Honorable Lord and Patron:

Yesterday, in accordance with the orders of His Holiness, I reported on Galileo's case to the Most Eminent Lords of the Holy Congregation by briefly relating its current state. Their Lordships approved what has been done so far, and then they considered various difficulties in regard to the manner of continuing the case and leading it to a conclusion; for in his deposition Galileo denied what can be clearly seen in the book he wrote, so that if he were to continue in his negative stance it would become necessary to use greater rigor in the administration of justice and less regard for all the ramifications of this business. Finally I proposed a plan, namely that the Holy Congregation grant me the authority to deal extrajudicially with Galileo, in order to make him understand his error and, once having recognized it, to bring him to confess it. The proposal seemed at first too bold, and there did not seem to be much hope of accomplishing this goal as long as one followed the road of trying to convince him with reasons; however, after I mentioned the basis on which I proposed this, they gave me the authority. In order not to lose time, yesterday afternoon I had a discussion with Galileo, and, after exchanging innumerable arguments and answers, by the grace of the Lord I accomplished my purpose: I made him grasp his error, so that he clearly recognized that he had erred and gone too far in his book; he expressed everything with heartfelt words, as if he were relieved by the knowledge of his error; and he was ready for a judicial confession. However, he asked me for a little time to think about the way to render his confession honest, for in regard to the substance he will hopefully proceed as mentioned above.

I have not communicated this to anyone else, but I felt obliged to inform Your Eminence immediately, for I hope His Holiness and Your Eminence will be satisfied that in this manner ⟨107⟩ the case is brought to such a point that it may be settled without difficulty. The Tribunal will maintain its reputation; the culprit can be treated with benignity; and, whatever the final outcome, he will know the favor done to him, with all the consequent satisfaction one wants in this. I am thinking of examining him today to obtain the said confession; after obtaining it, as I hope, the only thing left for me will be to question him about his intention and allow him to present a defense. With this done, he could be granted imprisonment in his own house, as Your Eminence mentioned. And to you I now express my humblest reverence.

Rome, 28 April 1633.

To Your Most Eminent and Most Reverend Lordship.

Your Humblest and Most Obliged Servant,
Fra Vincenzo da Firenzuola.

Galileo's Second Deposition (30 April 1633)

Called personally to the hall of the Congregations, in the presence and with the assistance of those mentioned above and of myself, the above-mentioned Galileo Galilei, who has since then petitioned to be heard, having sworn an oath to tell the truth, was asked by the Fathers the following:

Q: That he state whatever he wished to say.

A: For several days I have been thinking continuously and directly about the interrogations I underwent on the 16th of this month,[76] and in particular about the question whether sixteen years ago I had been prohibited, by order of the Holy Office, from holding, defending, and teaching in any way whatever the opinion, then condemned, of the earth's motion and sun's stability. It dawned on me to reread my printed *Dialogue*, ⟨343⟩ which over the last three years I had not even looked at. I wanted to check very carefully whether, against my purest intention, through my oversight, there might have fallen from my pen not only something enabling readers or superiors to infer a defect of disobedience on my part, but also other details through which one might think of me as a transgressor of the orders of Holy Church. Being at liberty, through the generous approval of superiors, to send one of my servants for errands, I managed to get a copy of my book, and I started to read it with the greatest concentration and to examine it in the most detailed manner. Not having seen it for so long, I found it almost a new book by another author. Now, I freely confess that it appeared to me in several places to be written in such a way that a reader, not aware of my intention, would have had reason to form the opinion that the arguments for the false side, which I intended to confute, were so stated as to be capable of convincing because of their strength, rather than being easy to answer. In particular, two arguments, one based on sunspots and the other on the tides, are presented favorably to the reader as being strong and powerful, more than would seem proper for someone who deemed them to be inconclusive and wanted to confute them, as indeed I inwardly and truly did and do hold them to be inconclusive and refutable. As an excuse for myself, within myself, for having fallen into an error so foreign to my intention, I was not completely satisfied with saying that when one presents arguments for the opposite side with the intention of confuting them, they must be explained in the fairest way and not be made out of straw to the disadvantage of the opponent, especially when one is writing in dialogue form. Being dissatisfied with this excuse, as I said, I resorted to that of the natural gratification everyone feels for his own subtleties and for showing himself to be cleverer than the average man, by finding ingenious and apparent considerations of probability even in favor of false propositions. Nevertheless—even though, to use Cicero's words, "I am more desirous of glory than is suitable"—if I had to write out the same arguments now, there is no doubt I would weaken them in such a way that they could not appear to exhibit a force which they really and essentially lack. My error then was, and I confess it, one of vain ambition, pure ignorance, and inadvertence. This is as much as I need to say on this occasion, and it occurred to me as I reread my book.

With this, having obtained his signature, and having sworn him to silence, the Fathers formally concluded the hearing.

I, Galileo Galilei, have testified as above.

⟨344⟩ And returning after a little, he said:

And for greater confirmation that I neither did hold nor do hold as true the condemned opinion of the earth's motion and sun's stability, if, as I desire, I am granted the possibility and the time to prove it more clearly, I am ready to do so. The occasion for it is readily available since in the book already published the speakers agree that after a certain time they should meet again to discuss various physical problems other than the subject already dealt with. Hence, with this pretext to add one or two other Days,[77] I promise to reconsider the arguments already presented in favor of the said false and condemned opinion and to confute them in the most effective way that the blessed God will enable me. So I beg this Holy Tribunal to cooperate with me in this good resolution, by granting me the permission to put it into practice.

And again he signed.

I, Galileo Galilei, affirm the above.

Galileo's Third Deposition (10 May 1633)

Summoned, there appeared personally at the hall of Congregations of the palace of the Holy Office in Rome, in the presence of the very Reverend Father Fra Vincenzo Maculano, O.P., Commissary General of the Holy Office, etc.

Galileo Galilei mentioned above; and, called before his Paternity, the same Father Commissary gave him a deadline of eight days to present his defense, if he wanted and intended to do it.

Having heard this, he said: I understand what Your Paternity has told me. In reply I say that I do want to present something in my defense, namely in order to show the sincerity and purity of my intention, not at all to excuse my having transgressed in some ways, as I have already said. I present the following statement, together with a certificate by the late Most Eminent Lord Cardinal Bellarmine, written with his own hand by the Lord Cardinal himself, of which I earlier presented a copy by my hand. For the rest I rely in every way on the usual mercy and clemency of this Tribunal.

After signing his name, he was sent back to the house[78] of the above-mentioned Ambassador of the Most Serene Grand Duke, under the conditions already communicated to him.

I, Galileo Galilei, with my own hand.

Galileo's Defense (10 May 1633)

In an earlier interrogation, I was asked whether I had informed the Most Reverend Father Master of the Sacred Palace about the private injunction issued to me sixteen years ago by order of the Holy Office—"not to hold, defend, or teach in any way whatever" the opinion of the earth's motion and sun's stability—and I answered No. Since I was not asked the reason why I did not inform him, I did not have the opportunity to say anything else. Now it seems to me necessary to mention it, in order to prove the absolute purity of my mind, always averse to using simulation and deceit in any of my actions.

⟨346⟩ I say, then, that at that time some of my enemies were spreading the rumor that I had been called by the Lord Cardinal Bellarmine in order to abandon some opinions and doctrines of mine, that I had had to abjure, that I had also received punishments for them, etc., and so I was forced to resort to His Eminence and to beg him to give me a certificate explaining why I had been called. I received this certificate, written by his own hand, and it is what I attach to the present statement.[79] In it one clearly sees that I was only told not to hold or defend Copernicus's doctrine of the earth's motion and sun's stability; but one cannot see any trace that, besides this general pronouncement applicable to all, I was given any other special order. Having the reminder of this authentic certificate, handwritten by the one who issued the order himself, I did not try later to recall or give any other thought to the words used to give me orally the said injunction, to the effect that one cannot defend or hold, etc.; thus, the two phrases besides "holding" and "defending" which I hear are contained in the injunction given to me and recorded, that is, "teaching" and "in any way whatever," struck me as very new and unheard. I do not think I should be mistrusted about the fact that in the course of fourteen or sixteen years I lost any memory of them, especially since I had no need to give the matter any thought, having such a valid reminder in writing. Now, when those two phrases are removed and we retain only the other two mentioned in the attached certificate, there is no reason to doubt that the order contained in it is the same as the injunction issued by the decree of the Holy Congregation of the Index. From this I feel very reasonably excused for not notifying the Father Master of the Sacred Palace of the injunction given to me in private, the latter being the same as the one of the Congregation of the Index.

Given that my book was not subject to more stringent censures than those required by the decree of the Index, I followed the surest and most effective way to protect it and purge it of any trace of blemish. It seems to me that this is very obvious, since I handed it over to the supreme Inquisitor ⟨347⟩ at a time when many books on the same subjects were being prohibited solely on account of the above-mentioned decree.

From the things I am saying, I think I can firmly hope that the idea of my having knowingly and willingly disobeyed the orders given me will be given no credence by the Most Eminent and Most Prudent Lord judges. Thus, those flaws that can be seen scattered in my book were not introduced through the cunning of an insincere intention, but rather through the vain ambition and satisfaction of appearing clever above and beyond the average among popular writers; this was an inadvertent result of my writing, as I confessed in another deposition of mine. I am ready to make amends and compensate for this flaw by every possible means, whenever I may be either ordered or allowed by Their Most Eminent Lordships.

Finally, I am left with asking you to consider the pitiable state of ill health to which I am reduced, due to ten months of constant mental distress, and the discomforts of a long and tiresome journey in the most awful season and at the age of seventy; I feel I have lost the greater part of the years which my previous state of health promised me. I am encouraged to do this by the faith I have in the clemency and kindness of heart of the Most Eminent Lordships, my judges; and I hope that if their sense of justice perceives anything lacking among so many ailments as adequate punishment for my crimes, they will, I beg them, condone it out of regard for my declining old age, which I humbly also ask them to consider. Equally, I want them to consider my honor and reputation against the slanders of those who hate me, and I hope that when the latter insist on disparaging my reputation, the Most Eminent Lordships will take it as evidence why it became necessary for me to obtain from the Most Eminent Lord Cardinal Bellarmine the certificate attached herewith.

Final Report to the Pope (May or June 1633) [80]

⟨293⟩ In the month of February 1615 Father Master Niccolò Lorini, a Dominican from Florence, forwarded here a composition by Galileo which was circulating in that city and which, following Copernicus's view that the earth moves and the sun stands still, contained many suspect or rash propositions. He reported that it was written for the purpose of contradicting certain sermons preached at the Church of Santa Maria Novella by Father Master Caccini, dealing with the tenth chapter of Joshua, where it says, "Sun, stand thou still." [81]

The composition is in the form of a letter written to Father Benedetto Castelli, a Benedictine Friar, then mathematician at Pisa, and it contains the following propositions:

That in the Holy Scripture there are many propositions which are false as regards the literal sense of the words;

That in natural disputations it should be saved for the last place;

That in order to cater to the incapacity of the masses, the Scripture has not refrained from perverting its most important dogmas, attributing even to God Himself properties contrary to and very far from His essence;

That in natural investigations philosophical argument should somehow prevail over the sacred;

That Joshua's command to the sun to stop should be understood as directed not to the sun but to the Prime Mobile, if one does not accept the Copernican system.

Despite diligent efforts one could not obtain the original of this letter.

Then Father Caccini was examined, and, besides the above-mentioned matters, he testified having heard Galileo utter other erroneous opinions: that God is an accident; that He really laughs, cries, etc.; and that the miracles attributed to the Saints are not true miracles.

He named some witnesses from whose examination one deduces that these propositions were not uttered by Galileo or his disciples in the manner of an assertion, but only in the context of a disputation.

⟨294⟩ Then, from the book on sunspots published in Rome by the same Galileo, two propositions were examined: "that the sun is the center of the world and wholly motionless regarding local motion; that the earth is not the center of the world and moves as a whole and also with diurnal motion." They were qualified as philosophically absurd.

Moreover, the first was also qualified as formally heretical, for expressly conflicting with Scripture and the opinion of the Saints; the second as at least erroneous in faith, considering the true theology.

Consequently, on 25 February 1616 His Holiness ordered the Lord Cardinal Bellarmine to summon Galileo[82] and give him the injunction that he must abandon and not discuss in any way the above-mentioned opinion of the immobility of the sun and the motion of the earth.

On the 26th the same Cardinal, in the presence of the Father Commissary of the Holy Office, notary, and witnesses, gave him the said injunction, which he promised to obey. Its tenor is that "he should abandon completely the said opinion, and indeed that he should not hold, teach, or defend it in any way whatever; otherwise the Holy Office would start proceedings against him."

In accordance with this the Holy Congregation of the Index issued a decree, which prohibited generally any book that treats of the said opinion of the earth's motion and sun's immobility.

In 1630 Galileo brought his book manuscript to the Father Master of the Sacred Palace in Rome in order to publish it. As mentioned in a previous report, he ordered that it be reviewed by an associate of his, though the relevant certificate is missing. In the same report it is stated that, for greater assurance, the Master of the Sacred Palace wanted to examine the book himself; so, to prevent delays, he agreed with the author that he would examine it page by page during the printing process, and, to enable the author to negotiate with the printer, he gave it the imprimatur for Rome.

Thereafter the author went to Florence, from where he petitioned the Father Master of the Sacred Palace for permission to publish it there, which was denied. The case was then transferred to the Inquisitor of Florence, and the Father Master of the Sacred Palace, removing himself from the proceedings, left him the task of whether or not to grant it and informed him of what had to be obeyed.

We have copies of a letter written by the Father Master of the Sacred Palace to the Inquisitor of Florence and of the Inquisitor's answer, in which the latter reported to have entrusted the correction of the book to Father Stefani, Consultant to the Holy Office. We also have copies of the work's preface or beginning and the suggestions concerning what the author was supposed to say at the end of the same work.

After this the Father Master of the Sacred Palace was not told anything, but then he saw the book had been printed in Florence, and published with the imprimatur of the Inquisitor there, and also with the imprimatur of Rome. By order of His Holiness he had all other copies seized, wherever ⟨295⟩ it was possible to do so with diligence. He examined the book and found that Galileo had transgressed the orders and the injunction given him by deviating from a hypothetical discussion.

After this and other offenses were referred to the Congregation of the Holy Office on 23 September 1632, His Holiness ordered that the Inquisitor of Florence be directed to issue an injunction to Galileo to come to Rome.

Having come, he was examined by the Holy Office on 12 April 1633. He said he thought he had been called to Rome because of a book written by him in dialogue form and printed in Florence in the year 1632, where he treats of the two chief world systems, that is, the arrangement of the heavenly orbs and the elements. He identified the book and he said he had been working on it for the previous ten or twelve years and had been principally but not continuously concerned with it for seven or eight years.

He said that in the year 1616 he came to Rome to hear what was appropriate to hold in regard to Copernicus's opinion on the earth's motion and sun's immobility. He discussed this topic several times with the Lord Cardinals of the Holy Office, and especially with the Lord Cardinals Bellarmine, Aracoeli, San Eusebio, Bonsi, and Ascoli. Eventually the Congregation of the Index declared that the above-mentioned opinion of Copernicus, taken absolutely, was contrary to Holy Scripture and that one could not hold or defend it, except as a supposition. Then the Lord Cardinal Bellarmine informed him of this declaration, as is evident from the certificate written for him in the Cardinal's own hand; in it he certifies that Galileo did not abjure but only was informed of the above-mentioned declaration, namely that the opinion that the earth moves and the sun stands still is contrary to Holy Scripture and therefore cannot be held or defended.

He admitted having been given an injunction, but he said that, based on this certificate in which the words "to teach in any way whatever" do not appear, he did not remember them.

To publish his book, he came to Rome and presented it to the Father Master of the Sacred Palace. The latter had it reviewed and gave him permission to print it in Rome. Forced to leave, he asked by letter for permission to publish it in Florence. The Father Master answered that he wanted to review the original again, but then, because of the risks of sending it to Rome due to the plague, he handed the case over to the Inquisitor of Florence. The latter had it reviewed by Father Stefani and then gave permission to publish it, in accordance with every order given by the Master of the Sacred Palace.

In asking for this authorization he did not mention the injunction to the Master of the Sacred Palace because he did not deem it necessary to tell him; for in his book he does not hold or defend the opinion of the sun's immobility and earth's motion, but rather he shows the opposite and that Copernicus's reasons are invalid.

On 30 April he asked for a hearing and said the following: [83] "Having reflected on the questions asked me about the injunction not to hold, defend, or teach in any way whatever the above-mentioned opinion, then condemned, I thought of rereading ⟨296⟩ my book, which I had not done over the last three years, to check whether against my purest intention I had inadvertently written anything that might indicate a defect of disobedience or other particulars that might suggest I was violating the orders of Holy Church. Having examined it very exactly, I perceived it, due to the long disuse, almost as a new work by another author. And I freely confess that it appeared to me in several places to be written in such a way that a reader, not aware of my intention, would have had reason to form the opinion that the arguments for the false side, which I intended to confute, were so stated as to be capable of convincing because of their strength, rather than being easy to answer. In particular, two arguments, one based on sunspots and the other on the tides, are presented favorably to the reader as being strong and powerful, more than would seem proper for someone who deemed them to be inconclusive and wanted to confute them, as indeed I inwardly and truly did and do hold them to be inconclusive and refutable. As an excuse of myself, within myself, I was not completely satisfied with saying that when one presents arguments for the opposite side with the intention of confuting them, they must be explained in the fairest way and not be made out of straw to the disadvantage of the opponent, especially when one is writing in dialogue form. Being dissatisfied with this excuse, as I said, I resorted to that of the natural gratification everyone feels for his own subtleties and for showing himself to be cleverer than the average man, by finding ingenious and apparent considerations of probability even in favor of false propositions. Nevertheless—even though, to use Cicero's words, "I am more desirous of glory than is suitable"—if I had to write out the same arguments now, there is no doubt I would weaken them in such a way that they could not appear to exhibit a force which they really and essentially lack. My error then was, and I confess it, one of vain ambition, pure ignorance, and inadvertence. And for greater confirmation that I did not and do not hold as true this opinion of the earth's motion and sun's immobility, I am ready to provide a better proof if I am allowed. There happens to be a very appropriate occasion, given that in the book already published the speakers agree that after some time they should meet to discuss various physical problems different from the topic already dealt with in their meetings; so, having to add one or two other Days, I promise to reconsider the arguments already presented in favor of this false and condemned opinion and to confute them in the most effective manner that God will suggest to me."

In his defense he presented the original copy of the Lord Cardinal Bellarmine's certificate to show that it does not contain the injunction's words "to teach in any way whatever" ⟨297⟩ and to have us accept his claim that in the course of fourteen or sixteen years he had no occasion to think about it and so lost any recollection of it.

He begged to be excused for having been silent about the injunction issued to him, since he did not remember the words "to teach in any way whatever," and so he thought the decree of the Congregation of the Index was sufficient. For it was public and agreed in every way with the words found in the certificate given him, namely that the said opinion should not be held or defended; and in publishing his book he obeyed whatever the decree of the Congregation stipulates. He said all this not to be excused from the error, but so that it be attributed to vain ambition rather than to malice and deception.

He humbly asked that due consideration be given to his advanced age of seventy years, his miserable indisposition, the mental affliction of the last ten months, the hardships suffered during the journey, and the slanders of his enemies to which his honor and reputation will be subject.

Galileo's Fourth Deposition (21 June 1633)

Called personally to the hall of Congregations in the palace of the Holy Office in Rome, fully in the presence of the Reverend Father Commissary General of the Holy Office, assisted by the Reverend Father Prosecutor, etc.

Galileo Galilei, Florentine, mentioned previously, having sworn an oath to tell the truth, was asked by the Fathers the following:

Q: Whether he had anything to say.

A: I have nothing to say.

Q: Whether he holds or has held, and for how long, that the sun is the center of the world and the earth is not the center of the world but moves also with diurnal motion.

A: A long time ago, that is, before the decision of the Holy Congregation of the Index, and before I was issued that injunction, I was undecided and regarded the two opinions, those of Ptolemy and Copernicus, as disputable, because either the one or the other could be true in nature. But after the above-mentioned decision, assured by the prudence of the authorities, all my uncertainty stopped, and I held, as I still hold, as very true and undoubted Ptolemy's opinion, namely the stability of the earth and the motion of the sun.

Having been told that he is presumed to have held the said opinion after that time, from the manner and procedure in which the said opinion is discussed and defended in the book he published after that time, indeed from the very fact that he wrote and published the said book, therefore he was asked to freely tell the truth whether he holds or has held that opinion.

A: In regard to my writing of the *Dialogue* already published, I did not do so because I held Copernicus's opinion to be true. Instead, deeming only to be doing a beneficial service, I explained the physical and astronomical reasons that can be advanced for one side and for the other; I tried to show that none of these, neither those in favor of this opinion or that, had the strength of a conclusive proof and that therefore to proceed with certainty one had to resort to the determination of more subtle doctrines, as one can see in many places in the *Dialogue*. So for my part I conclude ⟨362⟩ that I do not hold and, after the determination of the authorities, I have not held the condemned opinion.

Having been told that from the book itself and the reasons advanced for the affirmative side, namely that the earth moves and the sun is motionless, he is presumed, as it was stated, that he holds Copernicus's opinion, or at least that he held it at the time, therefore he was told that unless he decided to proffer the truth, one would have recourse to the remedies of the law and to appropriate steps against him.

A: I do not hold this opinion of Copernicus, and I have not held it after being ordered by injunction to abandon it. For the rest, here I am in your hands; do as you please.

And he was told to tell the truth, otherwise one would have recourse to torture.[84]

A: I am here to obey, but I have not held this opinion after the determination was made, as I said.

And since nothing else could be done for the execution of the decision,[85] after he signed he was sent to his place.

I, Galileo Galilei, have testified as above.

Sentence (22 June 1633)

We:

Gasparo Borgia, with the title of the Holy Cross in Jerusalem;

Fra Felice Centini, with the title of Santa Anastasia, called d'Ascoli;

Guido Bentivoglio, with the title of Santa Maria del Popolo;

Fra Desiderio Scaglia, with the title of San Carlo, called di Cremona;

Fra Antonio Barberini, called di Sant'Onofrio;

Laudivio Zacchia, with the title of San Pietro in Vincoli, called di San Sisto; ⟨403⟩

Berlinghiero Gessi, with the title of Sant'Agostino;

Fabrizio Verospi, with the title of San Lorenzo in Panisperna, of the order of priests;

Francesco Barberini, with the title of San Lorenzo in Damaso; and

Marzio Ginetti, with the title of Santa Maria Nuova, of the order of deacons;

By the grace of God, Cardinals of the Holy Roman Church, and especially commissioned by the Holy Apostolic See as Inquisitors-General against heretical depravity in all of Christendom.

Whereas you, Galileo, son of the late Vincenzio Galilei, Florentine, aged seventy years, were denounced to this Holy Office in 1615 for holding as true the false doctrine taught by some that the sun is the center of the world and motionless and the earth moves even with diurnal motion; for having disciples to whom you taught the same doctrine; for being in correspondence with some German mathematicians about it; for having published some letters entitled *On Sunspots*, in which you explained the same doctrine as true; for interpreting Holy Scripture according to your own meaning in response to objections based on Scripture which were sometimes made to you; and whereas later we received a copy of an essay in the form of a letter, which was said to have been written by you to a former disciple of yours and which in accordance with Copernicus's position contains various propositions against the authority and true meaning of Holy Scripture;

And whereas this Holy Tribunal wanted to remedy the disorder and the harm which derived from it and which was growing to the detriment of the Holy Faith, by order of His Holiness and the Most Eminent and Most Reverend Lord Cardinals of this Supreme and Universal Inquisition, the Assessor Theologians assessed the two propositions of the sun's stability and the earth's motion as follows:

That the sun is the center of the world and motionless is a proposition which is philosophically absurd and false, and formally heretical, for being explicitly contrary to Holy Scripture;

That the earth is neither the center of the world nor motionless but moves even with diurnal motion is philosophically equally absurd and false, and theologically at least erroneous in the Faith.

Whereas however we wanted to treat you with benignity at that time, it was decided at the Holy Congregation held in the presence of His Holiness on 25 February 1616 that the Most Eminent Lord Cardinal Bellarmine would order you to abandon this false opinion completely; that if you refused to do this, the Commissary of the Holy Office would give you an injunction to abandon this doctrine, not to teach it to others, not to defend it, and not to treat of it; and that if you did not acquiesce in this injunction, you should be imprisoned. To execute this decision, the following day at the palace of and in the presence of the above-mentioned Most Eminent Lord Cardinal Bellarmine, after being informed and warned in a friendly way by the same Lord Cardinal, you were given an injunction by the then Father Commissary of the Holy Office ⟨404⟩ in the presence of a notary and witnesses to the effect that you must completely abandon the said false opinion, and that in the future you could neither hold, nor defend, nor teach it in any way whatever, either orally or in writing; having promised to obey, you were dismissed.

Furthermore, in order to completely eliminate such a pernicious doctrine, and not let it creep any further to the great detriment of Catholic truth, the Holy Congregation of the Index issued a decree which prohibited books treating of such a doctrine and declared it false and wholly contrary to the divine and Holy Scripture.

And whereas a book has appeared here lately, printed in Florence last year, whose inscription showed that you were the author, the title being *Dialogue by Galileo Galilei on the two Chief World Systems, Ptolemaic and Copernican*; and whereas the Holy Congregation was informed that with the printing of this book the false opinion of the earth's motion and sun's stability was being disseminated and taking hold more and more every day, the said book was diligently examined and found to violate explicitly the above-mentioned injunction given to you; for in the same book you have defended the said opinion already condemned and so declared to your face, although in the said book you try by means of various subterfuges to give the impression of leaving it undecided and labeled as probable; this is still a very serious error since there is no way an opinion declared and defined contrary to divine Scripture may be probable.

Therefore, by our order you were summoned to this Holy Office, where, examined under oath, you acknowledged the book as written and published by you. You confessed that about ten or twelve years ago, after having been given the injunction mentioned above, you began writing the said book, and that then you asked for permission to print it without explaining to those who gave you such permission that you were under the injunction of not holding, defending, or teaching such a doctrine in any way whatever.

Likewise, you confessed that in several places the exposition of the said book is expressed in such a way that a reader could get the idea that the arguments given for the false side were effective enough to be capable of convincing, rather than being easy to refute. Your excuses for having committed an error, as you said so foreign from your intention, were that you had written in dialogue form, and everyone feels a natural satisfaction for one's own subtleties and showing oneself sharper than the average man by finding ingenious and apparently probable arguments even in favor of false propositions.

Having been given suitable terms to present your defense, you produced a certificate in the handwriting of the Most Eminent Lord Cardinal Bellarmine, which you said you obtained to defend yourself from the calumnies of your enemies, who were claiming that you had abjured and had been punished by the Holy Office. This ⟨405⟩ certificate says that you had neither abjured nor been punished, but only that you had been notified of the declaration made by His Holiness and published by the Holy Congregation of the Index, whose content is that the doctrine of the earth's motion and sun's stability is contrary to Holy Scripture and so can be neither defended nor held. Because this certificate does not contain the two phrases of the injunction, namely "to teach" and "in any way whatever," one is supposed to believe that in the course of fourteen or sixteen years you had lost any recollection of them, and that for this same reason you had been silent about the injunction when you applied for the license to publish the book. Furthermore, one is supposed to believe that you point out all of this not to excuse the error, but in order to have it attributed to conceited ambition rather than to malice. However, the said certificate you produced in your defense aggravates your case further since, while it says that the said opinion is contrary to Holy Scripture, yet you dared to treat of it, defend it, and show it as probable; nor are you helped by the license you artfully and cunningly extorted since you did not mention the injunction you were under.

Because we did not think you had said the whole truth about your intention, we deemed it necessary to proceed against you by a rigorous examination. Here you answered in a Catholic manner, though without prejudice to the above-mentioned matters confessed by you and deduced against you about your intention.

Therefore, having seen and seriously considered the merits of your case, together with the above-mentioned confessions and excuses and with any other reasonable matter worth seeing and considering, we have come to the final sentence against you given below.

Therefore, invoking the Most Holy name of Our Lord Jesus Christ and his most glorious Mother, ever Virgin Mary; and sitting as a tribunal, with the advice and counsel of the Reverend Masters of Sacred Theology and the Doctors of both laws, our consultants; in this written opinion we pronounce final judgment on the case pending before us between the Magnificent Carlo Sinceri, Doctor of both laws, and Prosecuting Attorney of this Holy Office, on one side, and you the above-mentioned Galileo Galilei, the culprit here present, examined, tried, and confessed as above, on the other side:

We say, pronounce, sentence, and declare that you, the above-mentioned Galileo, because of the things deduced in the trial and confessed by you as above, have rendered yourself according to this Holy Office vehemently suspected of heresy,[86] namely of having held and believed a doctrine which is false and contrary to the divine and Holy Scripture: that the sun is the center of the world and does not move from east to west, and the earth moves and is not the center of the world, and that one may hold and defend as probable an opinion after it has been declared and defined contrary to Holy Scripture. Consequently you have incurred all the censures and penalties imposed and promulgated by the sacred canons and all particular and general laws against such delinquents. We are willing to absolve you from them provided that first, with a sincere heart and unfeigned faith, in front of us you abjure, curse, and detest the above-mentioned errors and ⟨406⟩ heresies, and every other error and heresy contrary to the Catholic and Apostolic Church, in the manner and form we will prescribe to you.

Furthermore, so that this serious and pernicious error and transgression of yours does not remain completely unpunished, and so that you will be more cautious in the future and an example for others to abstain from similar crimes, we order that the book *Dialogue* by Galileo Galilei be prohibited by public edict.

We condemn you to formal imprisonment in this Holy Office at our pleasure. As a salutary penance we impose on you to recite the seven penitential Psalms once a week for the next three years. And we reserve the authority to moderate, change, or condone wholly or in part the above-mentioned penalties and penances.

This we say, pronounce, sentence, declare, order, and reserve by this or any other better manner or form that we reasonably can or shall think of.

So we the undersigned[87] Cardinals pronounce:

> Felice Cardinal d'Ascoli.
> Guido Cardinal Bentivoglio.
> Fra Desiderio Cardinal di Cremona.
> Fra Antonio Cardinal di Sant'Onofrio.
> Berlinghiero Cardinal Gessi.
> Fabrizio Cardinal Verospi.
> Marzio Cardinal Ginetti.

Galileo's Abjuration (22 June 1633)

I, Galileo, son of the late Vincenzio Galilei of Florence, seventy years of age, arraigned personally for judgment, kneeling before you Most Eminent and Most Reverend Cardinals Inquisitors-General against heretical depravity in all of Christendom, having before my eyes and touching with my hands the Holy Gospels, swear that I have always believed, I believe now, and with God's help I will believe in the future all that the Holy Catholic and Apostolic Church holds, preaches, and teaches. However, whereas, after having been judicially instructed with injunction by the Holy Office to abandon completely the false opinion that the sun is the center of the world and does not move and the earth is not the center of the world and moves, and not to hold, defend, or teach this false doctrine in any way whatever, orally or in writing; and after having been notified that this doctrine is contrary to Holy Scripture; I wrote and published a book in which I treat of this already condemned doctrine and adduce very effective reasons in its favor, without refuting them in any way; therefore, I have been judged vehemently suspected of heresy, namely of having held and believed that the sun is the center of the world and motionless and the earth is not the center and moves.

Therefore, desiring to remove from the minds of Your Eminences and every faithful ⟨407⟩ Christian this vehement suspicion, rightly conceived against me, with a sincere heart and unfeigned faith I abjure, curse, and detest the above-mentioned errors and heresies, and in general each and every other error, heresy, and sect contrary to the Holy Church; and I swear that in the future I will never again say or assert, orally or in writing, anything which might cause a similar suspicion about me; on the contrary, if I should come to know any heretic or anyone suspected of heresy, I will denounce him to this Holy Office, or to the Inquisitor or Ordinary of the place where I happen to be.

Furthermore, I swear and promise to comply with and observe completely all the penances which have been or will be imposed upon me by this Holy Office; and should I fail to keep any of these promises and oaths, which God forbid, I submit myself to all the penalties and punishments imposed and promulgated by the sacred canons and other particular and general laws against similar delinquents. So help me God and these Holy Gospels of His, which I touch with my hands.

I, the above-mentioned Galileo Galilei, have abjured, sworn, promised, and obliged myself as above; and in witness of the truth I have signed with my own hand the present document of abjuration and have recited it word for word in Rome, at the convent of the Minerva, this twenty-second day of June 1633.

I, Galileo Galilei, have abjured as above, by my own hand.

APPENDIXES

Chronology of Events

This chronology lists primarily the key events of the Galileo affair from 1613 to 1633, which is the period corresponding to the selected documents. It also includes other developments in Galileo's life, as well as important occurrences in the immediately preceding historical background and some relevant subsequent events. References not otherwise identified are to the various volumes of the National Edition of Galileo's collected works edited by Favaro (1890–1909). References preceded by an asterisk are to the documents translated in this volume. It should be mentioned that a 69-page biographical summary of Galileo's life appears in Favaro (1907b), which specialists will find extremely useful and informative. However, the references to the National Edition given by him are not always sufficient to justify fully his claims, which thus seem to be based on other sources as well. For example, Favaro (1907b, p. 33) gives 6 January 1615 as the publication date of Foscarini's *Letter*, citing as reference p. 150 of vol. 12; there we find a letter by Cesi to Galileo dated 7 March 1615 and stating merely that Foscarini's book had just been published.

1453   The Ottoman Turks conquer Constantinople.

1492   Italian navigator Christopher Columbus discovers America.

1517   German monk Martin Luther triggers the Protestant Reformation by posting ninety-five theses on the door of a church in Wittenberg.

1543   Polish astronomer Nicolaus Copernicus publishes his book *On the Revolutions of the Heavenly Spheres*; he dies the same year.

1545–1563   The Catholic church convenes the Council of Trent to deal with Protestantism, and the Catholic Counter-Reformation begins.

15 February 1564   Galileo is born in Pisa. ⟨19:23–24⟩

1571   Johannes Kepler is born.

1581–1585   Galileo is enrolled at the University of Pisa in medicine, studies mathematics privately, and leaves without a degree. ⟨19:32 and 602–605⟩

1589   Galileo begins teaching mathematics at the University of Pisa. ⟨19: 37–39⟩

1592   Galileo becomes a professor of mathematics at the University of Padua. ⟨19:111 and 117–119⟩

1600   Dominican Friar Giordano Bruno is burned at the stake by the Inquisition in Rome.

1601   Danish astronomer Tycho Brahe dies.

1609   Kepler publishes his *New Astronomy*, containing the first two of his famous three laws of planetary motion.

Fall 1609   Galileo builds an astronomically useful telescope and begins to observe the heavens. ⟨10:250–278⟩

13 March 1610   Galileo's *Starry Messenger* is published in Venice. ⟨10:288⟩

19 April 1610   Kepler sends Galileo his *Conversation with the Starry Messenger*, supporting the new discoveries. ⟨3:97–126⟩

April–June 1610   Galileo applies for and is granted the position of philosopher and chief mathematician to the grand duke of Tuscany; he resigns from his position at the University of Padua. ⟨10:307, 353, and 369; 19:125⟩

June 1610   Martin Horky publishes *A Very Short Excursion Against the Starry Messenger.* ⟨3:127–145⟩

November 1610   John Wedderburn publishes in Padua a reply to Horky and defense of Galileo. ⟨3:11 and 147–178⟩

1610 or 1611   Lodovico delle Colombe writes *Against the Earth's Motion*, containing religious objections to Galileo's views. ⟨3:12 and 251–290⟩

1611   Kepler publishes in Frankfurt an account of his observations of Jupiter's satellites, further supporting Galileo. ⟨3:11 and 179–190⟩

1611   Giovann'Antonio Roffeni publishes in Bologna another reply to Horky in defense of Galileo. ⟨3:11 and 191–200⟩

1611   Francesco Sizzi publishes in Venice his *Dianoia Astronomica, Optica, Physica*, containing religious objections to Galileo's views. ⟨3:12 and 201–250⟩

24 April 1611   Replying to an inquiry by Cardinal Bellarmine (19 April 1611), the four professors of mathematics at the Jesuit Collegio Romano (Fathers Christopher Clavius, Christopher Grienberger, Odo van Maelcote, and Gio.

Paolo Lembo) write a report confirming Galileo's telescopic observations, but avoiding commitment to any interpretation of them. ⟨11:87−88 and 92−93⟩

25 April 1611   Galileo is made a member of the Lincean Academy. ⟨19:265⟩

17 May 1611   At a regular meeting the Inquisition decides to check whether Galileo is mentioned in its ongoing proceeedings against Dr. Cesare Cremonini, professor of philosophy at the University of Padua, who was in trouble because of his philosophical writings; he was an old colleague and friend of Galileo, though the two were also well-known intellectual opponents. ⟨19:275; Pagano 1984, p. 219⟩

May 1611   The Jesuit Collegio Romano holds a special meeting at which, in the presence of Galileo, Father Maelcote delivers a lecture praising *The Starry Messenger*. ⟨3:13 and 293−298⟩

December 1611   A friend writes to Galileo that a group of malicious and envious persons are meeting regularly at the house of the archbishop of Florence and plotting to see whether they can fault him in regard to such matters as the earth's motion; one of them asked a preacher to attack Galileo from the pulpit, but the latter refused. ⟨11:241−242⟩

1612   Giulio Cesare Lagalla, professor of philosophy at the University of Rome, publishes in Venice a book, *On the Phenomena in the Orb of the Moon*, disputing Galileo's lunar discoveries. ⟨3:13 and 309−399⟩

November 1612   In a private conversation on 2 November, Dominican Friar Niccolò Lorini attacks Galileo for believing ideas (such as that the earth moves) which contradict the Bible and for thus being inclined to heresy; but on 5 November Lorini writes Galileo a letter of apology. ⟨11:427⟩

Fall 1612   The Lincean Academy decides to publish in Rome under its own auspices a number of writings on sunspots: primarily the long letters Galileo had written that year to Mark Welser, a German politician, businessman, and intellectual who was a member of the academy; but also the letters which Welser had written to Galileo and which had occasioned his writing; as well as a reprint of certain writings addressed to Welser by the German Jesuit astronomer Christopher Scheiner. ⟨5:12−14; 11:403−404; and 11:416⟩

Fall 1612−Winter 1613   The publication of the *Sunspot Letters* encounters difficulties and delays in part due to ecclesiastical censorship; some biblical references and anti-Aristotelian critical statements are deleted from the original text. ⟨5:12, 16, 93, 96, and 138−140; 11:431, 437−438, and 483 (no. 847)⟩

22 March 1613   Galileo's *Sunspot Letters* are published in Rome. ⟨5:12; 11:489−490⟩

Fall 1613   Ulisse Albergotti publishes in Viterbo (central Italy) the book *Dialogue ... in Which It Is Held ... That the Moon is Intrinsically Lumi-*

nous . . . , containing biblical criticism of Galileo's theories. ⟨11:598–599; Drake 1957, p. 190⟩

December 1613   Incited by Cosimo Boscaglia, special professor of philosophy at the University of Pisa, the Grand Duchess Dowager Christina questions Benedetto Castelli about the compatibility of Galileo's ideas with the Bible. ⟨*"Castelli to Galileo"; 11:605–606⟩

21 December 1613   In reply to a letter by Castelli informing him about the grand duchess's worries, Galileo writes a long letter to Castelli summarizing his views on the relationship between the Bible and scientific inquiry and discussing the passage from Joshua 10:12–13, which had been advanced as especially troublesome. ⟨*"Galileo to Castelli"; 5:281–288⟩

21 December 1614   At the Church of Santa Maria Novella in Florence, Dominican Friar Tommaso Caccini preaches a sermon against mathematicians in general, and Galileo in particular, because their beliefs and practices allegedly contradict the Bible and thus are heretical. ⟨*"Caccini's Deposition" (20 March 1615); 12:123; 19:307; Favaro 1907b, p. 33⟩

December 1614–January 1615   Galileo speaks and writes to a number of people about the best way to respond to Caccini. ⟨12:122–131⟩

10 January 1615   Dominican Friar Maraffi, Caccini's superior, writes to Galileo apologizing for his excessive zeal. ⟨12:127–128⟩

12 January 1615   Prince Cesi, founder and head of the Lincean Academy, answers Galileo's letter advising caution and informing Galileo that "in regard to Copernicus's opinion, Bellarmine himself, who often heads commissions investigating these matters, has told me that he regards it as heretical and that without doubt the earth's motion is against the Scripture; thus you should be careful. I have always thought that if the Congregation of the Index had been consulted at the time of Copernicus, they would have prohibited him; nor need I say anything else." ⟨12:129⟩

7 February 1615   Lorini files with the Inquisition a written complaint against Galileo, enclosing his letter to Castelli as incriminating evidence. ⟨*"Lorini's Complaint"; 19:297–298⟩

16 February 1615   Galileo writes to Monsignor Dini in Rome asking for support and advice about the charges and actions against him. ⟨*"Galileo to Monsignor Dini"; 5:291–295⟩

February or March 1615   Carmelite Friar Paolo Antonio Foscarini publishes a book entitled *Letter on the Pythagorean and Copernican Opinion of the Earth's Motion and Sun's Rest and on the New Pythagorean World System, in which are harmonized and reconciled those passages of the Holy Scripture and those theological propositions which could ever be adduced against this opinion* (Naples, 1615), in which he argues that Copernicanism is compatible with the Bible. ⟨12:150⟩

7 March 1615   From Rome Monsignor Dini replies to Galileo's letter, summarizing all that he has done so far in accordance with its requests and all that he has learned; he includes a report of a discussion with Bellarmine in which the cardinal, among other things, was particularly worried about the biblical passage in Psalm 19:5−6, which seems to attribute motion to the sun. ⟨*"Monsignor Dini to Galileo"; 12:151−152⟩

20 March 1615   Caccini gives a deposition to the Inquisition in Rome, charging Galileo with suspicion of heresy, based on the content of the letter to Castelli and the *Sunspot Letters* and on hearsay evidence of a general sort (allegedly known to everyone in Florence) and of a more specific type (involving two individuals named Ximenes and Attavanti). ⟨*"Caccini's Deposition"; 19:307−311⟩

23 March 1615   Galileo writes a long letter replying to the several points raised by Dini's own letter; the discussion focuses on the epistemological status of Copernicanism and on the elaboration of a speculative reinterpretation of the troublesome biblical passage (Psalm 19:5−6) consistent with Copernicanism. ⟨*"Galileo to Monsignor Dini"; 5:297−305⟩

12 April 1615   Cardinal Bellarmine writes to Father Foscarini, commenting on his book and stating explicitly that his remarks apply to Galileo as well. ⟨*"Cardinal Bellarmine to Foscarini"; 12:171−172⟩

1615   Galileo writes his "Considerations on the Copernican Opinion" and his "Letter to the Grand Duchess Christina" ⟨5:277 and 292⟩. An Inquisition consultant reports to the Holy Office that Galileo's letter to Castelli does not contain any significant errors but conforms essentially to Catholic doctrine. ⟨19:305⟩

13 November 1615   In a deposition before the Inquisition, Dominican Father Ferdinando Ximenes states that he has indeed discussed a number of heretical ideas with Giannozzo Attavanti, but only in the context of an argumentative disputation and in such a way as to point out clearly to him that these were indeed erroneous; Ximenes also states that he has heard that some of these heretical ideas are held by Galileo and his followers. ⟨*"Ximenes' Deposition"; 19:316−318⟩

14 November 1615   In a deposition before the Inquisition, the Reverend Giannozzo Attavanti confirms having discussed these matters with Ximenes but states that he has never heard Galileo express heresies. He also refers to the *Sunspot Letters* as his main source of information about Galileo's views. ⟨*"Attavanti's Deposition"; 19:318−320⟩

25 November 1615   The Inquisition decides to examine Galileo's *Sunspot Letters*. ⟨19:277−278 and 320⟩

December 1615   After a long delay due to illness, Galileo goes to Rome to try to clear his name and prevent the condemnation of Copernicanism; he lodges at the Tuscan embassy. ⟨12:207⟩

8 January 1616   At the request of Cardinal Alessandro Orsini, Galileo writes his "Discourse on the Tides." ⟨5:395⟩

24 February 1616   A committee of eleven consultants reports to the Roman Inquisition that they are of the unanimous opinion that the heliocentric and heliostatic thesis is philosophically absurd and formally heretical and that the geokinetic thesis is philosophically absurd and theologically erroneous. ⟨*"Consultants' Report on Copernicanism"; 19:320–321⟩

25 February 1616   The pope orders Cardinal Bellarmine to warn Galileo to abandon his Copernican views. ⟨*"Inquisition Minutes"; 19:321; Pagano 1984, pp. 100–101 and 222–223⟩

26 February 1616   Bellarmine calls Galileo to his house and gives him the warning. ⟨*"Special Injunction"; 19:321–322⟩

3 March 1616   Bellarmine reports to the Inquisition that Galileo has acquiesced and that the Congregation of the Index will suspend Copernicus's book. ⟨*"Inquisition Minutes"; 19:278⟩

5 March 1616   The Congregation of the Index publishes a decree prohibiting and condemning Foscarini's book, suspending until corrected Copernicus's book and Zuñiga's *On Job*, and ordering analogous censures for analogous works; Galileo is not mentioned at all. ⟨*"Decree of the Index"; 19:322–323⟩

6 March 1616   From Rome Galileo writes to the Tuscan secretary of state in Florence, giving his interpretation of the outcome of the Inquisition proceedings. ⟨*"Galileo to the Tuscan Secretary of State"; 12:243–245⟩

12 March 1616   Galileo reports to the Tuscan secretary of state that the day before he had an audience with Pope Paul V to discuss the recent happenings, spending three-quarters of an hour with him and being warmly received and reassured. ⟨*"Galileo to the Tuscan Secretary of State"; 12:247–249⟩

Spring 1616   From friends in Pisa and Venice Galileo receives letters reporting that rumors are circulating there to the effect that he had been personally put on trial and condemned by the Inquisition. ⟨12:254 and 257–259⟩

26 May 1616   Cardinal Bellarmine writes a declaration on Galileo's behalf, denying these rumors and clarifying how Galileo was affected by the Inquisition proceedings and the Index decree. ⟨*"Cardinal Bellarmine's Certificate"; 19:348⟩

June 1616   Galileo returns to Florence. ⟨12:261–266⟩

23 May 1618   Galileo writes to the Archduke Leopold of Austria, sending him a copy of the "Discourse on the Tides" written two years earlier, along with other gifts such as some telescopes. ⟨*"Galileo to Archduke Leopold of Austria"; 12:389–392⟩

15 May 1620   The Congregation of the Index issues the correction of Copernicus's book *On the Revolutions*, promised in the decree of 5 March 1616. ⟨*"Correction of Copernicus's *On the Revolutions*"; 19:400–401⟩

6 August 1623   Florentine Cardinal Maffeo Barberini, an admirer and patron of Galileo, is elected Pope Urban VIII. ⟨20:323⟩

October 1623   Galileo's *Assayer* is published in Rome, under the auspices of the Lincean Academy, and dedicated to the new pope. ⟨6:201; 13: 141–142⟩

Spring 1624   Galileo visits Rome for about six weeks and is warmly received by the pope and other Church officials; in particular he is granted papal audiences on six occasions. ⟨13:175 and 182–185⟩

8 June 1624   As he is about to leave Rome, Galileo writes to Prince Cesi reporting on his visit; in particular he mentions having been told the following detail by German Cardinal Hohenzollern: while discussing the subject of Copernicus with the pope, Hohenzollern told the latter that "all heretics accept his opinion and hold it as most certain, and so one must be very circumspect in arriving at any decision; to this His Holiness replied that the Holy Church had not condemned it, nor was about to condemn it, as heretical, but only as temerarious, and that one should not fear that it could ever be proved to be necessarily true." ⟨13:182⟩

Summer 1624   Galileo writes his "Reply to Ingoli," answering the anti-Copernican essay the latter had written in 1616. ⟨*"Galileo's Reply to Ingoli"; 6:503–508; 13:209.23⟩

Fall 1624   Galileo begins working on what was to become the *Dialogue* or, to be more exact, the "Dialogue on the Tides, Which Brings with It as a Consequence the Copernican System." ⟨13:236.23–25; cf. 13:209.25⟩

1624 or 1625   A written opinion is sent to a Church official by a person unknown to us, charging that the atomistic theory of matter in Galileo's *Assayer* conflicts with the Catholic doctrine of the Holy Sacrament. The exact date and the addressee are also unknown to us. ⟨*"Anonymous Complaint About *The Assayer*"; Pagano 1984, pp. 43–48 and 245–248⟩

April 1625   Galileo learns that a complaint against his *Assayer* has been investigated by the Inquisition, but the investigation led to his exoneration; nevertheless, he is advised that it is prudent at the moment not to deliver his "Reply to Ingoli," given that it explicitly defends Copernicanism. ⟨*"Guiducci to Galileo"; 13:265–266⟩

16 March 1630   Castelli writes to Galileo having heard from Prince Cesi that "in the last few days Father Campanella was speaking with His Holiness and told him that he had had the opportunity to convert some German gentlemen to the Catholic faith and they were very favorably inclined; however, having heard about the prohibition of Copernicus, etc., they had been scan-

dalized, and he had been unable to go further. His Holiness answered with the following exact words: 'It was never our intention, and if it had been up to us that decree would not have been issued'." ⟨14:87–88⟩

April 1630   Galileo completes work on the *Dialogue*. ⟨14:67.26–29, 92.43–47, and 96–97⟩

May–June 1630   For about two months Galileo is in Rome to obtain the imprimatur for the *Dialogue* from Church authorities and arrange for its publication by the Lincean Academy. ⟨14:97–98 and 121⟩

26 June 1630   Galileo leaves Rome without his book having been published there, but with a written endorsement from the Vatican secretary (who was the chief Roman censor) and with the impression that this is a final approval pending certain minor additions and finishing touches. ⟨14:121, 130, and 216⟩

1 August 1630   Prince Cesi dies, and the Lincean Academy is left with no one to replace him for leadership and support. ⟨14:126–127; 20:417⟩

Summer 1630   An outbreak of the plague begins to interrupt travel and commerce in Italy. ⟨*"Galileo's First Deposition"; 19:341.166–167⟩

Fall 1630   Difficulties and delays develop in executing the practical details of printing the *Dialogue* in accordance with the understanding reached in June between Galileo and the Vatican secretary. ⟨14:150–151, 156–157, 167, and 169⟩

7 March 1631   Galileo writes to the Tuscan secretary of state, indicating that he is tired of waiting for the Vatican secretary in regard to the publication of his book and that, in view of the death of Prince Cesi and the plague, he wants to publish his book in Florence. ⟨*"Galileo to the Tuscan Secretary of State"; 14:215–218⟩

Spring–Summer 1631   Intense negotiations take place with the result that the Vatican secretary transfers to the Florentine inquisitor jurisdiction over publication of the book in Florence, together with certain instructions (stemming from the pope) concerning its title, content, preface, and ending. ⟨*Correspondence 7 March–19 July 1631; cf. 14:215–285⟩

21 February 1632   Printing of the *Dialogue* is completed in Florence. ⟨7:8; 14:331⟩

Summer 1632   While the book is received with great enthusiasm and praise in many quarters, a number of questions, rumors, complaints, and criticisms emerge in Rome concerning its content, form, and manner of publication; these lead the pope to prohibit the printer from further distributing the book and to appoint a special commission to investigate the matter. ⟨14:333ff. and 375.14; 19:326; 20:571–572⟩

Early September 1632   The special papal commission on the *Dialogue* files its report, and as a result the pope decides to forward the case to the Inquisition.

⟨*"Diplomatic Correspondence (18 September 1632)"; *"Cardinal Barberini to the Florentine Nuncio"; 14:391–393 and 397–398⟩

23 September 1632   At a meeting of the Inquisition presided over by the pope, it is decided to summon Galileo to Rome for the whole month of October. ⟨19:279–280 and 330⟩

1 October 1632   The Florentine inquisitor conveys to Galileo the order from Rome, and Galileo signs a written statement promising to obey. ⟨19: 331–332⟩

13 October 1632   Galileo writes to Cardinal Francesco Barberini asking that the trial be transferred to Florence or that he be allowed to answer the charges in writing. ⟨14:406–410⟩

11 November 1632   At a meeting of the Inquisition presided over by the pope, Galileo's plea is refused and it is decided that he be compelled to obey if need be. ⟨19:280⟩

17 December 1632   Three Florentine physicians sign a statement that they have examined Galileo and have found him afflicted by a number of specific ills such that the least external problem poses a danger to his life. ⟨19:334–335⟩

30 December 1632   At an Inquisition meeting presided over by the pope, Galileo's medical excuses are discounted and it is decided that unless he comes on his own he will be arrested and brought to Rome. ⟨19:281–282 and 335⟩

15 January 1633   Galileo writes a very revealing letter to Elia Diodati in Paris ⟨*"Galileo to Diodati"; 15:23–26⟩. He makes his will ⟨19:520⟩. And he decides to leave for Rome ⟨15:27⟩.

20 January 1633   Galileo leaves Florence to go to Rome. ⟨15:27 and 29⟩

13 February 1633   Galileo arrives in Rome and is lodged at the Tuscan embassy. ⟨*"Diplomatic Correspondence (14 February 1633)"; 15:40–41⟩

15 February 1633   Galileo is ordered not to socialize and to remain in seclusion at the embassy. ⟨*"Diplomatic Correspondence (14 and 16 February 1633)"; 15:41⟩

Early April 1633   Cardinal Francesco Barberini notifies the Tuscan ambassador that the Inquisition plans to examine Galileo soon and that for this purpose he will have to go to the palace of the Inquisition and may have to be detained there. ⟨*"Diplomatic Correspondence (9 April 1633)"; 15:84–85⟩

12 April 1633   Galileo is formally interrogated by the Inquisition. ⟨*"Galileo's First Deposition"; 19:336–342⟩

12–30 April 1633   Galileo is detained at the Inquisition headquarters but is allowed to lodge in the chief prosecutor's apartment. ⟨*"Diplomatic Correspondence (16 April and 1 May 1633)"; 15:86–87, 94–95, and 109–110⟩

17 April 1633   Three Inquisition consultants file their reports about Galileo's *Dialogue*. ⟨*"Oreggi's Report, Inchofer's Report, and Pasqualigo's Report"; 19:348–360⟩

28 April 1633   The commissary general of the Inquisition reports to Cardinal Francesco Barberini about his success in arranging with Galileo an extrajudicial deal whereby he will admit some wrongdoing and then will be treated with leniency. ⟨*"Commissary General to Cardinal Barberini"; 15:106–107⟩

30 April 1633   Galileo signs a statement in which he admits some wrongdoing, but no malicious intent, in connection with the writing of the *Dialogue* ⟨*"Galileo's Second Deposition"; 19:342–344⟩. Galileo is allowed to return to the Tuscan embassy, with orders not to discuss the case with anyone, and to return to the Inquisition headquarters when summoned again. ⟨*"Diplomatic Correspondence (1 May 1633)"; 15:109–110; 19:344⟩

10 May 1633   Galileo appears before the Inquisition and makes a formal presentation of his defense, consisting of Bellarmine's certificate (26 May 1616) and a formal statement. ⟨*"Galileo's Third Deposition" and *"Galileo's Defense"; 19:345–347⟩

May or June 1633   A report is made to the pope by the Inquisition officials, summarizing the events and the proceedings since 1615. ⟨*"Final Report to the Pope"; 19:293–297⟩

16 June 1633   At a meeting of the Inquisition presided over by the pope, he decides that Galileo be examined on his intentions, even to the extent of torture; that if he sustains it, he is to abjure vehement suspicion of heresy before the full Holy Congregation, promising never again to speak or write about the motion of the earth or stability of the sun; that the *Dialogue* be prohibited; and that, as an example, his sentence be sent to apostolic nuncios and inquisitors to be read in public to gatherings of professors of mathematics. ⟨19:282–283 and 360–361⟩

21 June 1633   The Inquisition examines Galileo with the formal threat of torture to determine his intention. ⟨*"Galileo's Fourth deposition"; 19:361–362⟩

22 June 1633   Galileo is sentenced, with only seven of the ten cardinal-inquisitors signing the sentence; the punishment involves an abjuration, prohibition of the *Dialogue*, formal arrest at the pleasure of the Inquisition, and some religious penances ⟨*"Sentence"; 19:402–406⟩. Galileo recites in public a formal abjuration at the convent of the Minerva. ⟨*"Galileo's Abjuration"; 19:406–407⟩

24 June 1633   Galileo is allowed to be under house arrest at the Tuscan embassy. ⟨15:165⟩

30 June 1633   The pope refuses Galileo's petition to be allowed to go back to Florence but, as a step in that direction, permits him to stay under conditions

of house arrest at the residence of the archbishop of Siena. ⟨25:170–171; 19:362⟩

9 July 1633   Galileo arrives in Siena. ⟨19:364⟩

1 December 1633   The pope grants Galileo's petition to return to his villa in Arcetri near Florence, though the conditions of house arrest will continue. ⟨19:388–389⟩

17 December 1633   Galileo is home at Arcetri and writes to Cardinal Francesco Barberini to thank him. ⟨19:391⟩

1638   Galileo publishes in Holland his book entitled *Discourses and Mathematical Demonstrations on Two New Sciences Pertaining to Mechanics and to Local Motion.* ⟨8:41⟩

8 January 1642   Galileo dies at Arcetri; he is quietly buried at the Church of Santa Croce in Florence, in a grave without decoration or inscription. ⟨18:378–379; 19:558⟩

1729   English astronomer James Bradley discovers the aberration of starlight, providing direct observational evidence that the earth has translational motion.

16 June 1734   The Roman Inquisition agrees to a Florentine request to erect a mausoleum for Galileo in the Church of Santa Croce. ⟨19:398–399⟩

1744   The Church allows the *Dialogue* to be published in Padua as part of an edition of Galileo's collected works, but accompanied by a number of qualifications and disclaimers. ⟨19:292; Gebler 1879, p. 313⟩

16 April 1757   During the papacy of Benedict XIV, the Congregation of the Index decides to withdraw the decree which prohibited *all* books teaching the earth's motion, although the *Dialogue* and a few other books continue to be explicitly included. ⟨19:419; Gebler 1879, pp. 312–313⟩

16 August 1820   The Congregation of the Holy Office, with the pope's approval, decrees that Catholic astronomer Joseph Settele can be allowed to treat the earth's motion as an established fact. ⟨19:420⟩

11 September 1822   The Congregation of the Holy Office decides to allow in general the publication of books treating of the earth's motion in accordance with modern astronomy. ⟨19:421⟩

25 September 1822   Pope Pius VII ratifies this decision. ⟨19:421⟩

1835   The 1835 edition of the Catholic Index of Prohibited Books for the first time omits the *Dialogue* from the list. ⟨Gebler 1879, p. 315⟩

1893   In the encyclical letter *Providentissimus Deus*, Pope Leo XIII puts forth a view of the relationship between biblical interpretation and scientific investigation that corresponds to the one advanced by Galileo in the "Letter to the Grand Duchess Christina." ⟨Dubarle 1964, p. 25; Langford 1966, p. 66; Viganò 1969, p. 234; Martini 1972, p. 444; Poupard 1984a, p. 13⟩

10 November 1979   In a speech to the Pontifical Academy of Sciences commemorating the centenary of Albert Einstein's birth, Pope John Paul II admits
that Galileo suffered unjustly at the hands of the Church and praises Galileo's
religiousness and his views and behavior regarding the relationship between
science and religion. ⟨John Paul II (1979)⟩

Concordance to the Documents

Listed here are all the documents and writings translated and annotated in this documentary history, their dates, the exact page references to the National Edition of Galileo's *Opere* edited by Favaro (1890–1909) and to the Pontifical Academy of Sciences' edition of the Vatican manuscripts edited by Pagano (1984). They are listed in absolute chronological order, irrespective of source or topic. The documents labeled "Diplomatic Correspondence" are those translated in the chapter by this title, and they refer to letters by Niccolini (the Tuscan ambassador to the Holy See) to Cioli (the Tuscan secretary of state).

DATE	DOCUMENT	FAVARO	PAGANO
14 Dec. 1613	Castelli to Galileo	11:605–606	
21 Dec. 1613	Galileo to Castelli	5:281–288	
7 Feb. 1615	Lorini's Complaint	19:297–298	69–70
16 Feb. 1615	Galileo to Monsignor Dini	5:291–295	
7 Mar. 1615	Monsignor Dini to Galileo	12:151–152	
20 Mar. 1615	Caccini's Deposition	19:307–311	80–85
23 Mar. 1615	Galileo to Monsignor Dini	5:297–305	
12 Apr. 1615	Cardinal Bellarmine to Foscarini	12:171–172	
1615	Considerations on the Copernican Opinion	5:351–370	
1615	Letter to the Grand Duchess Christina	5:309–348	
1615	Consultant's Report on Letter to Castelli	19:305	68–69
13 Nov. 1615	Ximenes' Deposition	19:316–318	93–95

DATE	DOCUMENT	FAVARO	PAGANO
14 Nov. 1615	Attavanti's Deposition	19:318–320	95–98
8 Jan. 1616	Galileo's Discourse on the Tides	5:377–395	
24 Feb. 1616	Consultants' Report on Copernicanism	19:320–321	99–100
25 Feb. 1616	Inquisition Minutes	19:321	100–101
26 Feb. 1616	Special Injunction	19:321–322	101–102
3 Mar. 1616	Inquisition Minutes	19:278	223–224
5 Mar. 1616	Decree of the Index	19:322–323	102–103
6 Mar. 1616	Galileo to Tuscan Secretary of State	12:243–245	
12 Mar. 1616	Galileo to Tuscan Secretary of State	12:247–249	
26 May 1616	Cardinal Bellarmine's Certificate	19:348	138
23 May 1618	Galileo to Archduke Leopold of Austria	12:389–392	
15 May 1620	Correction of Copernicus's *On the Revolutions*	19:400–401	
1624	Galileo's Reply to Ingoli	6:509–561	
1624 or 1625	Anonymous Complaint About *The Assayer*		245–248
18 Apr. 1625	Guiducci to Galileo	13:265–266	
7 Mar. 1631	Galileo to Tuscan Secretary of State	14:215–218	
25 Apr. 1631	Vatican Secretary to Tuscan Ambassador	14:254–255	
3 May 1631	Galileo to Tuscan Secretary of State	14:258–260	
24 May 1631	Vatican Secretary to Florentine Inquisitor	19:327	108–109
31 May 1631	Florentine Inquisitor to Vatican Secretary	19:328	109–110
19 July 1631	Vatican Secretary to Florentine Inquisitor	19:330	113
19 July 1631	Tuscan Ambassador to Galileo	14:284–285	
1632	Preface to the *Dialogue*	7:29–31	
1632	Ending of the *Dialogue*	7:487–489	
15 Aug. 1632	Diplomatic Correspondence	14:372	
22 Aug. 1632	Diplomatic Correspondence	14:374–375	
28 Aug. 1632	Diplomatic Correspondence	14:377	
5 Sept. 1632	Diplomatic Correspondence	14:383–385	

DATE	DOCUMENT	FAVARO	PAGANO
11 Sept. 1632	Diplomatic Correspondence	14:388–389	
Sept. 1632	Special Commission Report on the *Dialogue*	19:324–327	105–108
18 Sept. 1632	Diplomatic Correspondence	14:391–393	
25 Sept. 1632	Cardinal Barberini to Florentine Nuncio	14:397–398	
24 Oct. 1632	Diplomatic Correspondence	14:418–419	
6 Nov. 1632	Diplomatic Correspondence	14:425	
13 Nov. 1632	Diplomatic Correspondence	14:428–429	
11 Dec. 1632	Diplomatic Correspondence	14:438–439	
26 Dec. 1632	Diplomatic Correspondence	14:443–444	
15 Jan. 1633	Diplomatic Correspondence	15:28 (no. 2387)	
15 Jan. 1633	Galileo to Diodati	15:23–26	
14 Feb. 1633	Diplomatic Correspondence	15:40–41	
16 Feb. 1633	Diplomatic Correspondence	15:41	
19 Feb. 1633	Diplomatic Correspondence	15:45	
27 Feb. 1633	Diplomatic Correspondence [I]	15:54–55	
27 Feb. 1633	Diplomatic Correspondence [II]	15:55–56	
6 Mar. 1633	Diplomatic Correspondence	15:61	
13 Mar. 1633	Diplomatic Correspondence	15:67–68	
19 Mar. 1633	Diplomatic Correspondence	15:73–74	
9 Apr. 1633	Diplomatic Correspondence	15:84–85	
12 Apr. 1633	Galileo's First Deposition	19:336–342	124–130
16 Apr. 1633	Diplomatic Correspondence	15:94–95	
17 Apr. 1633	Oreggi's Report on the *Dialogue*	19:348	139
17 Apr. 1633	Inchofer's Report on the *Dialogue*	19:349–356	139–148
17 Apr. 1633	Pasqualigo's Report on the *Dialogue*	19:356–360	148–153
23 Apr. 1633	Diplomatic Correspondence	15:103–104	
28 Apr. 1633	Commissary General to Cardinal Barberini	15:106–107	
30 Apr. 1633	Galileo's Second Deposition	19:342–344	130–132
1 May 1633	Diplomatic Correspondence	15:109–110	
3 May 1633	Diplomatic Correspondence	15:111–112	
10 May 1633	Galileo's Third Deposition	19:345	133–134
10 May 1633	Galileo's Defense	19:345–347	135–137
15 May 1633	Diplomatic Correspondence	15:124	
22 May 1633	Diplomatic Correspondence	15:132	

DATE	DOCUMENT	FAVARO	PAGANO
29 May 1633	Diplomatic Correspondence	15:140	
May or June 1633	Final Report to the Pope	19:293–297	63–68
19 June 1633	Diplomatic Correspondence	15:160	
21 June 1633	Galileo's Fourth Deposition	19:361–362	154–155
22 June 1633	Sentence	19:402–406	
22 June 1633	Galileo's Abjuration	19:406–407	

Biographical Glossary

Collected here are brief biographical sketches of the historical figures named or mentioned in this collection of documents (and in the Introduction). Since most of them are referred to in several different chapters, it seemed advisable to have the complete list in one convenient location rather than having this information scattered or duplicated in the Notes. Many of these biographical descriptions are digests of those found in Favaro (20:361–561), to which the scholar is referred for further information.

AGGIUNTI, Niccolò (1600–1635).   Tutor to the young Duke Ferdinando II, friend of Galileo, and successor to Castelli in 1626 as professor of mathematics at the University of Pisa.

ALDOBRANDINI, Pietro (1571–1621).   Cardinal, archbishop of Ravenna, and nephew of Pope Clement VIII.

ALIDOSI, Mariano.   Italian nobleman in the service of the Tuscan grand duke; became entangled in a complex territorial dispute with the pope; was charged with various crimes ranging from homicide to heresy.

ANAXAGORAS (c. 500–428 B.C.).   Ancient Greek thinker who was one of the originators of the atomic theory of matter, according to which all things are made up of tiny particles, too small to see directly, whose motions and properties account for all that is perceived.

ANTELLA, Niccolò (1560–1630).   Florentine nobleman who held numerous important positions in the Tuscan government, including that of reviewer and censor of books to be printed, at the time of the licensing of the *Dialogue*.

APELLES.   Pseudonym of Christopher SCHEINER.

AQUINAS, Thomas (1225–1274).   A saint and one of the greatest theologians and philosophers of Catholicism; his system of ideas, known as Thomism, was declared by Pope Leo XIII in 1879 to be the official philosophy of the Catholic church.

ARACOELI, Cardinal. See Agostino GALAMINI.

ARCHIMEDES of Syracuse (287–212 B.C.). Greek mathematician and physicist and a younger contemporary of Aristarchus; most famous for such achievements as the proof of theorems about the volume and surface of spheres and cylinders, a brilliant method for approximating the value of the number pi, and the principle of hydrostatics that bears his name. This principle states that a body immersed in a fluid loses weight by an amount equal to that of the fluid it displaces.

ARISTARCHUS of Samos (c. 310–250 B.C.). Ancient Greek astronomer who elaborated the theory that the earth moves.

ARISTOTLE (384–322 B.C.). Ancient Greek thinker, a pupil of Plato who often developed ideas opposite to those of his teacher. His writings ranged over all subjects from ethics and politics to astronomy, physics, and biology; they contain the earliest systematization of the geostatic and geocentric worldview.

ARRIGHETTI, Niccolò (1586–1639). Florentine nobleman, intellectual, and friend of Galileo.

ASCOLI, Cardinal d'. See Felice CENTINI.

ATTAVANTI, Giannozzo (c. 1582–1657). A minor cleric who was named in Caccini's deposition, and who then gave his own deposition, during the earlier Inquisition proceedings against Galileo.

AUGUSTINE of Hippo (354–430). Bishop of Hippo, theologian, philosopher, and one of the greatest Christian saints and Church Fathers. Author of one of the earliest and most authoritative treatises on the interpretation of the Bible, entitled *De Genesi ad litteram* (*On the Literal Interpretation of Genesis*).

AUGUSTINUS, Aurelius. See AUGUSTINE of Hippo.

BARBERINI, Antonio (1569–1646). Brother of Maffeo Barberini, appointed cardinal with the title of Sant'Onofrio when his brother became Pope Urban VIII, and one of the inquisitors conducting the trial in 1633.

BARBERINI, Francesco (1597–1679). Nephew of Maffeo Barberini (Pope Urban VIII), appointed cardinal by the latter in 1623, and a patron of Galileo and of culture in general; one of the inquisitors conducting the trial in 1633, but one of three who did not sign the final sentence.

BARBERINI, Maffeo (1568–1644). Member of an influential Florentine family; highly educated in such fields as philosophy, literature, and jurisprudence; held numerous diplomatic and ecclesiastic positions; cardinal since 1606; was elected Pope Urban VIII in 1623. He was at first a great admirer of Galileo, but their relationship soured after the publication of the *Dialogue* in 1632, which led to the trial and condemnation of the following year.

BARONIO, Cesare (1538–1607). A cardinal since 1596, reported by Galileo (in the "Letter to the Grand Duchess Christina") as the author of the aphorism that "the intention of the Holy Ghost is to teach us how one goes to heaven and not how heaven goes."

BELLARMINE, Robert (1542–1621). Jesuit theologian, perhaps the most influential Catholic churchman of his time, and now a saint. Besides the position of cardinal, at various times he was also a professor at the Collegio Romano, an archbishop, the pope's theologian, and a consultant to the Inquisition. In 1605, at the conclave which resulted in the election of Pope Paul V, he firmly refused to be elected pope himself, despite wide support among the cardinals.

BENTIVOGLIO, Guido (1577–1644). A cardinal since 1621, and one of the inquisitors who conducted the trial in 1633.

BOCCABELLA, Alessandro (d. 1639). A doctor of theology as well as of canon and civil law; held various offices with the Roman Inquisition, including that of assessor from 28 July 1632 to 26 January 1633.

BOCCHINERI, Geri (d. 1650). A private secretary to the Grand Duke Ferdinando II de' Medici.

BOETHIUS, Anicius (c. 475–525). Roman philosopher and politician, most famous as the author of two books, *The Consolation of Philosophy* and *On Music*; the latter was the unquestioned authority on music until the time of Galileo.

BOLOGNETTI, Giorgio (c. 1590–1680). Nuncio to the Grand Duchy of Tuscany from 1631 to 1634 and bishop of Ascoli.

BONSI, Giovanni Battista (c. 1555–1621). A cardinal since 1611 and a skillful diplomat on various special assignments.

BORGIA, Gaspare (1589–1645). Spanish ambassador to Rome, cardinal since 1611, and one of the three inquisitors conducting the trial in 1633 who did not sign the sentence.

BOSCAGLIA, Cosimo. Special professor of philosophy at the University of Pisa from 1610 to 1621.

BRAHE, Tycho (1546–1601). Danish astronomer, best known as an incomparable observer and collector of data and as the originator of the so-called Tychonic system, according to which the earth is motionless at the center of the universe, but the planets revolve around the sun and are carried by it around the earth.

BRECHT, Bertolt (1898–1956). Controversial, outspoken, and socially oriented German playwright and poet; an important innovator in the modern theater; and author of such plays as *Galileo* and *The Threepenny Opera*.

BRUNO, Giordano (1548–1600). Italian philosopher and martyr for freedom of thought; when put on trial by the Inquisition, he refused to recant and was burned alive at the stake. Little is known about his final trial, except that Bellarmine examined his books and his heresies included disbelief in transubstantiation and belief in a plurality of worlds. He began as a Dominican, but left the order in 1576 and fled Rome to escape the Inquisition; for the next fifteen years he wandered throughout Europe, lecturing at many universities and publishing many books; finally, in 1591 he returned to Italy, where he was soon arrested and tried by the Inquisition in Venice; he was then transferred to Rome and kept in prison there for another eight years until his final trial.

BUONARROTI, Michelangiolo (1568–1646). Nephew of the great artist by the same name and a well-known and influential Florentine intellectual and politician.

CACCINI, Tommaso (1574–1648). A Dominican and a well-sought-after preacher; held various administrative positions in his order and earned various academic degrees and positions in theology. He accused Galileo of heresy in a sermon in 1614 and testified against him with the Inquisition in 1615.

CAETANI, Bonifacio (1567–1617). Was born and died in Rome; from 1606 was cardinal with the title of St. Pudenziana.

CAMPANELLA, Tommaso (1568–1639). An important philosopher in his own right and a Dominican who had his own problems with the Inquisition; in 1616 he wrote a defense of Galileo entitled *Apologia pro Galileo*, which was not published until 1622 in Frankfurt.

CAPIFERREO (or CAPIFERREUS), Francesco Maddaleni (d. 1632). A Dominican, secretary of the Congregation of the Index from 1615.

CASTELLI, Benedetto (1578–1643). Benedictine monk, student of Galileo at the University of Padua, his successor at the University of Pisa, and friend and collaborator; also an important figure in his own right, mainly for his contributions to the science of hydraulics and as the teacher of many outstanding Italian scientists of the period.

CENTINI, Felice (1570–1641). A cardinal since 1611 and one of the inquisitors conducting the trial in 1633; also called Cardinal d'Ascoli, after his native town.

CESI, Federico (1585–1630). Wealthy and influential Italian aristocrat, patron of the arts and sciences, and himself interested in writing about scientific subjects and doing scientific research; most famous as the founder and head of the Lincean Academy, the first international scientific society in modern science.

CHIARAMONTI, Scipione (1565–1652). Professor of philosophy at the University of Pisa from 1627 to 1636, mentioned favorably in *The Assayer* (1623) but sharply criticized in the *Dialogue* (1632).

CHRISTINA of Lorraine. Mother of Cosimo II de' Medici and grand duchess dowager after the death of Cosimo's father, Ferdinando I de' Medici.

CIAMPOLI, Giovanni (1589 or 1590–1643). Florentine intellectual, friend of Galileo, clergyman from November 1614, later a member of the Lincean Academy; a confidant of Cardinal Maffeo Barberini and his correspondence secretary when he became pope. Ciampoli's role in the publication of the *Dialogue* and his involvement in the politics of the Thirty Years War in 1632 made him lose favor with the pontiff, who banished him from Rome and appointed him governor of a remote town in the Papal States.

CICERO, Marcus Tullius (106–43 B.C.). The greatest Roman orator and writer, as well as an important politician and philosopher.

CIOLI, Andrea (1573–1641). Secretary of state to the grand duke of Tuscany from 1627.

CLAVIUS, Christopher (1538–1612). A Jesuit, professor at the Collegio Romano, one of the leading mathematicians and astronomers of his time, and a friendly acquaintance of Galileo.

COPERNICUS, Nicolaus (1473–1543). Polish astronomer, author of *On the Revolutions of the Heavenly Spheres*, whose original title was *De revolutionibus orbium celestium* (Nuremberg, 1543). This was the first work to elaborate the technical details of the theory that the earth moves on its axis and around the sun, an idea originally conceived and discussed by the ancient Greeks.

COSIMO II. See Cosimo II de' MEDICI.

DARWIN, Charles Robert (1809–1892). English naturalist, whose *Origin of Species* elaborated the details of a theory of organic evolution according to which biological species are not immutable but have evolved from one into another over the course of millions of years. Normally regarded as an epochmaking contribution to science and an essential element of modern biology, his theory was originally attacked by biblical fundamentalists, and his approach continues to receive their criticism.

DEL MONTE, Francesco Maria (1549–1626). An influential churchman and a friend of the House of Medici; became cardinal in 1588 after Ferdinando I de' Medici resigned his position of cardinal to succeed his deceased brother Francesco I as grand duke of Tuscany.

DEMOCRITUS (c. 460–c. 370 B.C.). Ancient Greek thinker who was one of the originators of the atomic theory of matter, according to which all things are made up of tiny particles, too small to see directly, whose motions and properties account for all that is perceived.

DINI, Piero (1570–1625). An influential Florentine intellectual in the earlier part of his life and archbishop of the city of Fermo from 1621. In early 1615, at the time of his correspondence with Galileo, he was a minor Vatican official ("apostolic referendary") living in Rome with his uncle Cardinal Ottavio Bandini, and so he was well connected and well acquainted with goings-on in the eternal city.

DIODATI, Elia (1576–1661). Famous French lawyer and politician who carried out extensive correspondence with many of the greatest scientists, philosophers, and statesmen of his time.

DIONYSIUS the Areopagite. A disciple of St. Paul, and bishop of Athens.

DUHEM, Pierre (1861–1916). Frenchman who made important contributions to science, especially physical chemistry; to the history of science, by his monumental investigations into the medieval origins of modern science; and to the philosophy of science, where he stressed that scientific theories are to some extent arbitrary, not always descriptions of reality, and often mere instruments of calculation and prediction.

ECPHANTUS of Syracuse. Ancient Greek who lived at the beginning of the fifth century B.C.

EGIDI, Clemente. Inquisitor-general of Florence from 1626 to 1635.

EINSTEIN, Albert (1879–1955). German-born scientist who spent the last twenty years of his life in the United States; one of the greatest physicists who ever lived, whose discoveries and theories not only revolutionized physics but also affected other fields and activities. His contributions ranged over many topics, such as the equivalence between mass and energy, the constancy of the speed of light, mechanics (theory of relativity), atomic physics (quantum theory), and the explanation of the photoelectric effect and Brownian movement.

ELCI, Orso d' (d. 1636).   A secretary to Grand Duke Ferdinando II.

EUCLID of Alexandria.   Greek mathematician and physicist who lived about 300 B.C.; author of the *Elements* of geometry, one of the most influential books ever written.

FERDINANDO II.   See Ferdinando II de' MEDICI.

FIRENZUOLA, Vincenzo da.   See Vincenzo MACULANO.

FOSCARINI, Paolo Antonio (1580–1616).   Head of the Carmelites in the province of Calabria and professor of theology at the University of Messina; published in early 1615 a book that attempted to show the compatibility between the Bible and the earth's motion.

GALAMINI, Agostino (1552–1639).   A Dominican who held numerous ecclesiastic positions; in 1611 was appointed cardinal with the title of Santa Maria d'Aracoeli.

GALEN (c. 130–c. 200).   Greek physician and philosopher, physician to the Roman emperor Marcus Aurelius, and writer of many medical treatises; these made him the supreme authority on medicine until the sixteenth century.

GASSENDI, Pierre (1592–1655).   French philosopher, scientist, and priest who held numerous academic positions and ecclesiastic offices; best known for his anti-Aristotelianism, his revival of ancient Greek atomism, and his controversies with philosopher-scientist René Descartes.

GESSI, Berlinghiero (1564–1639).   A cardinal since 1626, one of the inquisitors conducting the trial in 1633, and governor of Rome at the time.

GIESE, Tiedemann (1480–1550).   Polish friend of Copernicus and bishop of Kulm.

GILBERT, William (1540–1603).   Englishman who was personal physician to Queen Elizabeth I; author of a pioneering book on magnetism; highly regarded by Galileo on account of his experimental approach and his Copernicanism.

GINETTI, Marzio (1585–1671).   A cardinal since 1626 and one of the inquisitors who conducted the trial in 1633; held the position of vicar of Rome under Urban VIII and his next four successors.

GIRALDI, Iacopo (1576–1630).   Florentine man of letters.

GRASSI, Orazio (1590–1654).   A Jesuit, professor of mathematics at the Collegio Romano, and engaged in a bitter polemic with Galileo over the nature of comets; in this polemic Grassi published under the pseudonym of Lothario Sarsi.

GRIENBERGER, Christopher (1561–1636).   A Jesuit, professor of mathematics at the Jesuit Collegio Romano in Rome, in which position he succeeded Christopher Clavius.

GUEVARA, Giovanni di (1561–1641).   A member of the Order of Clerks Regular Minor, bishop of Teano from 1627, author of a commentary on the pseudo-Aristotelian *Questions of Mechanics*, and consultant to the Inquisition in regard to a complaint against Galileo's *Assayer* in 1624–1625.

GUIDUCCI, Mario (1585–1646).   Florentine intellectual and friend of Galileo; became a member of the Lincean Academy in 1625.

GUSTAVUS, Adolphus (1594–1632).   King of Sweden from 1611 to 1632; led the Protestants to several victories in 1630–1632 during the Thirty Years War (1618–1648), but died in battle.

HERACLIDES of Pontus.   Ancient Greek who lived in the fourth century B.C.

HICETAS of Syracuse.   Ancient Greek who lived about 400 B.C.

INCHOFER, Melchior (1585–1648).   A Jesuit who for a long time taught philosophy, mathematics, and theology at the University of Messina; member of the special commission that reported on the *Dialogue* in September 1632; and consultant to the Inquisition during the trial of 1633 who wrote the most negative report on the *Dialogue*.

INGOLI, Francesco (1578–1649).   A clergyman, author in 1616 of an essay "On the Location and Rest of the Earth, Against the Copernican System," to which Galileo replied in 1624, by which time Ingoli had become secretary of the newly created Congregation for the Propagation of the Faith.

JEROME, Saint (c. 347–420?).   A contemporary of Augustine, one of the most important and influential early Christian writers, now a Father and a Doctor of the Church; most famous perhaps for his commentaries on and Latin translations of the Bible; these laid the foundations for the Church's official Latin version of the Bible, the so-called Vulgate.

JOSEPHUS, Flavius (A.D. 37–95?).   Jewish writer and military leader who wrote both about the war between Romans and Jews and about Jewish history since its earliest beginnings.

KEPLER, Johannes (1571–1630).   Mathematician to the Holy Roman Emperor and author of three books (published in 1596, 1609, and 1619) which elaborated Copernicanism in new and important ways. Most famous for the three laws of planetary motion named after him; the first of these states that the planets move around the sun in elliptical orbits with the sun located at one of the foci of these ellipses. One of the first and strongest supporters of Galileo's telescopic discoveries, though Galileo never did reciprocate in kind and did not even pay attention to Kepler's laws of planetary motion.

KOESTLER, Arthur (1905–1983).   Hungarian-born writer who spent the latter part of his life in England; his novel *Darkness at Noon* vividly depicts the Stalinist purges of the 1930s in the Soviet Union; his history of cosmology (*The Sleepwalkers*) contains a classic sympathetic portrayal of Kepler and a classic unfavorable (and unreliable) account of Galileo.

LACTANTIUS, Lucius (c. 260–340).   Writer and apologist of the early Christian Church; some of his theological views became influential, but others were rejected.

LEOPOLD of Austria (1586–1632).   Brother-in-law of Grand Duke Cosimo II, the latter having married Maria Maddalena of Austria in 1608; Leopold himself married Claudia de' Medici in 1625 or 1626.

LOMBARD, Peter (1095–1160).   Christian theologian, bishop of Paris, born in the Italian region of Lombardy.

LORINI, Niccolò (b. 1544).   Born in Florence and died at an uncertain date after 1617. A member of the Dominican order, he held various administrative posts in various convents, was highly regarded as a preacher, and became professor of Church history at the University of Florence. In 1615 he filed a written complaint against Galileo with the Roman Inquisition.

LUTHER, Martin (1483–1546). German theologian and religious leader who triggered the Protestant Reformation in 1517 by posting on a church door ninety-five theses critical of contemporary Catholic practices and ideas; also founded the Lutheran church and laid the basis for a reform of German society; his writings helped shape the modern German language.

LYSENKO, Trofim Denisovich (1898–1976). Russian biologist who held the basically incorrect view that environmentally acquired traits are inherited. His theories became predominant in the Soviet Union, where they were seen as essential to Marxist orthodoxy, and won the official sanction of political authorities. Though his ideas were later rejected, his influence retarded the progress of Soviet genetics.

MACULANO, Vincenzo (1578–1667). A Dominican also known as Vincenzo da Firenzuola; became general of his order, master of the Sacred Palace, and cardinal; best known as the commissary general of the Inquisition at the time of the 1633 trial.

MAGALHAENS, Cosme (1553–1624). Author of a commentary on Joshua published in 1612.

MAGALOTTI, Filippo (b. 1558). A member of a prominent Florentine family and a distant relative of Pope Urban VIII.

MAGINI, Giovanni Antonio (1555–1617). Mathematician, astronomer, and geographer; professor of mathematics at the University of Bologna from 1588 till his death.

MARIA MADDALENA of Austria. Wife of Cosimo II de' Medici from 1608.

MEDICI, Antonio de' (1576–1621). Adopted son of Francesco I de' Medici (uncle of Cosimo II) and a clergyman at the time of the incident at the ducal court described in Castelli's letter of 14 December 1613.

MEDICI, Carlo de' (1595–1666). Younger brother of Grand Duke Cosimo II; a cardinal from December 1615, he eventually became very influential, especially after the coronation of Pope Innocent X in 1644.

MEDICI, Cosimo II de' (1590–1621). Ruler of the Grand Duchy of Tuscany from 1609 to 1621.

MEDICI, Ferdinando II de' (1610–1670). Son of Cosimo II; grand duke of Tuscany from 1621 when Cosimo II died prematurely.

MERSENNE, Marin (1588–1647). French philosopher, scientist, theologian, and priest who made important contributions to mechanics and musicology; most famous for his extensive correspondence with other scientists, in regard to which he acted as a kind of clearinghouse.

MILLINI, Gio. Garzia (1572–1629). Held numerous positions in the ecclesiastic hierarchy and Vatican diplomatic corps; from 1606 a cardinal with the title "dei SS. Quattro Coronati"; secretary of the Holy Office at the time of earlier proceedings against Galileo.

MILTON, John (1608–1674). English poet; author of *Paradise Lost*, regarded as the greatest epic poem in the English language; also wrote on theological and political subjects, including *Areopagitica* (1644), one of the most famous arguments in favor of freedom of the press.

NICCOLINI, Caterina Riccardi (1598–1676). Wife of Francesco Niccolini and cousin of Niccolò Riccardi.

NICCOLINI, Francesco (1584–1650). Ambassador of the grand duke of Tuscany to the Holy See in Rome from 1621 to 1643.

OREGGI (or OREGIO, or OREGIUS), Agostino (1577–1635). Originally a member of Cardinal Maffeo Barberini's entourage, later appointed papal theologian when Barberini became pope, and then appointed cardinal in November 1633; member of the special commission reporting on the *Dialogue* in September 1632 and consultant to the Inquisition who reported on the book during the trial of 1633.

ORIGANUS. See David TOST.

ORSINI, Alessandro (1593–1626). Member of a powerful Roman family and related through his mother to the Florentine House of Medici; was made a cardinal on 22 December 1615.

ORSINI, Paolo Giordano (1591–1656). A relative of the House of Medici and a brother of the man (Alessandro) for whom Galileo wrote the "Discourse on the Tides" in 1616.

PASQUALIGO, Zaccaria (d. 1664). A member of the Order of the Theatines; taught philosophy to his confreres in Padua and theology in Rome; as consultant to the Inquisition during the trial in 1633, wrote a report on the *Dialogue*.

PAUL V (1552–1621). Pope from 1605 to his death.

PAUL of Burgos (d. 1435). Spanish Jew who converted to Christianity and became an influential scriptural theologian.

PAUL of Middelburg (1445–1533). Italian clergyman of Danish birth, bishop of Fossombrone, and professor of mathematics and astronomy at the University of Padua.

PERERIUS, Benedictus (1535–1610). A Jesuit who taught at the Collegio Romano, primarily philosophy, but occasionally theology and Sacred Scripture. His philosophical works seem to have influenced the composition of Galileo's early notebooks dealing with natural philosophy, logic, and epistemology.

PLATO (c. 427–347 B.C.). Ancient Greek thinker whose writings focused on questions of ethics and political theory and laid the foundations for much subsequent Western philosophy.

PLUTARCH (A.D. 46?–c. 120). Greek writer, the greatest biographer of antiquity, whose classic work *The Parallel Lives* was for a long time the best source of information about the lives of famous Greeks and Romans.

PTOLEMY, Claudius, of Alexandria (second century A.D.). Greek mathematician, astronomer, and geographer; author of a work entitled *Almagest*, containing a complete, detailed, mathematical, and systematic exposition of the geostatic worldview and considered the classical synthesis of ancient astronomy.

PYTHAGORAS (c. 580–c. 500 B.C.). Ancient Greek thinker famous for having been one of the earliest proponents of a theory of the moving earth and for the discovery of the geometrical theorem (named after him) that in a right triangle the square of the hypotenuse equals the sum of the squares of the other two sides.

RICCARDI, Niccolò (1585–1639). A Dominican; from 1629 master of the Sacred Palace at the Vatican.

SACROBOSCO, Johannes de (c. 1200–1256). English mathematician and astronomer, most famous as the author of an elementary astronomy textbook entitled *Treatise on the Sphere*, which was still circulating and being discussed at the time of Galileo. Also known as John of Hollywood.

SAGREDO, Giovanfrancesco (1571–1620). Venetian diplomat and public official who was Galileo's dearest friend when he lived in Padua; immortalized as one of the three speakers in the *Dialogue*.

SALVIATI, Filippo (1582–1614). Wealthy Florentine nobleman whose interest in science and philosophy earned him membership in the Lincean Academy in 1612; one of Galileo's closest friends in Florence; immortalized as one of the three speakers in the *Dialogue*.

SAN EUSEBIO, Cardinal. See Ferdinando TAVERNA.

SAN SISTO, Cardinal. See Laudivio ZACCHIA.

SANTA CECILIA, Cardinal. See Paolo SFONDRATI.

SANT'ONOFRIO, Cardinal. See Antonio BARBERINI.

SANTUCCI, Antonio (d. 1613). A professor of mathematics at the University of Pisa.

SARPI, Paolo (1552–1623). Venetian Servite (i.e., member of the Order of the Servants of Mary, O.S.M.), theologian, lawyer, historian, who also wrote on scientific subjects; most famous for his role in the struggle that developed in 1606 between the Republic of Venice and the Papacy (when he vigorously defended the right of the state to control the Church) and as the author of a history of the Council of Trent (which he interpreted in terms of papal absolutism and centralization).

SARSI, Lothario. Pseudonym of Orazio GRASSI.

SCAGLIA, Desiderio (1596–1639). A Dominican, at one time the commissary general of the Inquisition, a cardinal from 1621, and one of the inquisitors conducting the trial in 1633.

SCHEINER, Christopher (1573–1650). A Jesuit professor of mathematics at various universities who engaged in a dispute with Galileo over priority in the discovery of sunspots and over their interpretation; in this dispute Scheiner sometimes published under the pseudonym of Apelles.

SCHOENBERG, Nicolaus von (1472–1537). Cardinal, archbishop of Capua, and nuncio to France.

SEGIZZI (or SEGHIZZI, or SEGHITIUS), Michelangelo (1585–1625). A Dominican who, after being inquisitor in various Italian cities, held the office of commissary general of the Inquisition in Rome at the time of the earlier Inquisition proceedings against Galileo; beginning in November 1616, he was bishop of his native town of Lodi.

SELEUCUS. A Babylonian who lived about the middle of the second century B.C.

SENECA, Lucius Annaeus (c. 4 B.C.–A.D. 65). Ancient Roman philosopher and playwright.

SERARIUS (or SERURIER), Nicolaus (1555–1609). A French Jesuit who for twenty years was professor of theology and Sacred Scripture at the universities of Würzburg and Mainz (Germany).

SFONDRATI, Paolo (1561–1618). A cardinal (with the title of Santa Cecilia) and a member of the Congregation of the Holy Office at the time of the earlier Inquisition proceedings against Galileo (1615).

SIMPLICIO. One of the three speakers in the *Dialogue*, who takes the Aristotelian side; his name is supposed to refer to the Greek philosopher Simplicius.

SIMPLICIUS. Greek philosopher who lived in the sixth century A.D., famous as one of the greatest commentators of Aristotle.

SOLDANI, Iacopo (1579–1641). Florentine intellectual and politician.

STEFANI, Giacinto (1577–1633). A Dominican, a professor at the University of Florence, and a reviewer of the *Dialogue* during the process of its licensing.

TAVERNA (or TABERNA), Ferdinando (1558–1619). From 1604 cardinal with the title of San Eusebio and at one time governor of Rome.

TERTULLIAN, Quintus (c. 160–230). Early Christian theologian. Some of his views acquired proverbial status (e.g., his remark "I believe it because it is impossible"), but others differed from mainstream Christian doctrine.

THOMAS, Saint. See Thomas AQUINAS.

TOST, David. A Copernican, author of a book of astronomical tables published in Frankfurt in 1609.

TOSTADO, Alfonso (1400–1455). A Spaniard, bishop of Avila, and professor of theology and philosophy at the University of Salamanca.

TYCHO. See Tycho BRAHE.

URBAN VIII. See Maffeo BARBERINI.

VALERIO, Luca (1552?–1618). Professor of mathematics at the University of Rome; highly regarded by Galileo until 1616, at which time Valerio was expelled from the Lincean Academy because of his behavior in connection with the Copernican controversy.

VEROSPI, Fabrizio (1572–1639). A cardinal since 1627 and one of the inquisitors who conducted the trial in 1633.

VIO, Thomas de (1468–1534). Dominican friar, master general of his order, and author of a famous commentary on St. Thomas Aquinas's *Summa theologiae*. He was born in Gaeta, was made a cardinal, and served as bishop of Gaeta, on which account he is generally referred to as Cajetan in theological literature.

VISCONTI, Raffaello. A Dominican, a friend of Niccolò Riccardi, and a reviewer of Galileo's *Dialogue* during the process of its licensing.

XIMENES, Emanuele (b. 1542). A Jesuit named in Caccini's deposition; a consultant to the Inquisition in Florence at the time of the earlier proceedings against Galileo; died at an uncertain date soon thereafter.

XIMENES, Ferdinando (c. 1580–1630). A Dominican who was named in Caccini's deposition, and who then gave his own deposition, during the earlier Inquisition proceedings against Galileo.

XIMENES, Sebastiano.   Founder in 1583 of the Order of the Knights of Santo Stefano; named in Caccini's deposition.

ZACCHIA, Laudivio (c. 1560–1637).   From 1626 a cardinal with the title of San Sisto; in 1633 one of the inquisitors conducting the trial and one of the three who did not sign the final sentence.

ZUÑIGA, Diego de.   An Augustinian friar from Salamanca (Spain), author of a commentary on the Book of Job that was published in the sixteenth century and favored the earth's motion.

Notes

INTRODUCTION

1. This label is, of course, far from unprecedented, as may be seen from Apollonia (1964), Brophy (1961, 1964), Coyne et al. (1985), Daniel-Rops (1963), Gingerich (1982), Lenoble (1956), Namer (1975), Petit (1964), and Rochot (1956). For an explicit discussion of the proper label to use, especially the question of whether to speak of "trial" in the singular or "trials" in the plural, see Giacchi (1942), Paschini (1964, I:341), Morpurgo-Tagliabue (1981), Pagano (1984, p. 4), and D'Addio (1985).

2. This is meant in all seriousness, and thus general readers ought to be aware of the existence of a sizable paperback literature in English, which contains a good cross section of the available interpretations and evaluations: Coyne et al. (1985), Drake (1980), Duhem (1969), Feyerabend (1978, 1985), Geymonat (1965), Koestler (1959), Langford (1966), and Santillana (1955). To these one should add: Agassi (1971), which, while not easily available, and while being merely a journal article rather than a book, is nevertheless perhaps the most thoroughgoing attempt to articulate an account that tries to be charitable to all sides; Brecht (1966), which can be read with profit, especially in conjunction with Bentley (1966), despite the fact that it is a play rather than a historical work; Morpurgo-Tagliabue (1981), which is not available in English but discusses a number of crucial questions surrounding the concept of hypothesis and its role in the controversy; Poupard (1987), which is a semiofficial Vatican document and available only hardbound; and Redondi (1987), which is extremely valuable for reconstruction of the cultural and political milieu but also tries unconvincingly to cast the trial of 1633 in a completely new light. Finally, for a critical review of many of these interpretations, and a proposal for a new approach along the lines of this Introduction, see Finocchiaro (1986b, 1986c, 1988b).

3. See, for example, Coyne et al. (1985), Koestler (1959), Langford (1966), Poupard (1987), and Shea (1986).

4. See, for example, Drake (1980), Feyerabend (1978), Geymonat (1962, 1965), and Morpurgo-Tagliabue (1981).

5. See, for example, Feyerabend (1985), Lerner and Gosselin (1986), Redondi (1983, 1987), and Santillana (1955).

6. See also the "Diplomatic Correspondence" of 14 February 1633 in Chapter VIII.

7. Einstein (1967, p. vii).

8. See Koestler (1959) and Duhem (1969, pp. 113—17). Here I am following the traditional interpretation of Duhem, but it has recently been argued that when he asserted that "logic sides with Osiander, Bellarmine, and Urban VIII, not with Kepler and Galileo" (1969, p. 113), Duhem meant to show not that the Catholic church was therefore right in condemning him, but rather "the inadequacy of pure logic to account for the method of physics" (Martin 1987, p. 313).

9. See Gingerich (1982).

10. See, for example, Duhem (1969), Feyerabend (1978), Koestler (1959), Koyré (1978), Langford (1966), and McMullin (1967a).

11. See Finocchiaro (1980, 1986b).

12. See, for example, Duhem (1969) and Feyerabend (1985).

13. One is tempted to say that in 1610 Galileo went back to his native Tuscany to be the "scientific advisor" of the grand duke, but this description would be too modern and ultimately anachronistic; for a discussion of the nature of patronage and its importance, see Westfall (1984, 1985).

14. He began with a salary of 180 florins, and after about twenty years, as a result of his telescope, his salary of 500 was doubled to 1000 florins; even so, the senior professor of philosophy was earning 2000 florins. See Drake (1978, p. 160); Koestler (1959, p. 364); and Santillana (1955, p. 3).

15. See Westfall (1984, 1985).

16. Schroeder (1978, pp. 18—19).

17. Ibid., p. 25.

18. Ranke (1841, vol. II:98—125, especially pp. 116—19).

19. The following account is based on Masini (1621).

20. Masini (1621, pp. 16—17).

21. Ibid., pp. 17—18.

22. Ibid., pp. 166—67.

23. In principle there was also a third subcategory, namely, "violent suspicion of heresy"; however, the manual clarified that in practice this was equated to "vehement" suspicion of heresy (Masini 1621, p. 188).

24. For more details on this last point, see, for example, Kuhn (1957, pp. 160—65); this work may also be consulted for other details since Kuhn's classic remains the best account of the Copernican revolution.

25. For more details on what I am here calling explanatory coherence, see Lakatos and Zahar (1975, pp. 368—81), and Millman (1976); note, however, that their terminology is different.

26. An allusion to this objection is found in the letter by "Cardinal Bellar-

mine to Foscarini" in Chapter I; for more details, see Galilei (1967, pp. 32–38, and 247–56).

27. See section III, paragraphs 9 and 10, of "Galileo's Considerations on the Copernican Opinion" in Chapter II.

28. See Galilei (1967, pp. 38–105); see also section 18 of "Galileo's Reply to Ingoli" in Chapter VI.

29. See section 3 of "Galileo's Letter to the Grand Duchess Christina" in Chapter III (corresponding to Favaro 1890–1909, vol. 5, p. 328, hereafter cited as Favaro 5:328); see also Galilei (1967, pp. 318–40).

30. In many places—for example, in the "Letter to the Grand Duchess Christina" (Chapter III; see Favaro 5:328)—Galileo states that the expected variation in apparent size and brightness is of the order of 60; his discussion of this point in the *Dialogue* (Favaro 7:349–50; Galilei 1967, p. 321) indicates that this is meant to be approximately the square of 8, which he gives as the ratio of the maximum to the minimum distance between Mars and the earth. I am not sure how he arrives at this figure, given that the Copernican estimate of the mean distance between Mars and the sun is 1.52 times the mean distance between the earth and the sun (Dreyer 1953, p. 339), which would yield a ratio of about 4.85 to 1. Even if we add the eccentricity of Mars's orbit, which I estimate at about 0.1 (Dreyer 1953, p. 336), we would still have a ratio of 5.9. Adding the eccentricity of the earth's orbit would further increase this factor, but not by the required amount to make it correspond to 8. The reason why the observed variation in apparent brightness of Mars presented a serious difficulty for the Copernican system but not for the Ptolemaic system was that in the latter the relevant quantities (distance, epicycle, etc.) could be adjusted to correspond to the actual observations, whereas in the former the variation could be derived from other elements of the system, because of its greater coherence mentioned above.

31. See especially sections 2 and 11, but also section 3, of "Galileo's Reply to Ingoli," in Chapter VI; see also Galilei (1967, pp. 377–89).

32. For more details, including an answer, see sections 8 and 9 of "Galileo's Reply to Ingoli" in Chapter VI; for an even more extended analysis, see Galilei (1967, pp. 138–67).

33. For more details, see section 10 of "Galileo's Reply to Ingoli" in Chapter VI; for a full answer, see Galilei (1967, pp. 167–71).

34. See section 9 of "Galileo's Reply to Ingoli" in Chapter VI; see also Galilei (1967, pp. 143–49).

35. Chiaramonti (1633, p. 339).

36. See section 9 of "Galileo's Reply to Ingoli" in Chapter VI; see also Galilei (1967, pp. 143–49).

37. See Galilei (1967, pp. 132 and 188–218); see also section 7 of "Galileo's Reply to Ingoli" in Chapter VI.

38. See sections 6 and 17 of "Galileo's Reply to Ingoli" in Chapter VI; see also Galilei (1967, pp. 14–38 and 133–36).

39. See section 17 of "Galileo's Reply to Ingoli" in Chapter VI; see also Galilei (1967, pp. 256–64).

40. See the letters by "Galileo to Castelli" and by "Cardinal Bellarmine to

Foscarini" in Chapter I; and especially "Galileo's Letter to the Grand Duchess Christina" in Chapter III, which is devoted to a detailed refutation of this objection.

41. See the letter by "Cardinal Bellarmine to Foscarini" in Chapter I and the criticism in section 4 of "Galileo's Letter to the Grand Duchess Christina" in Chapter III.

42. For example, he discussed some of these objections in Book One of his great work; see, for example, Copernicus (1976).

43. For a general account, see Drake (1957, pp. 1–88; 1978, pp. 134–76); for an account that stresses the scientific and epistemological issues, see Ronchi (1958).

44. Favaro (10:369).

45. This also contrasts with the attitude of a true believer and that of zealous commitment, which one finds attributed to Galileo in such works as Feyerabend (1978), Koestler (1959), and Langford (1966). However, see Finocchiaro (1988b) for more details and documentation of my interpretation, which also corresponds in large measure to Drake (1978).

46. There seems to be no direct documentary evidence that the *Sunspot Letters* were examined. The *indirect* evidence for this is a note on the back of one of the folios containing Attavanti's deposition; the note is dated 25 November 1615 and says simply: "Those letters of Galileo, published in Rome under the title *Delle macchie solari*, should be looked at" (Favaro 19:320; Pagano 1984, p. 98).

47. Urban may also have had in mind the objection that God could have created a world in which the evidence suggested a moving earth despite its being motionless. There is no doubt that the exact content, structure, origin, and consequences of Urban's argument remain an open question which deserves further exploration. It should also be mentioned that if Urban's argument did indeed include the objection just mentioned, this would lend support to Morpurgo-Tagliabue's thesis that Descartes' "*Discourse on Method . . .* is essentially an answer to the argument of Urban VIII" (1981, p. 104), which is a connection also worthy of further investigation.

48. I realize, of course, that my interpretation here goes against the prevailing view that Galileo regarded the tidal argument as conclusive, that he therefore wanted to advertise this in the book's title, and that the Church did not want to endorse this argument; for an exposition of this view, see Shea (1972, pp. 172–89). For some criticism, see Finocchiaro (1980, pp. 16–18 and 76–78); see also Finocchiaro (1988b) and Drake (1986).

49. Here I am following in part the interpretation found in Morpurgo-Tagliabue (1981).

50. I do not mean to imply that *all* of the complaints had been voiced before the pope took this step, but only that some of them emerged before; in particular, I am not sure there is any way of dating exactly what I have called "the most serious complaint," involving the special injunction. However, it is certain that the ban on sales preceded the report of the special commission (Favaro 20:571–72; 14:391–93; and 14:397–98), and it seems to me that the first

written mention of the special injunction does not occur till September (Favaro 14:391–93 and 397–98).

51. See Masini (1621, pp. 120–51).

52. Insufficient attention has been paid to the double character of the heresy attributed to Galileo and to the probabilistic character of the proscribed methodological principle; a step toward the proper analysis may be found in Finocchiaro (1986b).

53. John Paul II (1979).

54. Ibid.

55. For this and other details, see Pagano (1984, pp. 1–4).

56. Ibid., p. 25.

57. See Gebler (1879, pp. 319–29), Favaro (1902a), and Pagano (1984, pp. 10–26).

58. Venturi (1818–1821), Marini (1850), and Epinois (1867).

59. Berti (1876, 1878), Gebler (1876, 1877), and Epinois (1877).

60. Favaro (19:293–399).

61. Favaro (1902b) is a special edition of thirty copies.

62. In Favaro (1907a) and in the various reprints of the National Edition of Galileo's works in 1929–1939 and 1968.

63. Favaro (19:275–92).

64. See, for example, Ferrone and Firpo (1985a, pp. 178–79; 1986) and Redondi (1985c).

65. See, for example, Redondi (1983, 1987).

66. Pagano (1984, pp. 41–42).

67. This is not to deny the value of other collections, but it is to say that they were conceived differently; see Del Lungo and Favaro (1915), Drake (1957), Favaro and Del Lungo (1911), Galilei (1957, 1968), and Namer (1975).

I. CORRESPONDENCE (1613–1615)

1. Cosimo II de' Medici (1590–1621), ruler of the Grand Duchy of Tuscany from 1609 to 1621.

2. "Medicean planets" was the name Galileo gave to the satellites of Jupiter he discovered in his first telescopic observations. The name, which first appeared in *The Starry Messenger* (1610), was meant to honor the House of Medici, ruling family of his native Tuscany, where he wanted to return after an eighteen-year residence at Padua, in the Republic of Venice.

3. Christina of Lorraine, mother of Cosimo II, and grand duchess dowager since the death (in 1609) of Cosimo's father Ferdinando I de' Medici.

4. Antonio de' Medici (1576–1621), adopted son of Francesco I de' Medici (uncle of Cosimo II) and a clergyman at the time of this incident.

5. Maria Maddalena of Austria, wife of Cosimo II since 1608.

6. Paolo Giordano Orsini (1591–1656), a relative of the ducal family and a brother of the man (Alessandro) for whom Galileo wrote the "Discourse on the Tides" in 1616.

7. See Castelli's letter to Galileo of 14 December 1613.

8. Joshua 10:12–13, "Then spake Joshua to the Lord in the day when the Lord delivered up the Amorites before the children of Israel, and he said in the sight of Israel, Sun, stand thou still upon Gibeon; and thou, Moon, in the valley of Ajalon. And the sun stood still, and the moon stayed, until the people had avenged themselves upon their enemies. Is not this written in the book of Jasher? So the sun stood still in the midst of heaven, and hasted not to go down about a whole day" (King James Version).

9. This is obviously from the point of view of the Aristotelian–Ptolemaic system, according to which the sun is regarded as a planet, and like all planets it has two motions superimposed on each other: the daily revolution from east to west that all heavenly bodies have and its own planetary motion from west to east relative to the fixed stars with its own special period of one year.

10. In Aristotelian–Ptolemaic cosmology this is the outermost sphere in the universe whose axial rotation every twenty-four hours carries along all the other heavenly spheres and bodies in their daily motion from east to west around the motionless earth at the center. It was descriptively called *primum mobile*, which literally means the first movable body. In translating this phrase as "Prime Mobile" I am following Stillman Drake's latest translation (1978, p. 228), which I think is preferable both to his earlier retention of the Latin phrase (1957, p. 211), as well as to "prime mover" (which would suggest an entity which is both first and itself unmoved, and which is equated with God in Aristotelian metaphysics).

11. Galileo discovered the axial rotation of the sun on the basis of the discovery of sunspots and from a study of the specific details of how sunspots appear through telescopic observations.

12. Note that this date is nine days after Fra Niccolò Lorini wrote his complaint to the Inquisition (see below). Although Galileo probably did not know the exact details, he had obviously learned that something was happening, as he states in the initial paragraph of this letter.

13. On 20 December 1614, at the church of Santa Maria Novella in Florence, Fra Tommaso Caccini had preached a sermon against mathematicians in general, and Galileo in particular, because their beliefs and practices allegedly contradicted the Bible and thus were heretical. One of the scriptural passages discussed in the sermon was the tenth chapter of Joshua, which we have encountered already in Castelli's letter to Galileo and Galileo's letter to Castelli. It is also said that Caccini exploited the biblical statement "Ye men of Galilee, why stand you gazing up into heaven?" (Acts 1:11), which refers to the inhabitants of Galilee who had just witnessed Jesus' ascension into heaven, and which is telling them to stop looking at the sky because he is not going to come back to earth; Caccini was making a pun and addressing the advice to Galileo and his followers, on the grounds that their observations of the heavens had supposedly led them to heretical beliefs. Some of the documentary evidence for Caccini having discussed the Joshua passage is in his deposition to the Inquisition (20 March 1615) given below; as for his reference to the men of Galilee in the first chapter of the Acts, I am not aware of any direct documentary evidence, nor is any reference given by the various authors who repeat the story (e.g., Gebler

1879, p. 51; Geymonat 1962, p. 100; Santillana 1955, p. 42; and Langford 1966, p. 55). Favaro (12:127, n. 2) states that the earliest mention of the story he was able to find was in *Lettere inedite di uomini illustri* (Florence: F. Mouecke, 1773), p. 47, n. 1.

14. This is a reference to Fra Niccolò Lorini, who on 2 November 1612 had expressed objections to Copernicanism and to Galileo on scriptural grounds. Galileo knew about this from a letter of apology written to him by Lorini himself and dated 5 November 1612 (Favaro 11:427).

15. This mathematician is Benedetto Castelli, and the letter is the one given above bearing the date of 21 December 1613.

16. Galileo's suspicions were correct, as one can see from the copy of the letter which Lorini enclosed with his written complaint to the Inquisition dated 7 February 1615. This copy is now found in the Galileo dossier of the Vatican Secret Archives on folios 8r–11r, immediately after Lorini's complaint on folio 7r–7v (Pagano 1984, pp. 69–77). Aside from minor differences of punctuation and spelling, it differs from the original primarily as follows: (1) It omits the salutation and the first two paragraphs, as well as each of the several references to the ducal family; (2) in the third paragraph, the second clause of the second sentence ("nevertheless some of its interpreters and expositors can sometimes err in various ways") reads instead "nevertheless its interpreters and expositors can err in various ways"; (3) in the same paragraph, the first clause of the last sentence ("thus in the Scripture one finds many propositions which look different from the truth if one goes by the literal meaning of the words") reads instead "thus in the Scripture one finds many propositions which are false if one goes by the literal meaning of the words"; and (4) in the fourth paragraph about two-thirds of the way down, the sentence "the Scripture has not abstained from somewhat concealing its most basic dogmas" reads instead "the Scripture has not abstained from perverting its most basic dogmas."

17. This is "Galileo's Letter to the Grand Duchess Christina," given in Chapter III.

18. In 1514 during the Fifth Lateran Council, Pope Leo X issued a general invitation to astronomers and theologians to help with the reform of the calendar by either coming to Rome or sending their comments in writing. Although Copernicus reported in writing, and although his system played a role in the reform, the new so-called Gregorian calendar was not instituted until 1582 during the reign of Gregory XIII, on the basis of non-Copernican ideas.

19. Paul of Middelburg (1445–1533), Italian bishop of Danish birth and professor of mathematics and astronomy at the University of Padua.

20. Actually, the reform was accomplished without basing it on Copernican ideas. Galileo's statements about Copernicus, on this and other matters that need not concern us here, are not always exactly correct; see Rosen (1958).

21. Dreyer (1953, pp. 317–18) states that in 1536 this cardinal "wrote to Copernicus urging him to make his discovery known to the learned world and begging for a copy of whatever he had written, together with the tables belonging thereto, all to be copied at the cardinal's expense."

22. Actually Copernicus was Polish (though of German parents).

23. See Matthew 5:29 and 18:9.

24. Again, this refers to what is now known as "Galileo's Letter to the Grand Duchess Christina."

25. The mathematician is Benedetto Castelli, and the letter is the one Galileo wrote to him on 21 December 1613 (see above).

26. Galileo had visited Rome in the spring of 1611, soon after his first telescopic discoveries and his publication of *Sidereus Nuncius* (Venice, 1610; now in Favaro 3:51–96), translated as *The Starry Messenger* and given in part in Drake (1957, pp. 21–58) and in its entirety and interspersed with commentary in Drake (1983, pp. 12–90). He had been very well received by Church officials, by the Jesuit professors of the Collegio Romano, and by Bellarmine in particular.

27. At that time, and in such contexts, the phrase "to save the appearances" meant to explain observed natural phenomena by means of assumptions which are not taken to describe real physical processes, but rather to be merely convenient means for making calculations and predictions.

28. Psalm 18:6 (Douay); cf. Psalm 19:5 (King James).

29. "His going out is from the end of heaven, and his circuit even to the end thereof: and there is no one that can hide himself from his heat" (Psalm 18:7, Douay); cf. Psalm 19:6 in the King James Version.

30. This is a reference to the essay Galileo had mentioned in his letter to Dini, which corresponds to the "Letter to the Grand Duchess Christina" (see below).

31. Maffeo Barberini (1568–1644), who was to become Pope Urban VIII in 1623.

32. Again, this is Galileo's "Letter to the Grand Duchess Christina" (see below).

33. This seems to be, at least in part, a reference to Galileo's "Considerations on the Copernican Opinion" (see below). The sentence also seems to refer to what was later to become his *Dialogue on the Two Chief World Systems*, first published as *Dialogo* (Florence, 1632; now in Favaro 7:21–520) and available in English in Galilei (1953, 1967).

34. See Genesis 1:2 (King James), "The earth was without form and void, and darkness was upon the face of the deep; and the Spirit of God moved upon the face of the waters."

35. Galileo is obviously referring to the Catholic Bible for which this reference is correct. This corresponds to Psalm 74:16 in the King James Version.

36. Psalm 73:16 (Douay).

37. Dionysius the Areopagite, a disciple of St. Paul, was bishop of Athens. Modern scholarship no longer considers him the author of this and other theological works, as was held at the time of Galileo. They are now thought to have been written in the fifth century by an author now referred to as Pseudo-Dionysius.

38. Galileo gives no precise reference for this Latin quotation. Here and elsewhere, translations from the Latin are my own unless otherwise noted.

39. See Psalm 18:6 (Douay) and Psalm 19:4 (King James).

40. Psalm 18:6 (Douay); cf. Psalm 19:5 (King James).

41. Galileo is referring to the next verse, each clause of which he comments

upon in turn: "His going out is from the end of heaven, and his circuit even to the end thereof: and there is no one that can hide himself from his heat" (Psalm 18:7, Douay Version; cf. Psalm 19:6 in the King James Version).

42. "The law of the Lord is unspotted, converting souls: the testimony of the Lord is faithful, giving wisdom to little ones" (Psalm 18:8 in the Douay Version; cf. Psalm 19:7 in the King James Version).

43. Psalm 18:8 (Douay); cf. Psalm 19:7 (King James).

44. Besides this position, Foscarini (1580–1616) was also professor of theology at the University of Messina. In late February or early March 1615 (cf. Favaro 12:149–50) he published in Naples a 64-page book (Langford 1966, p. 59, n. 19) whose full title conveys a good idea of its form and content: *Lettera sopra l'opinione de' Pittagorici e del Copernico della mobilità della terra e stabilità del sole e del nuovo Pittagorico sistema del mondo, nella quale si accordano ed appaciono i luoghi della Sacra Scrittura e le proposizioni teologiche, che giammai potessero addursi contro tale opinione* (cf. Geymonat 1962, p. 96); this translates as *Letter on the Pythagorean and Copernican opinion of the earth's motion and sun's rest and on the new Pythagorean world system, in which are harmonized and reconciled those passages of the Holy Scripture and those theological propositions which could ever be adduced against this opinion.* Galileo was immediately sent a copy of this booklet by Prince Federico Cesi (1585–1630), founder and head of the Lincean Academy and a patron and friend of Galileo. Foscarini sent a copy to Cardinal Bellarmine, and this elicited the present letter. In his chronological summary, Favaro (1907b, p. 33) gives 6 January 1615 as the publication date of Foscarini's book, although the reference he gives (12:150) is inconclusive; if the January date is indeed correct, this would be significant since it would be prior to Lorini's complaint (7 February 1615).

45. The term here is *ex suppositione*, which is left in Latin in the Italian text and which I translate as "suppositionally" to preserve the flavor of the original, though I am aware that other translators frequently render it as "hypothetically." Here and elsewhere, I use "supposition" for the Latin *suppositio* and "hypothesis" for the Latin *hypothesis*. For more details, see Wallace (1981a, 1981b, 1984, 1986), Wisan (1984), and Chapter II (n. 4).

46. "In the third heaven" just means in the third orbit around the sun.

47. The Council of Trent (1545–1563); see section 3 of the Introduction for the exact wording of its most relevant decrees.

48. See Galileo's "Letter to the Grand Duchess Christina" (below) about whether this is really so.

49. Ecclesiastes 1:5.

II.   GALILEO'S CONSIDERATIONS ON THE COPERNICAN OPINION (1615)

1. Although the manuscripts of these essays bear no date and no explicit indication of their author, scholars generally accept the conclusion of the editor of Galileo's collected works that they were written by Galileo sometime in 1615 or the first two months of 1616 at the latest (Favaro 5:277). Note, in fact, that

besides referring to his "Letter to the Grand Duchess Christina," in his second letter to Dini (at the end of the sixth paragraph) Galileo states that he is in the process of collecting and simplifying the arguments pertaining to the controversy. Note also that the third of these considerations addresses many of the points mentioned in Bellarmine's letter to Foscarini. Finally, it should be mentioned that it may be useful to read these Galilean considerations in conjunction with Copernicus's remarks in the preface and dedication to his work *On the Revolutions of the Heavenly Spheres* (Copernicus 1976); for an enlightening analysis of this dedication, see Westman (1987).

2. Here and at the beginning of the two other considerations the roman numerals have been added by the present editor, and they are enclosed in brackets to signal the fact that they are not part of the original text.

3. This description is reminiscent of the expression that would be used in the Inquisition Consultants' Report on Copernicanism (24 February 1616): "foolish and absurd in philosophy" (see below).

4. "In a suppositional manner" is my translation of *ex suppositione*, a Latin phrase which is left untranslated in the original Italian document and which is the same phrase that occurs at the beginning of the fourth paragraph of Bellarmine's letter to Foscarini (see above). The importance of this phrase has been underscored by Wallace, first in an essay originally published in 1976 but later revised and expanded (1981a, pp. 129–59) and then more fully revised (1981b); in his later writings (1984, 1986), he has made further clarifications on the meaning and usage of the expression in Galileo's writings and in those of his opponents; for fuller documentation one is referred to Wallace's introduction, notes, and commentary to Galileo's logical notebooks in Galilei (1988). Wisan (1984) offers a critique of Wallace (1981a) but takes no account of his revision (1981b) and perforce of his later writings.

5. Hicetas of Syracuse lived about 400 B.C. Galileo misspells this name (*Iceta* in Italian) by adding an initial N, which yields *Niceta* in Italian or *Nicetas* in English, thus repeating Copernicus's own misspelling in the dedication to *On the Revolutions*.

6. Seneca, *Quaestiones Naturales*, book 7, chapter 2. Galileo gives no exact reference for this passage, which he leaves in Latin; it is here translated directly from his quotation.

7. This is *On the Revolutions of the Heavenly Spheres*, first published as *De revolutionibus orbium celestium* (Nuremberg, 1543).

8. William Gilbert (1540–1603), *De Magnete magneticisque corporibus et de magno magnete Tellure physiologia nova* (London, 1600); the full title translates as "New physics of the lodestone, of magnetic bodies, and of the great lodestone the Earth"; published as *On the Loadstone and Magnetic Bodies*, Mottelay translation (New York, 1893).

9. That is, *Origani Novae Coel. Motuum Ephemerides* (Frankfurt, 1609); *ephemerides* (plural of *ephemeris*) are astronomical tables showing in a systematic way the positions of heavenly bodies at various times.

10. This is probably a reference to Aristotle.

11. The phrase used by Galileo here is *ex hypothesi*, not the *ex suppositione* he used in the second paragraph above. Wallace (1981a, 1981b, 1984) has ar-

gued that in late medieval and early modern Aristotelianism there was an important epistemological difference between these two notions and contends that Galileo was aware of this distinction and used it. I am not sure any difference can be discerned in the documents directly related to the Galileo affair, and the present context seems to be a good example of this indifference. Rather I am more inclined to accept the thesis of Morpurgo-Tagliabue (1981, 1985) that a crucially important distinction and ambiguity is that between the instrumentalist and fallibilist senses of hypothesis: in its instrumentalist meaning a hypothesis is not a description of physical reality (which could be either true or false) but merely an instrument for prediction and calculation (which can only be more or less useful useful or convenient); in its fallibilist meaning a hypothesis is a description of physical reality which is not *known* to be true but for which we have reasons for believing that it *may be* true, with varying degrees of probability, likelihood, or plausibility. I believe that this latter distinction cuts across both terms, *ex suppositione* and *ex hypothesi*.

12. Cardinal Schoenberg, archbishop of Capua.

13. Tiedemann Giese (1480–1550), Polish friend of Copernicus.

14. Copernicus (1976, p. 79); I have emended this text, as translated by Duncan, by using quotation marks rather than displaying the last two clauses.

15. "Medicean planets" was the name Galileo gave to the satellites of Jupiter he discovered in his first telescopic observations. The name, which first appeared in *The Starry Messenger* (1610), was meant to honor the House of Medici, ruling family of his native Tuscany, where he wanted to return after an eighteen-year residence at Padua, in the Republic of Venice.

16. This preface was in fact written by Andreas Osiander (1498–1552), a Lutheran theologian who supervised the last phase of the printing of Copernicus's book at Nuremberg. The action was soon discovered by Copernicus's friends and followers, causing a controversy, but it did not become generally known for some time, perhaps not until Kepler announced it in his book on Mars of 1609.

17. This phrase did not mean then what it most often means today; Galileo is saying that Copernicus was speaking in such a way as to agree with some of his opponents' statements for the purpose of the argument, and thereby to derive from these same statements some conclusions critical of his opponents.

18. Begging the question is the fallacy of assuming, in the course of a dispute, the truth of what is being questioned; thus, for example, if part of the dispute is about what is the correct meaning of a particular scriptural passage, then to argue against the earth's motion on the basis of a given meaning of that passage would be to beg the question.

19. This statement is indeed correct, but it has recently been discovered that as soon as Copernicus's book was published there was some concern for its potential heretical implications and some talk of its condemnation. Between 1544 and 1547 a Florentine Dominican theologian and astronomer by the name of Giovanni Maria Tolosani wrote a lengthy criticism of the book (cf. Garin 1971, 1975; Rosen 1975), which concludes by saying that "the Master of the Sacred and Apostolic Palace had planned to condemn this book, but, prevented first by illness and then by death, he could not fulfill this intention.

However, I have taken care to accomplish it in this little work for the purpose of preserving the truth to the common advantage of the Holy Church" (quoted in Westman 1986, p. 89).

20. The Council of Trent (1545–1563).

21. Note that in his letter to Foscarini, Bellarmine had used the same word *irritate* [*irritare*] when he stated that to interpret Copernicanism as a description of physical reality (rather than "suppositionally") "is . . . likely . . . to irritate all scholastic philosophers and theologians" (see above).

22. This point using this same language had been made by Bellarmine in his letter to Foscarini (see above).

23. This example had been used by Bellarmine in his letter to Foscarini (see above).

24. Again this statement is explicitly found in Bellarmine's letter to Foscarini (see above).

25. To "save the appearances" meant to explain observed natural phenomena by means of assumptions which are not taken to describe real physical processes, but rather to be merely convenient means for making calculations and predictions.

26. This distinction is also found in Bellarmine's letter to Foscarini (see above).

27. Solomon had been explicitly mentioned by Bellarmine in his letter to Foscarini (see above).

28. This phenomenon had been explicitly mentioned by Bellarmine in his letter to Foscarini (see above).

III. GALILEO'S LETTER TO THE GRAND DUCHESS CHRISTINA (1615)

1. This is Christina of Lorraine, mother of the Grand Duke Cosimo II. Note that she was directly involved in one of the initial incidents which triggered the Galileo affair, as described in Castelli's letter to Galileo of 14 December 1613 (see above).

2. This and subsequent section numbers are not in Galileo's text but are additions by the present editor. The forty pages of text in the National Edition of Galileo's works (Favaro 5:309–48) do not have any subdivisions, and so it is useful to have some for orientation and reference. Moreover, the six sections into which the text is being subdivided will be seen to correspond to six distinct though interrelated points of discussion; see Finocchiaro (1986b).

3. See Galileo Galilei, *Sidereus Nuncius* (Venice, 1610; now in Favaro 3:53–96), translated with the title *The Starry Messenger* in part in Drake (1957, pp. 21–58) and in its entirety and interspersed with commentary in Drake (1983, pp. 12–90). In it Galileo had described his discovery, through telescopic observation, of lunar mountains, four satellites of Jupiter (which he named "Medicean planets"), the stellar composition of the Milky Way and of nebulas, and the existence of thousands of previously invisible fixed stars. Within a few years, Galileo added to these his observations of sunspots, the phases of Venus, and Saturn's rings.

4. These are propositions such as that the earth is motionless, that the earth is located at the center of the universe, that all heavenly bodies are intrinsically luminous and physically unchangeable, and that there is an essential difference between terrestrial bodies and heavenly bodies.

5. Galileo is not exaggerating here, and the following may be mentioned as perhaps the principal writings against him: Martin Horky, *Brevissima Peregrinatio contra Nuncium Sidereum* (1610; now in Favaro 3:127–46), meaning *A Very Short Excursion Against the Starry Messenger*; Francesco Sizzi, *Dianoia Astronomica, Optica, Physica* (Venice, 1611; now in Favaro 3:201–50); Lodovico delle Colombe, *Contro il Moto della Terra*, meaning *Against the Motion of the Earth*, written in late 1611 or early 1612 and apparently unpublished (now in Favaro 3:251–90); and Giulio Cesare Lagalla, *De Phoenomenis in Orbe Lunae* (Venice, 1612; now in Favaro 3:309–93), meaning *On the Phenomena in the Orb of the Moon*.

6. Sizzi and Colombe specifically used biblical objections; it should be noted that they were not clergymen and were apparently the first to do so, a point often stressed by Drake (1957, 1978, 1980, 1983).

7. This reference is given by Galileo in the margin to his text; the numerous references to various Latin authors and writings which he similarly gives throughout this essay are here always inserted in parentheses in the text. In the case of references to the Bible and to other works for which there exists an English translation I have usually quoted the corresponding passage from such translations; otherwise Galileo's Latin quotations are translated directly from the wording as he has it, even when modern editions of the works from which he is quoting have a slightly different text. The present passage is from book 2, chapter 18 (see Augustinus 1894, p. 62.11–62.16).

8. The work in question is Augustine's *De Genesi ad litteram*, which is not available in English translation in its entirety, as one may infer from Przywara (1958, p. xi), Bourke (1964, pp. 251, 262), O'Meara (1973, pp. 546–52), and Meagher (1978, p. 299); the most readily available translation is one for book 12 of this work, from which none of Galileo's quotations are taken. Thus, as indicated in a previous note, his Latin quotations from this work are here translated directly.

9. By speaking of his "first announcement" (*primo avviso*) Galileo is making something of a play on words, since the Latin title of his *Sidereus Nuncius* could be taken to mean either a messenger or a message from the stars.

10. Caccini had indeed done this on 21 December 1614. See, for example, his deposition to the Inquisition (below).

11. For a sketch of the prehistory of Copernicanism, see the first of Galileo's "Considerations on the Copernican Opinion" (above).

12. As mentioned in a previous note, when Galileo first gives this story (second letter to Dini), Copernicus did not respond by actually going to Rome but sent a written report.

13. Actually Poland.

14. Paul of Middelburg (1445–1533).

15. As previously noted, though the Copernican system did play a role in the reform of the calendar, the new Gregorian calendar was constructed on the basis of non-Copernican ideas.

16. Cardinal Nicolaus von Schoenberg (1472–1537), archbishop of Capua.

17. Tiedemann Giese (1480–1550), Polish friend of Copernicus.

18. Of course, Galileo had no way of knowing that one Giovanni Maria Tolosani had had quite a few scruples about it; see Garin (1971, 1975), Rosen (1975), Westman (1986), and n. 19 to Chapter II.

19. For an analysis of the rhetorical import of Galileo's frequent use of the expression "necessary demonstration," see Moss (1986); but compare Finocchiaro (1986b) for an implicit criticism of Moss's interpretation and for a more methodological analysis.

20. This translation of Galileo's Latin quotation is taken from Copernicus (1976, pp. 26–27).

21. For example: Joshua 10:12–13, already mentioned several times and discussed at great length below; Psalm 18:6–7 (Douay) or Psalm 19:5–6 (King James), mentioned in Dini's letter to Galileo and discussed in his second letter to Dini; and Ecclesiastes 1:5, mentioned in Bellarmine's letter to Foscarini.

22. Here I have emended Evans' translation of this passage, which reads: "For my part I postulate that a god ought first to be known by nature, and afterwards further known by doctrine—by nature through his works, by doctrine through official teaching" (Tertullian 1972, p. 47).

23. In geostatic astronomy, both the sun and the moon are regarded as planets—a term that originally meant "wandering stars," namely heavenly bodies which appear to move relative to the fixed stars as well as to the earth.

24. See Augustinus (1894, pp. 45.20–46.10).

25. See ibid., pp. 47.22–48.11.

26. This translation of Galileo's Latin quotation is taken from Mourant (1964, p. 110). This letter is labeled Letter 143 in most editions of Augustine's works.

27. Ecclesiastes 3:11 (Douay Version).

28. Galileo misspells this name (in Italian *Iceta*) by adding an initial *N*, which yields *Niceta* in Italian or *Nicetas* in English, thus repeating Copernicus's error in the dedication to *On the Revolutions*.

29. Flora (1953, p. 1019, n. 4) gives the following reference: Cicero, *Academia*, II, 39, 123.

30. For example, Francesco Sizzi, *Dianoia Astronomica, Optica, Physica* (Venice, 1610; now in Favaro 3:201–50).

31. This may refer to Giulio Cesare Lagalla, *De Phoenomenis in Orbe Lunae* (Venice, 1612; now in Favaro 3:309–99); or it could refer to Ulisse Albergotti, *Dialogo . . . nel quale si tiene . . . la luna essere da sè luminosa* (Viterbo, 1613), concerning which see Favaro (11:599; 12:60, 65), and Drake (1957, p. 190).

32. Schaff and Wace (1893, p. 99). Galileo indicates the number of this letter as 103, but there is no doubt that his quotation is from what modern scholars and editors now designate as Letter 53. Further, I have slightly altered the punctuation and spelling in Schaff and Wace's translation for the sake of uniformity and easier comprehension.

33. These disciplines, especially geometry, had already reached the status of

model sciences in antiquity, primarily as a result of the work of Euclid, a Greek from Alexandria who lived about 300 B.C. and whose *Elements* is one of the most influential books ever written.

34. See Augustinus (1894, p. 31.2–31.12).

35. Antonio Santucci, who died in 1613.

36. Both the variation in apparent magnitude of Mars and Venus, and the phases of Venus, had been previously undetected, but they became observable with the telescope.

37. This seems to refer to Psalm 103:2 (Douay), which reads in part "who stretchest out the heaven like a pavilion," corresponding to Psalm 104:2 in the King James Version, which reads "who stretchest out the heavens like a curtain." Another relevant passage is Isaiah 40:22. Russo (1968, p. 346, nn. 1 and 2) comments that "neither St. Augustine nor Galileo seems to have understood that the hide concerned the hide of a tent" and that "the 'hide' in question is not a hide stretched out flat 'but the hide of a tent'."

38. This presumably corresponds to Psalm 103:2 (Douay), Psalm 104:2 (King James), and Isaiah 40:22; however, I have translated the word *pellem* in this sentence as "hide" because this is how Galileo understands it here.

39. See Augustinus (1894, p. 46.11–46.19).

40. Again, the exact reference for this passage, part of which Galileo quoted at the beginning of this essay, is *De Genesi ad litteram*, book 2, chapter 18; cf. Augustinus (1894, p. 62.8–62.16).

41. Job 26:7 (Douay).

42. Saint Thomas Aquinas (1225–1274), one of the greatest theologians and philosophers of Catholicism, whose system of ideas (Thomism) was declared by Pope Leo XIII in 1879 to be the official philosophy of the Catholic church. Though Aquinas had not yet reached such importance at the time of Galileo, even then he was sufficiently authoritative that one should not miss the point that Galileo is here basing his views on those of what are perhaps the three greatest theological authorities available to him (Augustine, Jerome, and Aquinas).

43. In mathematical astronomy, prosthaphaeresis is "the correction necessary to find the 'true', i.e. actual apparent, place of a planet, etc. from the mean place" (*Oxford English Dictionary*).

44. This is *Didaci a Stunica Salmaticensis Eremitae Augustiniani in Job Commentaria*, first published in Toledo (Spain) in 1584 and reprinted in Rome (1591); Galileo had apparently been given this reference by Cardinal Carlo Conti in a letter dated 7 July 1612 (Favaro 11:354–55). Zuñiga was an Augustinian friar from Salamanca, and his book was to be "suspended until corrected" (together with Copernicus's *Revolutions*) by the decree of the Index of 6 March 1616 (see below).

45. See Augustinus (1894, p. 48.6–48.11), previously cited by Galileo at note 25 above.

46. Galileo is here referring to interpretations of the miracle described in Joshua 10:12–13, in which God stopped the sun in order to prolong daylight. This is discussed at great length a few pages later, where more precise references are also given.

47. Alfonso Tostado (1400–1455), professor of theology and philosophy at the University of Salamanca (Spain).

48. Isaiah 38:8.

49. See Augustinus (1894, p. 27.12–27.20). Of the several quotations in this paragraph and the next, this is the only one that comes from chapter 18 of book 1. The next six quotations all come from chapter 19, and in fact they reproduce that entire chapter with the exception of its first five lines.

50. See Augustinus (1894, p. 28.1–28.4).

51. See ibid., p. 28.4–28.9.

52. See ibid., p. 28.9–28.12.

53. See ibid., p. 28.12–28.15.

54. See ibid., pp. 28.16–29.11.

55. See ibid., p. 29.11–29.20.

56. Joshua 10:12–13.

57. The ecliptic is the imaginary circle on the celestial sphere defined by the apparent path of the sun among the fixed stars.

58. I have not been able to confirm the precise identity of this work which Galileo (Favaro 5:344) cites as *De mirabilibus Sacrae Scripturae*. See, for example, Portalie (1960, pp. 401–6).

59. See *Istoria e Dimostrazioni Intorno alle Macchie Solari e Loro accidenti* (Rome, 1613; now in Favaro 5:71–249; translated in part in Drake 1957, pp. 87–144).

60. Joshua 10:12 (King James).

61. Thomas de Vio (1468–1534), author of a commentary on St. Thomas Aquinas's *Summa Theologiae*.

62. Joshua 10:13.

63. This is the version given in *The English Hymnal with Tunes*, p. 89. These are the first two of five stanzas of the hymn whose first Latin line is "Caeli Deus sanctissime" and which derives from the fourth or fifth century; see Julian (1892, p. 241).

64. See Proverbs 8:26. I have translated Galileo's Latin quotation literally in order to appreciate his point, which would certainly be lost with the King James Version and might still be obscured with the Douay Version.

IV. GALILEO'S DISCOURSE ON THE TIDES (1616)

1. Alessandro Orsini (1593–1626), member of a powerful Roman family, related through his mother to the Florentine House of Medici, was made cardinal on 22 December 1615.

2. Section numbers in brackets are not in the original and have been added by the present editor.

3. Galileo did not complete this work until his *Dialogo* (Florence, 1632; now in Favaro 7:21–520). For an English translation, which may be referred to simply as the *Dialogue*, see Galilei (1953, 1967). Much of the present discourse is included (at times almost verbatim) in Day IV of the *Dialogue*; the refutations in question are found in Favaro 7:442–49, corresponding to Galilei (1967, pp. 416–23). For some of the most reliable and enlightening accounts of Galileo's theory of tides, see Aiton (1954, 1963, 1965), Burstyn (1962, 1963,

1965), and Drake (1970, pp. 200–213; 1979; 1983); for a review of some of these works and others, see Finocchiaro (1980, pp. 74–79).

4. This is a reference to the precession of the equinoxes—the phenomenon of the apparent motion of the celestial pole around the pole of the ecliptic, whose period is about 26,000 years. First noticed by Hipparchus in the second century B.C., the phenomenon was explained in geostatic astronomy in terms of a celestial sphere which rotated at that rate. The Copernican explanation was in terms of a motion of the earth's axis, which is the oscillation mentioned by Galileo. In Newtonian physics this axial motion is in turn explained as an effect of the moon's attraction for the earth's equatorial bulge. See Kuhn (1957, pp. 268–71).

5. The original term here is *ex hypothesi.*

6. Copernicus had attributed to the earth a third motion, consisting of an annual rotation by the terrestrial axis around the axis of the earth's orbit in a direction opposite (east-to-west) to that of its orbital revolution, to account for the fact that the terrestrial axis always stays parallel to itself. As Galileo argued in the *Dialogue,* this phenomenon is simply an instance of rest, and thus no additional terrestrial motion needs to be postulated. See Favaro (7:424–25), Galilei (1967, pp. 398–99), and section [15] of "Galileo's Reply to Ingoli" (below).

7. Galileo is here thinking of 1 out of 360 degrees in a circle, so that if the circle were the equator (which is about 24,000 miles), 1 degree would be about 67 miles.

8. Lisbon, Portugal, on the western side of the Iberian peninsula, is obviously on its far side when viewed from Italy.

9. The Hellespont, which is nowadays more commonly known as the Dardanelles, is actually a strait between Asia and Europe that connects the Aegean Sea and the Sea of Marmara.

10. A seaport on the same (western) coast of Italy as Rome, but some distance to the northwest.

11. This channel is commonly known as the Strait of Messina.

12. According to ancient Greek mythology, Scylla and Charybdis were two monsters which lived on either side of the Strait of Messina.

13. Today Constantinople is commonly known as Istanbul and is located in the European part of Turkey.

14. This was the ancient mythological name for the Strait of Gibraltar, separating Europe and Africa and connecting the Mediterranean Sea and the Atlantic Ocean; the pillars were the two mountains on each side of the strait when approached from the Mediterranean.

V. THE EARLIER INQUISITION PROCEEDINGS
(1615–1616)

1. On 5 November 1612 Lorini had written a letter to Galileo apologizing for charging him with heresy in a private conversation a few days earlier. After writing the present complaint, he does not seem to have been further involved in the affair.

2. This letter as found in the Vatican manuscripts bears no date, though

there is a note added by some Inquisition clerk indicating that it was received in February 1615 (folio 12v; Pagano 1984, p. 71). Gebler (1879, p. 54, n. 2) states that "according to Epinois this letter was of the 5th, but Gherardi publishes a document which shows it to have been of the 7th"; here he is referring to Epinois (1867) and Gherardi (1870), two of the first scholars to have had access to the Vatican manuscripts.

3. This letter was sent from Florence to Cardinal Paolo Sfondrati (1561–1618), who was a member of the Congregation of the Holy Office in Rome. Soccorsi (1947, pp. 66–67, n. 14) states that at the time Sfondrati was the head of this congregation, as well as of the Congregation of the Index.

4. This was Galileo's letter to Castelli of 21 December 1613 (see above).

5. The copy was not in fact that faithful, as one can see by examining the text of Galileo's letter to Castelli enclosed by Lorini, now found in the special Vatican Galileo file immediately after Lorini's own letter (Pagano 1984, pp. 69–77; Favaro 19:297–305). For a description of the major differences, see note 16 to Chapter I.

6. *Saccenti* in the original.

7. Such an explicit statement of one's motivation was a formal requirement in complaints filed with the Inquisition. In fact, the manual for inquisitional procedure stated that the plaintiff had to declare that he was filing the complaint "for the sake of unburdening his own conscience, out of zeal for the Holy Faith, in order not to be excommunicated, or because his confessor had ordered him to do so" (Masini 1621, pp. 25–26).

8. In the original this parenthetical remark reads "*io non dico la scrittura*" (Favaro 19:298.37), which means literally "I am not referring to the essay." My translation is justified by the fact that the term *scrittura* here is the same one used at the beginning of the second paragraph following the salutation above (Favaro 19:297.6).

9. A Dominican who on 20 December 1614 had preached a sermon against mathematicians in general, and Galileo in particular, because of their allegedly heretical beliefs. For more information on his role in the affair, see Ricci-Riccardi (1902).

10. Today the last three words of this sentence are missing from the Vatican manuscripts, so that the text appears to end abruptly with the word *of* (*delle*). However, scholars who consulted these manuscripts in the last century were able to read the additional words "*sue sante orazioni*," which I have translated as "your holy prayers"; see Pagano (1984, p. 70, n. 4).

11. In the Vatican manuscripts this report bears no date and is curiously placed (folio 6r; Pagano 1984, pp. 68–69) *before* Lorini's complaint, which in turn precedes the copy of Galileo's letter to Castelli. Since for at least two months after Lorini's complaint the Inquisition attempted unsuccessfully to obtain the original letter (Pagano 1984, pp. 77–80 and 86–87), the date of this report is presumably no earlier than April 1615; and since the matter was resolved in late February and early March 1616, the report must have been written before then.

12. Note that this consultant had examined Lorini's copy of Galileo's letter to Castelli, which differs from the original in a number of ways which were

mentioned earlier (note 16 to Chapter I) and which will be repeated as they become relevant below.

13. The original letter said instead that "in the Scripture one finds many propositions which look different from the truth if one goes by the literal meaning of the words" (see above; cf. Pagano 1984, p. 72, or Favaro 19:299 and 5:282); thus Galileo had not used the word *false*, which is precisely one of those that the consultant finds objectionable.

14. The original letter said instead that "the Scripture has not abstained from somewhat concealing its most basic dogmas" (see above; cf. Pagano 1984, p. 72, or Favaro 19:300 and 5:283), which again might have prevented one of the consultant's qualms.

15. This was the same in the original.

16. Agostino Galamini (1552–1639), a Dominican who had been appointed cardinal with the title of Santa Maria d'Aracoeli in 1611.

17. Joshua 10:12.

18. For the exact wording of the relevant decrees issued by this council, see section 3 of the Introduction.

19. Ferdinando Ximenes (c. 1580–1630), a Dominican who will be called for a deposition to the Inquisition as a result of being mentioned here.

20. This was the title of Cardinal Paolo Sfondrati, to whom Lorini's letter was addressed.

21. Giannozzo Attavanti (c. 1582–1657), a minor cleric who had not yet been ordained priest and would be examined by the Inquisition on 14 November 1615 (see below).

22. This was the *Istoria e Dimostrazioni Intorno alle Macchie Solari e Loro Accidenti* (Rome, 1613; now in Favaro 5:71–249), commonly known in English as the *Sunspot Letters* (translated in part in Drake 1957, pp. 87–144).

23. Note that, here and in subsequent depositions, the letter Q is meant as an abbreviation for the sentence "He was asked," which yields, together with the expression that follows, an *indirect* rather than a direct question.

24. Emanuele Ximenes, born in 1542, at the time a consultant to the Inquisition in Florence, died soon after this incident in 1614. This Jesuit is not to be confused with either the earlier-named Dominican Ferdinando Ximenes or with a third individual (Sebastiano Ximenes) by the same surname mentioned below.

25. Paolo Sarpi (1552–1623), Venetian lawyer, theologian, and historian who also wrote on scientific subjects. Galileo and Sarpi were indeed friends, especially during the eighteen years that Galileo taught at the University of Padua, which is near Venice and was a public institution financially supported by the republic.

26. Sebastiano Ximenes, founder in 1593 of the order of the Knights of Santo Stefano.

27. The Lincean Academy, or Academy of the Linceans (in Italian *Accademia dei Lincei*, which means literally "Academy of the Lynx-eyed"), was the first international scientific society in modern science, having been founded in 1603 by Prince Federico Cesi (1585–1630), though it fell apart soon after his death. Galileo was made a member in 1611, became a friend of Cesi, and re-

ceived support from the academy for the publication of many of his subsequent works.

28. As its title indicates, the *Sunspot Letters* (1613) originated as an exchange of letters between German Jesuit astronomer Christopher Scheiner and German intellectual and politician Mark Welser and between Welser and Galileo.

29. See the letter by Cardinal-Inquisitor Fabrizio Veralli in Rome to Father Lelio Marzari, inquisitor of Florence, bearing this date (Favaro 20:569).

30. As previously mentioned, note that the letter *Q* stands for "He was asked," thus yielding *indirect* rather than direct questions.

31. See Psalm 104:5 (King James), "*Who* laid the foundations of the earth, *that* it should not be removed for ever"; or Psalm 103:5 (Douay), "Who hast founded the earth upon its own bases: it shall not be moved for ever and ever."

32. Again, by Cardinal-Inquisitor Fabrizio Veralli in Rome to Father Lelio Marzari, inquisitor of Florence, dated 7 November 1615 (Favaro 20:569).

33. Again, note that the questions are phrased as *indirect* ones, so that the *Q* should be read as something like "He was asked."

34. See Joshua 10:12–13.

35. At this point, between the word "philosophy" (*philosophia*) and the phrase "and formally heretical" (*et formaliter haereticam*), the original text in the Vatican manuscripts (folio 42r) shows a semicolon; Favaro (19:321) has a comma; and Pagano (1984, p. 99) has no punctuation. While rules and practices for punctuation were not as developed and strict in the seventeenth century as they are today, and while the Pagano volume is extremely valuable and generally reliable, in this particular instance there seems to be no justification for Pagano's transcription, which conveys the impression that biblical contradiction is being given as a reason for ascribing both philosophical-scientific falsehood and theological heresy.

36. Armagh is a town in Ireland which has been its ecclesiastical capital since St. Patrick founded his church there in the fifth century.

37. The original text in the Vatican manuscripts has this surname abbreviated "Jus.nus," which is kept by Favaro (19:321) in its abbreviated form but transcribed as "Justinus" by Pagano (1984, p. 100). However, an eighteenth-century summary of the affair in Italian (Favaro 19:417–19) lists this name as "Benedetto Giustiniani," and Favaro's Index (19:206) makes the same identification. So it is more likely that this individual was Benedetto Giustiniani (c. 1550–1622), born in Genoa, who taught rhetoric at the Collegio Romano and theology at Toulouse, Messina, and Rome.

38. Besides being found in the Vatican manuscripts (folio 43v), and thus published by Favaro (19:321) and by Pagano (1984, pp. 100–101), the latter has also discovered another essentially identical copy of these minutes in the Roman Archives of the Congregation for the Doctrine of the Faith, the new name for the Holy Office; cf. Pagano (1984, pp. 42 and 222–23). As Pagano discusses, this is one of the few additions his research was able to make to the documents already known.

39. Gio. Garzia Millini (or Mellini), who at the time was secretary of the Holy Office (Favaro 19:298, 306, and 311–13; and Viganò 1969, p. 120).

40. "And thereafter, indeed immediately" is my translation of the phrase *et successive, ac incontinenti* (folio 43v), which both Favaro (19:322.7) and Pagano (1984, p. 101) transcribe without the comma. The fact is that at this point the Latin text is somewhat ambiguous, and the sequence of events is far from clear; but the whole issue is too complex to be discussed here, let alone resolved. See Drake (1965), Gebler (1879, pp. 76–84), Geymonat (1962, pp. 112–15), Morpurgo-Tagliabue (1981, pp. 19–27), Langford (1966, pp. 92–97), and Santillana (1955, pp. 125–31; 1965; 1972).

41. See the next document below.

42. An "ordinary" would usually be a bishop.

43. Most of the works and individuals in this list are difficult to identify, but Usserius is likely to be James Ussher (1581–1656).

44. The original word here is *serpat*, which literally means "creep" or "crawl" in the manner of a snake. The Italian equivalent of this same term is used in the preface to the standard manual of inquisitional procedure (Masini 1621, p. 5) to describe how heresy moves, and it is also used in the Sentence (below).

45. Paolo Sfondrati.

46. Galileo had gone to Rome of his own initiative, in December 1615, to defend himself personally and try to prevent the condemnation of Copernicanism; lodging at the Tuscan embassy, he remained for about six months. The secretary of state of the Grand Duchy of Tuscany since 1613 was Curzio Picchena (1553–1626), who had come to that position after being private ducal secretary since 1601, before which time he had held a number of diplomatic posts.

47. Caccini had preached his famous sermon in Florence on 20 December 1614 and then had given a deposition to the Roman Inquisition on 20 March 1615 (see above).

48. Obviously this statement and the ones that follow constitute either Galileo's own *interpretation* of the outcome or perhaps the interpretation he wants the Tuscan court to believe.

49. See Job 9:6.

50. Such corrections were made a few years thereafter, and they were published by the Congregation of the Index on 15 May 1620; see the "Correction of Copernicus's *On the Revolutions*" (below).

51. Carlo de' Medici (1595–1666), younger brother of the reigning Grand Duke Cosimo II, had been appointed cardinal in December 1615. Galileo lingered in Rome for a few more months after 6 March, in part with the pretext of wanting to welcome the new cardinal upon his visit to Rome.

52. Galileo had earlier indicated his desire to go to Naples (Favaro 12:234), but he did not explicitly say why; I am inclined to agree with Drake (1957, p. 218) that it was probably to see Foscarini and perhaps Tommaso Campanella (1568–1639). A very controversial figure and an important thinker in his own right, Campanella was a strong supporter who at about this time wrote a defense of Galileo which was not published until 1622 in Frankfurt with the title *Apologia pro Galileo*, available in English in Campanella (1937), concerning which one of the best recent studies is Bonansea (1986). He was still in prison

(where he had been since 1603) as a result of a condemnation by the Inquisition, but later he was freed by Pope Urban VIII and tried unsuccessfully to help Galileo again in the later trial (1632–1633). Because of this, it is unclear that the meeting with Campanella could have really taken place; then, after Bellarmine's warning and the decree of the Index, it is likely that Galileo was in no mood or position to go through with his earlier plan.

53. As already mentioned, this did not happen until 15 May 1620; see the "Correction of Copernicus's *On the Revolutions*" (below).

54. Copernicus (1976, p. 51). For the exact corrections, see the document itself (below).

55. Paul V, born in 1552, pope from 1605 to his death in 1621.

56. Federico Cesi (1585–1630), who at this time was looking for a wife after the death of his first one, whom he had married in 1614.

57. The negotiations were successful, and in 1616 Cesi did marry Isabella Salviati, a relative of the grand duke.

58. In the Vatican manuscripts, this dated document is found among the proceedings of the later trial (1633) since it was at that time that Galileo introduced it in his defense (see Favaro 19:342 and 348; Pagano, 1984, pp. 134–35 and 138).

59. Having lingered in Rome for some time after the decree of the Index, Galileo soon received at least two letters from friends in Pisa and in Venice informing him that such rumors were circulating in these cities (Favaro 12:254 and 257–59). He showed these letters to Bellarmine (Favaro 12:257, n. 2; Baldini and Coyne 1984, p. 6), and so in this first sentence the cardinal is essentially reporting their content. This certificate makes it obvious that the two were still on good personal terms, despite their theological, philosophical, and scientific disagreement. The extent of Bellarmine's respect and goodwill is further supported by the autograph draft copy of this certificate, which is found in the Roman Archive of the Society of Jesus and has recently been published by Baldini and Coyne (1984, p. 25); the autograph draft copy shows that Bellarmine deleted two sentences and inserted one other, to make the end result more favorable to Galileo. In this context one should also mention the recent efforts toward a reinterpretation of Bellarmine undertaken by Baldini (1984) and by Coyne and Baldini (1985), which, while being more controversial, are likely to lead to fruitful discussions and new important insights; see Finocchiaro (1986a, 1988a).

60. The content of this sentence should be compared, on the one hand, with that of the decree of the Index (5 March 1616) and, on the other, with that of the Inquisition Minutes (25 February and 3 March 1616) and the Special Injunction (26 February 1616) document; many of the issues in the affair, and disputes about it, hinge on them (see, for example, Finocchiaro 1986b).

VI. GALILEO'S REPLY TO INGOLI (1624)

1. These convenient section numbers are not in the original and have been put in square brackets to indicate that they have been added by the present editor. The principle of subdivision has been to denote by a different section

number Galileo's reply to a different argument advanced by Ingoli. Besides sections [1] and [8], we thus have sixteen other sections which may be summarized and grouped as indicated by the following outline:

A. [1] INTRODUCTION: origin of essay, reasons for delay
B. ASTRONOMICAL ARGUMENTS FOR GEOCENTRISM
   [2] the nature and the facts of parallax
   [3] the constancy of apparent stellar magnitudes
   [4] the simultaneous visibility of exactly half of the celestial sphere
   [5] the eccentricities of Mars and Venus and the apogee of Venus
C. PHYSICAL ARGUMENTS FOR GEOCENTRISM
   [6] the lower position of heavy and dense bodies
   [7] the properties of the sieve and process of sieving
D. PHYSICAL ARGUMENTS AGAINST THE EARTH'S DIURNAL MOTION
   [8] general considerations
   [9] the vertical fall of bodies from a tower and from a ship's mast
   [10] the motion of projectiles
E. ASTRONOMICAL ARGUMENTS AGAINST THE EARTH'S ANNUAL MOTION
   [11] the constant latitude of the rising and setting of fixed stars
   [12] the constancy of the elevation of the celestial pole
   [13] the inequality of night and day
   [14] Tycho and the motion of comets
F. [15] COPERNICUS'S THIRD MOTION OF THE EARTH
G. PHYSICAL ARGUMENTS AGAINST THE EARTH'S MOTION IN GENERAL
   [16] the disinclination of heavier bodies to move
   [17] natural motion
   [18] the luminosity of bodies

2. In 1616, after Galileo and Ingoli had an oral dispute about Copernicanism in the presence of others, they decided to put their arguments in writing, with the latter doing so first and the former then replying (Favaro 5:399). The result was Ingoli's essay "On the Location and Rest of the Earth, Against the Copernican System" ("De Situ et Quiete Terrae Contra Copernici Systema," now in Favaro 5:403–12), which, in view of the decisions reached by the Inquisition and by the Index at the time, was left both unpublished and unanswered.

3. Obviously Galileo's silence had to do also with the warning he received from the Inquisition and with the decree of the Index; similarly, his breaking the silence now has to do not only with the things he mentions below but also with the fact that in 1623 Florentine Cardinal Maffeo Barberini had been elected Pope Urban VIII, and the climate was then more favorable toward freedom of discussion.

4. Galileo did not get around to doing this until his *Dialogue* of 1632.

5. In the National Edition (Favaro 6:515–16) these diagrams are *not* both on the same page; rather, the first one is on p. 515 and the second on p. 516.

6. This refers to Euclid's classic mathematical treatise, the *Elements*.

7. See the relevant passage in the *Dialogue* (Favaro 7:385–99; Galilei 1967, pp. 358–72).

8. In this argument the word *earth* is ambiguous: sometimes it means the "element" earth (e.g, rocks, sand, soil, and minerals) by contrast to other "elements" like water and air; and sometimes it means the terrestrial globe as a whole (certainly including the elements earth and water, and perhaps even air) by contrast to other bodies like the moon, sun, planets, and stars. To avoid confusion in section 6, the initial letter will be capitalized when the word has the latter meaning; thus, "Earth" refers to the terrestrial globe (or what we would today call the planet Earth), whereas "earth" refers to the solid element.

9. These are the four satellites of Jupiter which Galileo discovered and named the "Medicean planets," after the ruling family in Florence.

10. In classical logic, a syllogism is a special type of argument in which a conclusion is deduced from two premises and the three propositions involve exactly three terms arranged in various patterns, some of which turn out to be valid and others invalid; for example, "all A are B, all C are A, therefore all C are B" and "all A are B, some C are A, therefore some C are B" are valid; whereas "all A are B, all C are B, therefore all C are A" and "all A are B, some C are B, therefore some C are A" are invalid. Syllogisms with four terms would be arguments where such conclusions are deduced from two premises in such a way that, instead of three terms, four or more terms are involved; all such arguments are invalid—for example, "all A are B, all C are A, therefore all C are D." In some cases, terms like the B and the D of this last example are not explicitly different, but rather involve two different meanings of the same word; then we have a syllogism with four terms which is also an equivocation, and this is what Galileo seems to have in mind. The actual analysis of this passage in accordance with these concepts is beyond the scope of these notes; however, the study of the historical, philosophical, and theoretical connections between Galileo and logic is a fascinating and increasingly fruitful line of investigation, as one can see from such works as Barone (1972), Carugo and Crombie (1983), Crombie (1975), Finocchiaro (1980), Mertz (1980), Pizzorno (1979), and Wallace (1984).

11. A term used to refer to the Aristotelians and a nickname deriving from a Greek word meaning a place for walking and referring to the covered walk existing in the school founded by Aristotle in Athens.

12. Galileo's full analysis and critique of all these arguments is in the *Dialogue* of 1632.

13. It should be noted that in what follows there are at least three things at issue: whether this experiment had been made, what the results were, and how to interpret the results. See the elaboration in Day II of Galileo's *Dialogue* (Favaro 7:169–75) and the comments in Koyré (1966, pp. 223–38; 1978, pp. 164–75) and Finocchiaro (1980, pp. 116–17 and 210–12).

14. See the relevant sections of the *Dialogue* (Favaro 7:193–209; Galilei 1967, pp. 167–83).

15. The other things Galileo has in mind are such effects as the tides and the trade winds (for which see the "Discourse on the Tides" above), the motion of bodies in general (which he had researched extensively in his pretelescopic career, though he did not publish this work until the *Two New Sciences* of 1638), as well as heavenly phenomena like the periods of Jupiter's satellites (Favaro

7:144–45; Galilei 1967, pp. 118–19), the phases of Venus and the variations in apparent size and brightness of Venus and Mars (Favaro 7:346–68; Galilei 1967, pp. 318–40), and the annual period of the apparent motion of sunspots (Favaro 7:372–83; Galilei 1967, pp. 345–56). At this time Galileo may not have been fully aware of the precise details and implications of this last phenomenon (Drake 1978, pp. 209–10 and 310–11), but the *Sunspot Letters* (1613) certainly show a general awareness, and the assertion in this passage obviously refers to future expectations as well as past observations or previously established facts.

16. See the relevant section of the *Dialogue* (Favaro 7:403–16; Galilei 1967, pp. 376–89).

17. See the relevant section in the *Dialogue* (Favaro 7:399–416; Galilei 1967, pp. 372–89).

18. Besides attributing to the earth a diurnal axial rotation and an annual heliocentric revolution (both from west to east), Copernicus had felt it necessary to let the terrestrial axis rotate annually in the opposite direction around the orbital axis (in order to account for the fact that the terrestrial axis remains parallel to itself as the earth revolves around the sun).

19. See the relevant passage in the *Dialogue* (Favaro 7:424–25; Galilei 1967, pp. 398–99).

20. In a controversial passage in the *Dialogue* (Favaro 7:188–93; Galilei 1967, pp. 162–67) Galileo speculated that the actual path of a falling body on a rotating earth might be along the arc of a circle having as diameter the line going from the earth's center to the point from which the body was dropped.

21. This conclusion and this argument were repeated by Galileo in the *Dialogue* (Favaro 7:38–59 and 188–93; Galilei 1967, pp. 14–34 and 162–67) and have led some scholars to attribute to him a theory of "circular inertia"; see, for example, Koyré (1966, pp. 205–90; 1978, pp. 154–236), Shea (1972, pp. 116–38), and Shapere (1974, pp. 87–121). I believe a more correct interpretation is one along the lines of Coffa (1968), Drake (1970, pp. 240–78; 1978, pp. 126–32), Finocchiaro (1980, pp. 33–34, 87–92, and 349–53), and Sosio (1970, pp. l–li).

22. The traditional Aristotelian definition of nature reads: "*Nature is a source or cause of being moved and of being at rest in that to which* [*one*] *belongs primarily*, in virtue of itself and not in virtue of a concomitant attribute" (Aristotle, *Physics* 192b21–24, italics in the original).

23. This does *not* refer to the essay Galileo wrote in 1616 at the request of Cardinal Orsini (see above), but rather to what eventually became the *Dialogue on the Two Chief World Systems*, which he had wanted to entitle "Dialogue, or Discourse, on the Tides." For some of the background to this title see Drake (1986); for a discussion of its import see Finocchiaro (1980, pp. 12–18).

VII. MISCELLANEOUS DOCUMENTS
   (1618–1633)

1. See Galileo's "Discourse on the Tides" above.

2. Compare this interpretation of the anti-Copernican decree with the one

found in Galileo's letter (dated 6 March 1616) to the Tuscan secretary of state (see above).

3. Actually the decree of the Index of 5 March 1616 (see above) stated less severely that Copernicus's work was to be "suspended until corrected."

4. This amounts to deleting the first several sentences of the last paragraph of that preface, namely the following passage: "There may be triflers who though wholly ignorant of mathematics nevertheless abrogate the right to make judgements about it because of some passage in Scripture wrongly twisted to their purpose, and will dare to criticise and censure this undertaking of mine. I waste no time on them, and indeed I despise their judgement as thoughtless. For it is well known that Lactantius, a distinguished writer in other ways but no mathematician, speaks very childishly about the shape of the Earth when he makes fun of those who reported that it has the shape of a globe. Mathematics is written for mathematicians, to whom this work of mine . . ." (Copernicus 1976, pp. 26–27). The paragraph would then begin with the following sentence: "For the rest, this work of mine, if my judgement does not deceive me, will be seen to be of value to the ecclesiastical Commonwealth over which your Holiness now holds dominion" (Copernicus 1976, p. 27). Note that this deleted passage is also the one Galileo quoted approvingly at the beginning of his "Letter to the Grand Duchess Christina" (see above, paragraph 4).

5. This and subsequent references are to the original edition of Copernicus's book (Nuremberg, 1543). However, as Favaro (19:400, n. 2) points out, this first reference is erroneous and should read "book 1, chapter 5, page 3."

6. The original sentence reads: "However, if we consider it more closely the question will be seen to be still unsettled, and so decidedly not to be despised" (Copernicus 1976, p. 40). The sentence immediately preceding introduces the "question" by asserting: "Among the authorities it is generally agreed that the Earth is at rest in the middle of the universe, and they regard it as inconceivable and even ridiculous to hold the opposite opinion" (Copernicus 1976, p. 40).

7. The original passage says: "Why therefore do we still hesitate to concede movement to that which has a shape naturally fitted for it, rather than believe that the whole universe is shifting, although its limit is unknown and cannot be known? And why should we not admit that the daily revolution itself is apparent in the heaven, but real in the Earth; and the case is just as if Virgil's Aeneas were saying 'We sail out from the harbor, and the land and cities recede'?" (Copernicus 1976, p. 44).

8. The original sentence says: "I also add that it would seem rather absurd to ascribe motion to that which contains and locates, and not rather to that which is contained and located, that is the Earth" (Copernicus 1976, p. 46).

9. The deleted passage says: "You see then that from all these arguments the mobility of the Earth is more probable than its immobility, especially in the daily revolution, as that is particularly fitting for the Earth" (Copernicus 1976, p. 46).

10. The original initial sentence says: "Since, then, there is no objection to the mobility of the Earth, I think it must now be considered whether several motions are appropriate for it, so that it can be regarded as one of the wandering stars" (Copernicus 1976, p. 46).

11. The original sentence is: "Consequently we should not be ashamed to admit that everything that the Moon encircles, including the centre of the Earth, passes through that great sphere between the other wandering stars in an annual revolution round the Sun, and the centre of the universe is in the region of the Sun" (Copernicus 1976, p. 49).

12. The original sentence reads: "That the Sun remains motionless and whatever apparent motion the Sun has is correctly attributed to the motion of the earth" (Copernicus 1976, p. 49).

13. Copernicus (1976, p. 51).

14. The original title is "Derivation of the triple motion of the Earth" (Copernicus 1976, p. 51).

15. The original title is "The size of these three stars, the Sun, the Moon, and the Earth, and a comparison of them with each other" (Copernicus 1976, p. 217).

16. The original of this document is in Italian and found in the Roman Archives of the Sacred Congregation for the Doctrine of the Faith (formerly Holy Office), filed under "Acta et Documenta," vol. EE, folios 292r–293r. We do not know the exact date when it was written, but it obviously must have been after the fall of 1623, when *The Assayer* was published, and probably before 18 April 1625, when a friend wrote Galileo informing him in a general way about such a complaint (see next document below). Nor do we know the identity of the writer, though it is obvious from the content of the letter that it was *not* sent anonymously, but rather that the author expected some kind of personal response from the recipient. The document was first published by Redondi (1983, 1985a, 1987) and used by him as the essential basis of his reinterpretation of the trial of 1633; he attributes this complaint to Orazio Grassi, a Jesuit professor at the Collegio Romano with whom Galileo was involved in a bitter controversy that produced several books, including *The Assayer*; Redondi speculates that it was antiatomistic complaints such as those expressed in this document which provided the real though hidden basis for the trial, and he tries to explain away the official charges, the actual trial, and the public condemnation. Redondi's book has itself become the subject of a controversy, and his central thesis has been generally rejected by scholars (e.g., D'Addio 1984; Pagano 1984, pp. 43–48; and Ferrone and Firpo 1985a, 1985b, 1986), though defended by some (e.g., Morpurgo-Tagliabue 1984). The document was also published by Pagano (1984, pp. 245–48) in an improved transcription which I have used for the purpose of this translation.

17. This obviously refers to the original edition of *The Assayer* (1623), but it corresponds to section 48, to Favaro (6:347), to Drake (1957, pp. 273–74), and to Drake and O'Malley (1960, pp. 308–14).

18. See Aristotle, *On the Heavens*, book 2, chapter 7.

19. This is one of Galileo's most famous epistemological doctrines, which philosophers sometimes call the distinction between primary and secondary qualities; for a discussion of its significance, see, for example, O'Connor (1952, pp. 60–66) and Danto and Morgenbesser (1960, pp. 21–32), which also contains a translation of the relevant Galilean passage.

20. See Favaro (6:349), Drake and O'Malley (1960, pp. 310–11), and

Danto and Morgenbesser (1960, p. 29). The translation of *The Assayer* in Drake (1957, pp. 231–80) is abridged, and this particular passage is one of those that are truncated (p. 275).

21. See Favaro (6:350), Drake (1957, pp. 276–77), and Danto and Morgenbesser (1960, p. 30).

22. According to the Catholic doctrine of the Eucharist, during a mass after bread and wine are consecrated they become the body and the blood of Jesus Christ; such consecrated bread and wine are called the Eucharist. This mystery of the faith is in part explained by saying that the Eucharist contains the *substance* of Christ's body while retaining the usual accidents and qualities of bread. For further explanations of the connections between atomism and this doctrine, see Redondi (1983, pp. 257–87; 1987, pp. 203–26).

23. See Favaro (6:350–51) and Danto and Morgenbesser (1960, p. 31).

24. This canon reads: "If anyone says that in the sacred and holy sacrament of the Eucharist the substance of the bread and wine remains conjointly with the body and blood of our Lord Jesus Christ, and denies that wonderful and singular change of the whole substance of the bread into the body and the whole substance of the wine into the blood, the appearances only of bread and wine remaining, which change the Catholic Church most aptly calls transubstantiation, let him be anathema" (Schroeder 1978, p. 79). The same session defined transubstantiation as follows: "But since Christ our Redeemer declared that [i.e., the consecrated bread and wine] to be truly his own body which He offered under the form of bread, it has, therefore, always been a firm belief [of] the Church of God, and this holy council now declares it anew, that by the consecration of the bread and wine a change is brought about of the whole substance of the bread into the substance of the body of Christ our Lord, and of the whole substance of the wine into the substance of His blood. This change the holy Catholic Church properly and appropriately calls transubstantiation" (Schroeder 1978, p. 75).

25. Francisco Suarez, *Metaphysicarum Disputationum in quibus et universa naturalis theologia ordinate traditur . . . tomi duo* (Venice, 1610), vol. 2, p. 340.

26. See Favaro (6:347–48) and Drake (1957, p. 274).

27. Federico Cesi, head of the Lincean Academy.

28. *The Assayer* does not in fact praise Copernicanism or even discuss it directly. It does discuss a wide range of astronomical, physical, and philosophical topics, under the pretext of answering an opponent in the controversy about the nature of comets; but it steers clear of the earlier controversy about the motion and location of the earth and the sun. This, of course, raises doubts about whether here Guiducci is referring to the anonymous complaint contained in the preceding document.

29. Here Guiducci identifies Guevara somewhat incorrectly, for the latter was indeed a member of the Order of Clerks Regular Minor, but not of the Theatines; for a discussion of the possible significance of this confusion, see Redondi (1987, p. 143).

30. Francesco Barberini (1597–1679), nephew of the reigning Pope Urban VIII (Maffeo Barberini). He had been sent by his uncle to France to arrange a peace in connection with the Thirty Years War, but the attempt failed.

31. This involved a phase of the Thirty Years War (1618–1648), which was a general European war fought primarily in central Europe over territorial, dynastic, and religious issues, the latter being the struggle between Catholics and Protestants. The particular episode at the time involved a small valley in northern Italy near Switzerland called Valtellina, which had great strategic importance as a highway linking southern and central Europe.

32. Lothario Sarsi was the pseudonym used in the controversy about comets by Orazio Grassi (1590–1654), a Jesuit and professor of mathematics at the Collegio Romano. The work mentioned here was not published until 1626 and was meant to be an answer to *The Assayer*; see Favaro (6:3–500), Drake and O'Malley (1960), and Redondi (1983, pp. 223–56; 1987, pp. 176–202).

33. Sarsi (Grassi) was a native of Savona in the Republic of Genoa, which was involved in this phase of the Thirty Years War.

34. Alessandro Orsini, for whom Galileo had written the "Discourse on the Tides" in 1616.

35. Pseudonym of Christopher Scheiner, the German Jesuit astronomer with whom Galileo had been involved in a dispute about sunspots, which in part resulted in the *Sunspot Letters* and was to continue after this time.

36. Andrea Cioli (1573–1641), who had held this position since 1627.

37. Ever since his visit to Rome in the spring of 1624, Galileo had been working on a book containing a physical explanation of the tides in terms of the earth's motion, together with a discussion of the problems associated with assuming such a cosmological hypothesis. In January 1630 he had completed the work, and in May–June of that year he had gone to Rome to obtain the imprimatur from the Church authorities and arrange for its publication by the Lincean Academy (Drake 1978, pp. 311–12).

38. This office included the responsibility for issuing imprimaturs to books being published in Rome, and since 1629 it had been held by Niccolò Riccardi (1585–1639). He was a Dominican who in 1623 had been assigned by his predecessor in that office to review *The Assayer* and had praised it highly (Favaro 6:200).

39. Like Riccardi, Visconti was a Dominican.

40. This was due to an outbreak of the plague.

41. Niccolò Antella (1560–1630), Florentine nobleman who held numerous government offices; he was one of four regents when the Grand Duke Cosimo II died in 1621 and his son Ferdinando II was still a minor, as well as ducal reviewer and censor of books to be printed. His endorsement, dated 12 September 1630, appears on the imprimatur page of the published *Dialogue* of 1632 (Favaro 7:26).

42. Caterina Riccardi Niccolini (1598–1676), who happened to be a cousin of the master of the Sacred Palace.

43. Castelli was now living in Rome, having earlier left the University of Pisa to teach at the University of Rome and to supervise water projects there at the request of Pope Urban VIII.

44. Ferdinando II de' Medici (1610–1670), who had become grand duke upon his father's premature death in 1621; in view of his minor age, at first the government was administered by a Council of Regents consisting of his mother,

grandmother, and other notables, and it was not until 1627 that he assumed control of the state.

45. Geri Bocchineri (d. 1650), a private secretary to the grand duke.

46. In 1617 Galileo had moved to a house located in an area of the Florentine hills known as Bellosguardo (Drake 1978, p. 262).

47. The year is written in the original letter as 1630, concerning which Favaro (14:218) states that 1630 is merely a "Florentine custom" equivalent to 1631.

48. This is the short common description I am using to designate the person whose official title was master of the Sacred Palace and who was then Niccolò Riccardi.

49. Francesco Niccolini (1584–1650), the grand duke's ambassador to Rome, who held the office from 1621 to 1643.

50. Remember that Ciampoli was a strong supporter of Galileo and that at this time he was correspondence secretary to the pope.

51. Orso d'Elci (d. 1636), a secretary of the Grand Duke Ferdinando II and a member of the Council of Regents when the previous duke died and Ferdinando was still a minor (1621–1627).

52. Galileo's preferred title for the work he began writing after Urban VIII became pope had been *Dialogue on the Ebb and Flow of the Sea*; why this became *Dialogue . . . on the Two Chief World Systems*, and how the latter came to be interpreted as *Dialogue on the Two Chief World Systems*, is a subject of controversy. See, for example, Drake (1986), Finocchiaro (1980, pp. 112–18), and Shea (1972, pp. 172–86).

53. The Italian expression is *Nostro Signore*, which literally means "Our Lord" and which in such correspondence was a description referring to the pope.

54. See the decree of the Index (5 March 1616) above.

55. Clemente Egidi, inquisitor-general of Florence from 1626 to 1635.

56. The preface which the Vatican secretary enclosed with this letter corresponds verbatim to the one found in the published *Dialogue* of 1632; see the "Preface to the *Dialogue*" below, and see Favaro 7:29–31 and 19:328–30 (or Pagano 1984, pp. 110–12). The only significant difference occurs in the second sentence of the third paragraph. Here the published preface speaks of "mobility," whereas the copy attached to this letter has the word *mobility* corrected to *immobility* by inserting the prefix *im* between the lines. Thus in the latter copy the sentence reads: "First, I shall attempt to show that all experiments feasible on the earth are insufficient to prove its immobility but can be adapted indifferently to a moving as well as to a motionless earth." It seems to me that this version is more accurate in its emphasis, though the published one is not literally incorrect.

57. No suggested text for the ending was enclosed with this letter; rather, the following instructions followed the text of the Preface: "At the end one must have a peroration of the work in accordance with this preface. Mr. Galilei must add the reasons pertaining to divine omnipotence which Our Master gave him; these must quiet the intellect, even if there is no way out of the Pythagorean arguments" (Favaro 19:330; Pagano 1984, p. 112). See the ending Ga-

lileo came up with in our document entitled "Ending of the *Dialogue*" below.

58. Printing of the book was done in Florence and completed on 21 February 1632 (Favaro 7:8). This preface corresponds verbatim to the latest version sent by the Vatican secretary with his letter of 19 July 1631 (see above), the only exception being that the word *mobility* in the second sentence of the third paragraph here had been corrected to *immobility* in that copy.

59. See Galileo's "Discourse on the Tides" above.

60. This individual was probably Cesare Cremonini (1550–1631), professor of philosophy at the University of Padua and a colleague with whom Galileo was on friendly terms despite their philosophical and scientific opposition.

61. It should be mentioned, however, that the Italian name "Simplicio" which Galileo used also has the connotation of simpleton.

62. In a number of places in the *Dialogue* Galileo has the speakers refer to himself as the Academician, meaning member of the Lincean Academy. Here we also have a reference to a work on motion the research for which Galileo had essentially completed at Padua but not published yet; eventually the book was published in Holland in 1638 as the *Two New Sciences* (Favaro 8:39–318; Galilei 1974).

63. As soon as the *Dialogue* was published, various complaints and difficulties were voiced about its content and the circumstances of its publication (see "Diplomatic Correspondence" below)—so much so that the pope suspended further distribution and appointed a special commission to see what the problems were (Favaro 20:571–72).

64. There appear to have been three members of this special commission: the master of the Sacred Palace (Riccardi), the papal theologian Agostino Oreggi (alias Augustinus Oregius), and a Jesuit who is presumably Melchior Inchofer (see the letter dated 11 September 1632 in the "Diplomatic Correspondence" below, also in Favaro 14:388–89). However, it seems that the commission produced neither a single unified report, nor three individual ones, but rather a report consisting of two largely overlapping but distinct parts. This is initially suggested by the content from even a cursory reading, and it is confirmed by an examination of the Vatican manuscripts, which reveals the following. The first part is written on folios 52r and 52v; the second on folios 53r, 53v, 54r, and 54v; they are separated by a blank space of about one-quarter page at the bottom of folio 52v; moreover, the handwriting of the first part is less smooth than that of the second part, which seems to be the same handwriting as in the four attachments (A, B, C, and D) to the report; and finally, the watermarks of the paper on which the two parts are written are different, the first showing an inverted bird, the second showing none on one folio and a different watermark on the second folio. It thus seemed proper to indicate by numbers in square brackets that this report does have two parts.

65. See the document "Special Injunction (26 February 1616)" above.

66. Note that in the seventeenth century the expression "*ad hominem*" did not mean what it does today, but rather referred to an argument designed to examine the correctness of an opponent's views by showing that they imply consequences not acceptable to the opponent.

67. We have already come across this letter as the document entitled "The Vatican Secretary to the Florentine Inquisitor (24 May 1631)" above.

68. See the document entitled "The Florentine Inquisitor to the Vatican Secretary (31 May 1631)" above.

69. See the "Preface to the *Dialogue*" above.

70. See the document entitled "The Vatican Secretary to the Florentine Inquisitor (19 July 1631)" above.

71. Here, to avoid confusion between the two sets of numbers, the arabic numerals of the original have been replaced by roman numerals.

72. In the original edition of the *Dialogue* (Florence, 1632), the text in the body of the book is printed in italics whereas the preface, the dedication, and the lengthy index are printed in (what is for us) normal type.

73. That is, the pope's favorite argument; see the "Ending of the *Dialogue*" above.

74. That is, the last two pages of the text (Galilei 1632, pp. 457–58), continuous with the rest of the discussion in Day IV of the book.

75. Favaro (7:124–31) or Galilei (1967, pp. 98–105).

76. Favaro (7:154–55) or Galilei (1967, pp. 128–29).

77. Francesco Barberini, the pope's nephew.

78. Giorgio Bolognetti (c. 1590–1680), nuncio to the Grand Duchy of Tuscany from 1631 to 1634.

79. Here the original text reads "*opere*" in the plural (Favaro 14:397); this is somewhat puzzling, and it has led some scholars to speculate that indeed other works were examined and provided the basis for the trial of 1633. This is the sort of thing on which Redondi (1983, e.g., pp. 313–14; 1987) builds his interpretation, another being a similar reference to Galileo's "books" in the plural by Campanella in a letter dated 21 August 1632 (Favaro 14:373).

80. See the letter dated 18 September 1632 in the "Diplomatic Correspondence" below.

81. Jean Baptiste Morin, *Famosi et antiqui problematis de Telluris motu vel quiete hactenus optata solutio* (Paris, 1631), concerning which see Favaro (7:15 and 549–61) and L. Froidmont, *Ant-Aristarchus, sive Orbis-terrae immobilis* (Antwerp, 1631).

82. It may seem puzzling that Galileo would want to "praise" two books critical of the geokinetic theory, and in fact he may be writing with tongue in cheek. However, part of his meaning must be that these are intelligent books deserving serious criticism; for example, in an earlier letter to Diodati, Galileo had stated that "among the opponents of Copernicus, Froidmont seems to me more sensible and capable than anyone else I have seen so far" (Favaro 14:341); see also Favaro (16:60.40).

83. See Galileo's "Letter to the Grand Duchess Christina" above; the preceding paragraph is, of course, a summary of one of the central arguments in that essay.

84. Pierre Gassendi, *Epistolica Exercitatio, in qua principia philosophiae Roberti Fluddi medici reteguntur, . . . , cum Appendice aliquot observationum caelestium* (Paris, 1630).

VIII. DIPLOMATIC CORRESPONDENCE (1632–1633)

1. All the letters in this chapter were written by the Tuscan ambassador at the Holy See (Francesco Niccolini) to the Tuscan secretary of state (Andrea Cioli). They are translated here as they are printed in the National Edition of Galileo's collected works edited by Favaro; thus I have followed Favaro's practice of reproducing some letters in their entirety and excerpting from others only what is relevant to the affair. In the latter case, the omissions are indicated by ellipses.

2. Francesco Barberini, the pope's nephew.

3. Scipione Chiaramonti (1565–1652) actually taught philosophy rather than mathematics at the University of Pisa from 1627 to 1636, though he had earlier taught mathematics at Perugia. While he had been mentioned favorably in *The Assayer*, he was sharply criticized in the *Dialogue*, to which he was quick to reply with his *Difesa . . . al suo Antiticone e Libro delle tre stelle nuove* (Florence, 1633).

4. Besides being Tuscan secretary of state, Andrea Cioli was at this time "*Balì*" of Arezzo, that is, a magistrate appointed by the duke and responsible for executing the laws and administering justice in a given town.

5. Mariano Alidosi.

6. Tommaso Campanella, who had long been a supporter of Galileo, but who had himself had his share of troubles with Church authorities, had suggested in a letter to Galileo (21 August 1632) that he and Castelli be appointed to the special commission investigating the case (Favaro 14:373).

7. Benedetto Castelli.

8. The Master of the Sacred Palace (Niccolò Riccardi) was a cousin of the ambassador's wife.

9. Agostino Oreggi (1577–1635).

10. Melchior Inchofer (1585–1648); it is strange and ironical that in the same year his work on a supposed letter by the Virgin Mary to the inhabitants of Messina was placed on the Index.

11. In his letter dated 13 October 1632, Galileo made a plea to Cardinal Francesco Barberini to allow him to defend himself in writing or, failing that, to move the trial to Florence, in part because of his old age, ill health, and dangers of a trip to Rome; nevertheless, he ends by saying that if this is not granted then "I will make the trip since I would rather obey than live" (Favaro 14:410.131). Galileo also says in the letter that one of the things that had encouraged him to undertake the writing of the *Dialogue* was "hearing, almost like an echo of the Holy Spirit, a very short but admirable and most holy assertion suddenly come out of the mouth of a person who is most eminent in learning and venerable for the holiness of his life; that is, an assertion which, in no more than ten words cleverly and beautifully combined, summarizes what one can gather from the lengthy discussions scattered in the books of the sacred doctors" (Favaro 14:408–9); but he adds that it would not be proper to disclose the name of such a person.

12. Here and elsewhere Niccolini overstates Galileo's age; at this time he was in his sixty-ninth year.

13. Recall that Giovanni Ciampoli was the pope's correspondence secretary and a strong supporter of Galileo.

14. Galileo's last attempt to avoid having to go to Rome was to obtain a medical certificate (dated 17 December 1632) signed by three physicians, testifying that he had a number of specific health problems which made a journey to Rome dangerous to his life (Favaro 19:334–35; Pagano 1984, p. 121).

15. A church in Rome next to the Tuscan embassy.

16. See the "Ending of the *Dialogue*" above.

17. Sant'Onofrio was the title of Antonio Barberini (1569–1646), the pope's brother; the named Barberini was Francesco Barberini, the pope's nephew; Borgia was Gaspare Borgia (1589–1645), Spanish ambassador to Rome, cardinal since 1611; Gessi was Berlinghiero Gessi (1564–1639), governor of Rome and cardinal since 1626; Ginetti was Marzio Ginetti (1585–1671), whom we have encountered earlier; San Sisto was the title of Laudivio Zacchia (c. 1560–1637), cardinal since 1626; and Verospi was Fabrizio Verospi (1572–1639), cardinal since 1627. These seven, plus the other two mentioned above (Bentivoglio and Scaglia), correspond to the list in the "Sentence" below, with the exception of the second one in that list, Felice Centini (1570–1641), cardinal since 1611.

18. See Galileo's first deposition (12 April 1633) below.

19. Antonio Barberini, the pope's brother, not to be confused with Francesco Barberini, the pope's nephew.

20. Geri Bocchineri, a personal secretary of the grand duke, to whom Galileo wrote frequently since his arrival in Rome. In his 16 April letter, Galileo states that "I have had to stay in seclusion, though under conditions of unusual room and convenience, in three chambers which are part of the residence of the prosecutor of the Holy Office, and with unrestricted opportunity and ample space for walking about. My health is good . . ." (Favaro 15:88).

21. See the letter by the commissary to Barberini (28 April 1633) and Galileo's second deposition (30 April 1633) below.

22. On 4 May the Tuscan secretary of state had written the ambassador that Galileo had to pay his own expenses beyond the first month (Favaro 15:112).

23. Note in the final "Sentence" that only seven of the ten cardinals signed the document; though the reasons are a matter of speculation, the existence of some kind of dissent is a distinct possibility.

IX. THE LATER INQUISITION PROCEEDINGS
   (1633)

1. Remember that, here and in other depositions, the questions are recorded as *indirect* queries, so that the letter Q ought to be taken to mean "He was asked," rather than simply "Question."

2. The full title of the original edition (Florence, 1632) means: *Dialogue of Galileo Galilei, Lincean Academician, Extraordinary Mathematician at the University of Pisa, and Philosopher and Chief Mathematician to the Most Se-*

rene Grand Duke of Tuscany; where in meetings over the course of four days one discusses the two Chief World Systems, Ptolemaic and Copernican, proposing indeterminately the Philosophical and Natural reasons for the one as well as for the other side. For an account of the vicissitudes of the titling of the book, see Drake (1986); for an analysis of its logical and rhetorical import, see Finocchiaro (1980, esp. pp. 12–18).

3. Aside from Bellarmine and Bonsi, the identity of these cardinals is as follows: "Aracoeli" was Agostino Galamini, who had the title of Santa Maria d'Aracoeli; "San Eusebio" was the title of Ferdinando Taverna; and "d'Ascoli" was Felice Centini.

4. As previously mentioned, this word is my translation of the phrase *ex suppositione*, which Galileo leaves in Latin even though he is answering the inquisitor in Italian; see Wallace (1981a, 1981b, 1984) for a discussion of the importance of this notion and Wisan (1984) for a criticism of Wallace's interpretation; see also Chapter I, note 45, and Chapter II, note 4.

5. See the letter by Bellarmine to Foscarini (12 April 1615) above.

6. For the text of Bellarmine's certificate (26 May 1616), see above.

7. See the text and the form of the special injunction (26 February 1616) above.

8. This individual has been mentioned before, and it should be noted that his last name may also be spelled Oregio or (in Latin) Oregius. Here and elsewhere I am following the spelling given in Favaro's biographical index (Favaro 20:361–561), even though he himself sometimes uses slightly different spellings in other volumes of Galileo's *Opere*.

9. I presume that this is the special commission report of September 1632 (see above).

10. Unlike Oreggi's report, this one bears no date but is placed immediately after it in the Vatican manuscripts.

11. This and subsequent part numbers are not in the original, though the existence of the four parts identified here is suggested by the repetitiveness of their content and by Inchofer's signature being present four times; a direct examination of the Vatican manuscripts confirms this by revealing that each part begins at the top of a folio page (ff. 96r, 97r, 99r, and 100r, respectively) and is separated from the previous part by one or more blank folio pages. It will also be noted below that part [IV] has itself quite a complex structure, but an extensive analysis is beyond the scope of this documentary history.

12. The page numbers found in the report refer to the original edition of the *Dialogue*, a facsimile edition of which is in print with Culture et Civilisation, 115 Avenue Gabriel Lebon, Brussels, Belgium. The page correspondence to other editions or translations may be obtained by consulting Finocchiaro (1980, pp. 440–53); here Galilei (1632, p. 25) corresponds to Favaro (7:58) and to Galilei (1967, p. 33).

13. Inchofer's numerous quotations are generally accurate, though not verbatim; I have translated the wording as he gives it, rather than from Galileo's own text.

14. See Favaro (7:347–48) or Galilei (1967, p. 319).

15. See Favaro (7:349–56) or Galilei (1967, pp. 321–28).

16. See Favaro (7:368–70) or Galilei (1967, pp. 340–42).

17. I presume Inchofer means Galileo's "Letter to the Grand Duchess Christian" (see above).

18. The original text here speaks of Augustine's commentary to Psalm 108, but Pagano (1984, p. 142, n. 48) says that the number should be 118; compare, below, Pasqualigo's reference to the correct number 118.

19. In the *Dialogue* Galileo is never mentioned explicitly, but the speakers (especially Salviati) often refer to him indirectly when they speak of the ideas, works, and discoveries of the "Academician," meaning Lincean Academician.

20. At this point the original manuscript has the reference "L. 1. C. de Test., et ibi Baldus" (Favaro 19:351.86), which is probably a reference to the then current Code of Canon Law, chapter on witnesses, and Baldus's commentary thereon.

21. Christopher Scheiner (1573–1650), the German Jesuit astronomer whose correspondence with Mark Welser occasioned Galileo's letters to the latter, which were then published in the *Sunspot Letters* (1613); Scheiner and Galileo were involved in a scientific disagreement about the interpretation of sunspots, as well as in a priority dispute about their discovery. While Galileo was writing the *Dialogue*, the controversy continued with Scheiner's publication of his *Rosa Ursina* (Bracciano, 1626–1630); later he wrote a book against the *Dialogue* which was published only posthumously in 1651 (*Prodomus Pro Sole Mobile, et Terra Stabili contra Academicum Florentinum Galilaeum a Galilaeis*).

22. That is, against the decree of the Index (6 March 1616); throughout the *Dialogue* Galileo has the Copernican spokesman (Salviati) express frequent qualifications to the effect that the discussion should not be taken as deciding the issue in question, but only as presenting the relevant evidence and arguments.

23. See Favaro (7:299–300) or Galilei (1967, pp. 276–77).

24. See Favaro (7:141) or Galilei (1967, p. 115).

25. See Favaro (7:146) or Galilei (1967, p. 120). Beginning with this passage, the references in this list are given in the margin rather than the text of the manuscript.

26. See Favaro (7:142–43) or Galilei (1967, pp. 116–17).

27. See Favaro (7:155) or Galilei (1967, p. 129).

28. The manuscript numbers this item no. 3, thus assigning this number twice; similarly, the next item is numbered no. 4. I have eliminated these repetitions by numbering all items in a single consecutive order.

29. See Favaro (7:404–5) or Galilei (1967, pp. 377–78).

30. See Favaro (7:401) or Galilei (1967, p. 374).

31. Neither this sentence nor anything close to it is found on page 399 of the original edition of the *Dialogue*; it may, however, be regarded as a paraphrase of Sagredo's speech on page 302, corresponding to Favaro (7:434) or Galilei (1967, p. 408).

32. See Favaro (7:81–82) or Galilei (1967, pp. 56–57).

33. See Favaro (7:348) or Galilei (1967, p. 320).

34. See Favaro (7:348) or Galilei (1967, p. 320).

35. See Favaro (7:349–50) or Galilei (1967, pp. 320–21).

36. See Favaro (7:349) or Galilei (1967, p. 321).

37. See Favaro (7:349) or Galilei (1967, p. 321).

38. See Favaro (7:350) or Galilei (1967, p. 322).

39. See Favaro (7:355.8–12); this remark gets lost in the English translation found in Galilei (1967, p. 327). What Galileo is saying is that certain anti-Copernican arguments are worthy of men whose definition ("rational animals") has only the genus ("animals") but lacks the species ("rational"); cf. Finocchiaro (1980, pp. 60 and 244).

40. See Favaro (7:355) or Galilei (1967, pp. 327–28).

41. See Favaro (7:355–56) or Galilei (1967, p. 328).

42. See Favaro (7:367) or Galilei (1967, p. 339).

43. See Favaro (7:368) or Galilei (1967, p. 340).

44. See Favaro (7:369) or Galilei (1967, p. 341).

45. See Favaro (7:370) or Galilei (1967, p. 342).

46. See Favaro (7:371–72) or Galilei (1967, p. 344).

47. This parenthetical remark is found in the margin rather than within the text of the original; page 27 corresponds to Favaro (7:60) or Galilei (1967, p. 35).

48. See Favaro (7:372–73) or Galilei (1967, pp. 344–45).

49. See Favaro (7:379) or Galilei (1967, p. 352).

50. See Favaro (7:383) or Galilei (1967, p. 356).

51. See Favaro (7:383) or Galilei (1967, p. 356); this parenthetical remark appears in the margin.

52. See Favaro (7:429) or Galilei (1967, pp. 402–3).

53. See Favaro (7:437) or Galilei (1967, pp. 410–11).

54. Zaccaria Pasqualigo (d. 1664), of the Order of Theatines, taught philosophy to his confreres in Padua and theology in Rome; it is curious that some time after passing such a severe judgment on the *Dialogue* he had two works put on the Index until corrected, one of them being a work in moral theology (*Decisiones morales*).

55. Like Inchofer's report, and unlike Oreggi's, this one bears no date, but it is placed immediately after Inchofer's report in the Vatican manuscripts.

56. Again, these part numbers are not in the original; but, as in the case of Inchofer's report, they are suggested by the fact that each begins at the top of a folio page (even when the preceding page is partly or wholly blank) and the signature is repeated at the end of each.

57. Here and in the rest of this report Pasqualigo's term is *sub hypothesi*, left in Latin even though he writes this third part of his report in Italian.

58. These parenthetical page references to the 1632 edition of the *Dialogue* are found in the margin to the text; page 109 corresponds to Favaro (7:142) or Galilei (1967, p. 116).

59. See Favaro (7:143–44) or Galilei (1967, pp. 117–18).

60. See Favaro (7:144–45) or Galilei (1967, pp. 118–19).

61. See Favaro (7:146) or Galilei (1967, p. 120).

62. See Favaro (7:349) or Galilei (1967, p. 321).

63. See Favaro (7:370) or Galilei (1967, p. 342).

64. See Favaro (7:374, 381, 382) or Galilei (1967, pp. 346–47, 353–54, 354–55).
65. See Favaro (7:443) or Galilei (1967, p. 417).
66. See Favaro (7:452) or Galilei (1967, p. 426).
67. This is a reference to St. Thomas Aquinas's classic *Summa theologiae.*
68. See Favaro (7:244) or Galilei (1967, p. 218).
69. See Favaro (7:142) or Galilei (1967, p. 116).
70. See Favaro (7:349) or Galilei (1967, p. 321).
71. See Favaro (7:374–75) or Galilei (1967, p. 347).
72. See Favaro (7:443) or Galilei (1967, p. 417).
73. See Favaro (7:349) or Galilei (1967, p. 321).
74. See Favaro (7:374–75) or Galilei (1967, p. 347).
75. Vincenzo Maculano da Firenzuola to Francesco Barberini.
76. The actual date was 12 April (see above).
77. The *Dialogue* is divided into four chapters called "Days" to reflect the dialogical nature of the work.
78. Galileo had been detained at the Inquisition palace (in the prosecutor's quarters) from the time of his first deposition (12 April) to the time of his second one (30 April); at this latter date he had been sent back to the Tuscan embassy with instructions not to go out or socialize.
79. See this document among the earlier Inquisition proceedings presented above.
80. This report bears no date, and its original is placed at the *beginning* of the Vatican manuscripts as a summary of the affair. Favaro (19:293–97), who did some rearranging of the documents, also prints it at the beginning. However, its content makes it evident that it was written sometime between the date of Galileo's third deposition and defense (10 May) and the Inquisition meeting of 16 June 1633 when the pope decided to have Galileo interrogated a fourth time to determine his intention (Favaro 19:282–83 and 360–61; Pagano 1984, pp. 154 and 229).
81. Here the original document refers one to "folio 2." This is only the first of a series of references given by the writer of this report, who also numbered the 103 folios of the documents up to 10 May 1633 in the lower right corner of the recto side; today these numbers are only one of four sets found on the manuscripts, the latest set consisting of printed numerals that were stamped onto the manuscripts in 1923 when these loose manuscripts were bound into a volume; for more details, see Pagano (1984, pp. 2–3 and 55–60). At any rate, this reference corresponds to Galileo's letter to Castelli (21 December 1613), an edited and inaccurate copy of which (by Lorini) is found in the Vatican manuscripts and concerning which see above. In my translation I have simply omitted these references.
82. Here Santillana (1955, p. 280) comments that "this leaves it implied that Galileo came to Rome under summons in 1615, which he had denied explicitly" (in his first deposition); but this inference is questionable since the next sentence of this report states that the warning was given by Bellarmine the next day, which would have been impossible if Galileo had had to come from Florence.

The important point is that the earlier Inquisition proceedings do speak of Galileo being summoned by Cardinal Bellarmine in 1616; see the "Inquisition Minutes (25 February 1616)" and the "Special Injunction (26 February 1616)" above. However, Galileo was already in Rome at the time, and therefore Bellarmine did not have to summon him from Florence.

83. What follows is an almost verbatim transcription of Galileo's second deposition (30 April 1633); the original does not have quotation marks, but I have used them in my translation for the sake of clarity.

84. The exact status, import, and propriety of such a verbal threat is, of course, one of those things that make the Galileo affair such a controversial cause célèbre; see Garzend (1913), Giacchi (1942), Langford (1966, pp. 150–51), and Santillana (1955, pp. 292–301). The Inquisition manual had a whole chapter "On the Manner of Interrogating Culprits by Torture" (Masini 1621, pp. 120–51), which began by saying: "The culprit having denied the crimes with which he has been charged, and the latter not having been fully proved, in order to learn the truth it is necessary to proceed against him by means of a rigorous examination; in fact, the function of torture is to make up for the shortcomings of witnesses, when they cannot adduce a conclusive proof against the culprit" (Masini 1621, p. 120).

85. That is, the pope's decision, at the Inquisition meeting of 16 June, that Galileo be interrogated about his intention, under the formal threat of torture, and that even if he answered satisfactorily he should make a public abjuration (Favaro 19:282–83 and 360–61; Pagano 1984, pp. 154 and 229–30; Santillana 1955, pp. 292–93; Langford 1966, pp. 150–51).

86. "Vehement suspicion of heresy" was a technical term which meant much more than it may sound to our modern ears; in fact, it seems that it was a specific category of crime, second in seriousness only to formal heresy. See Masini (1621) and section 3 of the Introduction.

87. Note that only seven of the ten cardinals in the commission signed the sentence; the three that did not are Borgia, Francesco Barberini, and Zacchia. This fact is certainly worthy of further reflection and speculation; see, for example, Santillana (1955, pp. 310–11), Langford (1966, p. 153), and Redondi (1983, p. 328).

Selected Bibliography

Agassi, Joseph. 1971. "On Explaining the Trial of Galileo." *Organon* 8:137–66.

Aiton, E. J. 1954. "Galileo's Theory of the Tides." *Annals of Science* 10:44–57.

Aiton, E. J. 1963. "On Galileo and the Earth-Moon System." *Isis* 54:265–66.

Aiton, E. J. 1965. "Galileo and the Theory of the Tides." *Isis* 56:56–61.

Apollonia, L. d'. 1964. "L'affaire Galilée." *Relations* 24:168–70.

Augustinus, Aurelius. 1894. *De Genesi ad Litteram Libri Duodecimi.* In *Corpus Scriptorum Ecclesiasticorum Latinorum*, edited by Iosephus Zycha. Vol. 28, pt. 1, pp. 1–456. Vienna: Tempsky.

Baldini, Ugo. 1984. "L'Astronomia del Cardinale Bellarmino." In Galluzzi (1984), pp. 293–305.

Baldini, Ugo, and George V. Coyne, editors. 1984. *The Louvain Lectures (Lectiones Lovanienses) of Bellarmine and the Autograph Copy of His 1616 Declaration to Galileo.* Vatican Observatory Publications, Studi Galileiani, Research Studies Promoted by the Study Group Constituted by John Paul II, vol. 1, no. 2. Vatican City: Specola Vaticana.

Barone, Francesco. 1972. "Galileo e la logica." In Maccagni (1972), pp. 52–70.

Bentley, Eric. 1966. "Introduction: The Science Fiction of Bertolt Brecht." In Brecht (1966), pp. 7–42.

Berti, Domenico. 1876. *Il processo originale di Galileo Galilei pubblicato per la prima volta.* Rome.

Berti, Domenico. 1878. *Il processo originale di Galileo Galilei. Nuova edizione accresciuta, corretta e preceduta da un'avvertenza.* Rome.

Bonansea, Bernardino M. 1986. "Campanella's Defense of Galileo." In Wallace (1986), pp. 205–39.

Bourke, Vernon J., editor. 1964. *The Essential Augustine.* New York: New American Library.

Brecht, Bertolt. 1966. *Galileo*. Edited by Eric Bentley. New York: Grove Press.

Brodrick, James. 1961. *Robert Bellarmine, Saint and Scholar*. Westminster, Md.: Newman.

Brophy, Liam. 1961. "The Galileo Affair." *Apostle* 39 (February 1961):26–27.

Brophy, Liam. 1964. "The End of the Galileo Affair." *Social Justice Review* 57:79–82.

Burstyn, Harold L. 1962. "Galileo's Attempt to Prove That the Earth Moves." *Isis* 53:161–85.

Burstyn, Harold L. 1963. "Galileo and the Earth-Moon System." *Isis* 54:400–401.

Burstyn, Harold L. 1965. "Galileo and the Theory of the Tides." *Isis* 56:61–63.

Campanella, Thomas. 1937. *The Defense of Galileo*. Translated by Grant McColley. Smith College Studies in History, vol. 22, nos. 3–4.

Carugo, Adriano, and A. C. Crombie. 1983. "The Jesuits and Galileo's Ideas of Science and Nature." *Annali dell'Istituto e Museo di Storia della Scienza* 8:3–68.

Chiaramonti, Scipione. 1633. *Difesa . . . al suo Anticone e Libro delle tre nuove Stelle*. Florence: Landini.

Coffa, José A. 1968. "Galileo's Concept of Inertia." *Physis* 10:261–81.

Copernicus, Nicolaus. 1976. *On the Revolutions of the Heavenly Spheres*. Translated by A. M. Duncan. Newton Abbot: David & Charles.

Coyne, G. V., and U. Baldini. 1985. "The Young Bellarmine's Thoughts on World Systems." In Coyne et al. (1985), pp. 103–11.

Coyne, G. V., M. Heller, and J. Zycinski, editors. 1985. *The Galileo Affair: A Meeting of Faith and Science*. Proceedings of the Cracow Conference, 24–27 May 1984. Vatican City: Specola Vaticana.

Crombie, A. C. 1975. "Sources of Galileo's Early Natural Philosophy." In Righini Bonelli and Shea (1975), pp. 157–175 and 303–5.

D'Addio, Mario. 1984. "Alcune fasi dell'istruttoria del processo a Galileo." *L'Osservatore Romano*, 2 March 1984, p. 2.

D'Addio, Mario. 1985. *Considerazioni sui processi a Galileo*. Rome: Herder Editrice e Libreria.

Daniel-Rops, Henri. 1963. "Que faut-il penser de l'affaire Galilée?" *Historia* 33:502–5.

Danto, Arthur, and Sidney Morgenbesser, editors. 1960. *Philosophy of Science*. New York: World Publishing Company.

Del Lungo, Isidoro, and Antonio Favaro, editors. 1915. *Dal carteggio e dai documenti. Pagine di vita di Galileo*. Florence: Sansoni.

Drake, Stillman, editor. 1957. *Discoveries and Opinions of Galileo*. Garden City: Doubleday.

Drake, Stillman. 1965. "The Galileo–Bellarmine Meeting: A Historical Speculation." In Geymonat (1965), pp. 205–20.

Drake, Stillman. 1970. *Galileo Studies*. Ann Arbor: University of Michigan Press.

Drake, Stillman, editor. 1976. *Galileo Against the Philosophers*. Los Angeles: Zeitlin & Ver Brugge.

Drake, Stillman. 1978. *Galileo at Work: His Scientific Biography.* Chicago: University of Chicago Press.

Drake, Stillman. 1979. "History of Science and Tide Theories." *Physis* 21:61–69.

Drake, Stillman. 1980. *Galileo.* New York: Hill & Wang.

Drake, Stillman. 1983. *Telescopes, Tides, and Tactics: A Galilean Dialogue About the Starry Messenger and Systems of the World.* Chicago: University of Chicago Press.

Drake, Stillman. 1986. "Reexamining Galileo's *Dialogue*." In Wallace (1986), pp. 155–75.

Drake, Stillman, and C. D. O'Malley, editors and translators. 1960. *The Controversy on the Comets of 1618.* Philadelphia: University of Pennsylvania Press.

Dreyer, J. L. E. 1953. *A History of Astronomy from Thales to Kepler.* Revised by W. H. Stahl. New York: Dover.

Dubarle, Dominique. 1964. "Le dossier Galilée." *Signes du Temps* 14:21–26.

Duhem, Pierre. 1969. *To Save the Phenomena: An Essay on the Idea of Physical Theory from Plato to Galileo.* Translated by E. Dolan and C. Maschler. Chicago: University of Chicago Press.

Einstein, Albert. 1967. "Foreword." In Galilei (1967), pp. vi–xx.

The English Hymnal with Tunes. London: Oxford University Press, 1933.

Epinois, Henri de L'. 1867. *Galilée, son procès, sa condemnation d'après des documents inédits.* Extrait de la *Revue des questiones historiques.* Paris.

Epinois, Henri de L'. 1877. *Les pièces du procès de Galilée précédées d'un avant-propos.* Rome-Paris.

Fabris, Rinaldo. 1986. *Galileo Galilei e gli orientamenti esegetici del suo tempo.* Vatican City: Pontificia Academia Scientiarum.

Favaro, Antonio, editor. 1890–1909. *Le Opere di Galileo Galilei.* 20 vols. National Edition of Galileo's collected works. Florence: Barbèra.

Favaro, Antonio. 1902a. "I documenti del processo di Galileo." *Atti del Reale Istituto Veneto di Scienze, Lettere ed Arti,* vol. 61, pt. 2, pp. 757–806. Venice: Ferrari.

Favaro, Antonio. 1902b. *Il processo di Galileo estratto dal volume XIX della Edizione Nazionale delle Opere di Galileo Galilei.* Edition of thirty copies. Florence: Barbèra.

Favaro, Antonio. 1907a. *Galileo e l'Inquisizione. Documenti del processo galileiano esistenti nell'Archivio del S. Uffizio e nell'Archivio Segreto Vaticano, per la prima volta integralmente pubblicati.* Florence: Barbèra.

Favaro, Antonio. 1907b. *Regesto biografico galileiano dalla Edizione Nazionale delle Opere.* Florence: Barbèra.

Favaro, Antonio, and Isidoro Del Lungo, editors. 1911. *La prosa di Galileo.* Florence: Sansoni.

Ferrone, Vincenzo, and Massimo Firpo. 1985a. "Galileo tra inquisitori e microstorici." *Rivista storica italiana* 97:177–238.

Ferrone, Vincenzo, and Massimo Firpo. 1985b. "Replica." *Rivista storica italiana* 97:957–68.

Ferrone, Vincenzo, and Massimo Firpo. 1986. "From Inquisitors to Microhistorians." *Journal of Modern History* 58:485–524.

Feyerabend, Paul K. 1978. *Against Method*. New York: Schocken Books.

Feyerabend, Paul K. 1985. "Galileo and the Tyranny of Truth." In Coyne et al. (1985), pp. 155–66.

Finocchiaro, Maurice A. 1973. *History of Science as Explanation*. Detroit: Wayne State University Press.

Finocchiaro, Maurice A. 1980. *Galileo and the Art of Reasoning*. Boston Studies in the Philosophy of Science, vol. 61. Dordrecht: Reidel.

Finocchiaro, Maurice A. 1986a. "Review of Coyne et al.'s *The Galileo Affair*." *Isis* 77:192.

Finocchiaro, Maurice A. 1986b. "The Methodological Background to Galileo's Trial." In Wallace (1986), pp. 241–72.

Finocchiaro, Maurice A. 1986c. "Toward a Philosophical Reinterpretation of the Galileo Affair." *Nuncius: Annali di Storia della Scienza* 1:189–202.

Finocchiaro, Maurice A. 1988a. "Review of Baldini and Coyne's *Louvain Lectures (Lectiones Lovanienses) of Bellarmine* and of Coyne et al.'s *The Galileo Affair*." *Journal of the History of Philosophy* 26:149–51.

Finocchiaro, Maurice A. 1988b. "Galileo's Copernicanism and the Acceptability of Guiding Assumptions." In *Scrutinizing Science: Empirical Studies of Scientific Change*, edited by Arthur Donovan, Larry Laudan, and Rachel Laudan, pp. 49–67. Dordrecht: Reidel.

Flora, Ferdinando, editor. 1953. *Galileo Galilei: Opere*. Milan: Riccardo Ricciardi Editore.

Galilei, Galileo. 1632. *Dialogo . . . sopra i due Massimi Sistemi del mondo, tolemaico e copernicano*. Florence: Landini.

Galilei, Galileo. 1890–1909. *Opere*. 20 vols. National Edition edited by Antonio Favaro. Florence: Barbèra.

Galilei, Galileo. 1953. *Dialogue on the Great World Systems*. Salusbury's translation revised by G. de Santillana. Chicago: University of Chicago Press.

Galilei, Galileo. 1957. *La prosa*. Edited by I. Del Lungo and A. Favaro. New preface by Cesare Luporini. Florence: Sansoni.

Galilei, Galileo. 1967. *Dialogue Concerning the Two Chief World Systems*. Translated by Stillman Drake. 2nd ed. Berkeley: University of California Press.

Galilei, Galileo. 1968. *Dal carteggio e dai documenti. Pagine di vita di Galileo*. New preface by E. Garin. Florence: Sansoni.

Galilei, Galileo. 1974. *Two New Sciences*. Translated with introduction and notes by Stillman Drake. Madison: University of Wisconsin Press.

Galilei, Galileo. 1988. *Tractatio de praecognitionibus et praecognitis* and *Tractatio de demonstratione*. Translated from the Latin autograph by William F. Edwards, with an introduction, notes, and commentary by William A. Wallace. Padua: Editrice Antenore.

Galluzzi, Paolo, editor. 1984. *Novità celesti e crisi del sapere: Atti del Convegno Internazionale di Studi Galileiani*. Florence: Barbèra.

Garin, Eugenio. 1971. "A Proposito del Copernico." *Rivista critica di storia della filosofia* 26:83–87.

Garin, Eugenio. 1975. "Alle origini della polemica anticopernicana." In *Colloquia Copernicana*, vol. 2 (Studia Copernicana, vol. 6), pp. 31–42. Wroclaw: Ossolineum.

Garzend, Leon. 1913. *L'Inquisition et l'hérésie*. Paris: Desclée de Bouwer.

Gebler, Karl von. 1876. *Galileo Galilei und die römische Curie nach den autentischen Quellen*. Stuttgart.

Gebler, Karl von. 1877. *Die Acten des Galileischen Processes nach der vaticanischen Handschrift*. Stuttgart.

Gebler, Karl von. 1879. *Galileo Galilei and the Roman Curia*. Translated by Mrs. George Sturge. London: C. K. Paul. Reprint. Merrick, N.Y.: Richwood Publishing Co., 1977.

Geymonat, Ludovico. 1962. *Galileo Galilei*. 2nd ed. Turin: Einaudi.

Geymonat, Ludovico. 1965. *Galileo Galilei: A Biography and Inquiry into His Philosophy of Science*. Translated by Stillman Drake. New York: McGraw-Hill.

Gherardi, Silvestro. 1870. *Il processo Galilei riveduto sopra documenti di nuova fonte*. Estratto dalla *Rivista Europea* 3(1):3–37 and 3(3):398–410. Florence.

Giacchi, Orio. 1942. "Considerazioni giuridiche sui due processi contro Galileo." In *Nel terzo centenario della morte di Galileo Galilei*, edited by the Università Cattolica del Sacro Cuore, pp. 383–406. Milan: Vita e Pensiero.

Gingerich, Owen. 1982. "The Galileo Affair." *Scientific American* (August 1982):132–43.

John Paul II. 1979. "Deep Harmony Which Unites the Truths of Science with the Truths of Faith." *L'Osservatore Romano* (weekly edition in English), 26 November 1979, p. 9. (Reprinted in Poupard 1984b, pp. 271–77, and in Poupard 1987, pp. 195–200.)

Julian, John, editor. 1892. *A Dictionary of Hymnology*. New York: Charles Scribner's Sons.

Kempfi, Andrzej. 1982. "Tolosani versus Copernicus." *Organon*, number 16/17, 1981/1982, pp. 239–54.

Koestler, Arthur. 1959. *The Sleepwalkers: A History of Man's Changing Vision of the Universe*. New York: Macmillan.

Koyré, Alexandre. 1966. *Etudes galiléennes*. Reprint of 1939 edition. Paris: Hermann.

Koyré, Alexandre. 1978. *Galileo Studies*. Translated by John Mepham. Hassocks, England: Harvester Press.

Kuhn, Thomas S. 1957. *The Copernican Revolution*. Cambridge: Harvard University Press.

Lakatos, Imre, and E. Zahar. 1975. "Why Did Copernicus' Research Program Supersede Ptolemy's?" In *The Copernican Achievement*, edited by Robert S. Westman, pp. 354–83. Berkeley: University of California Press.

Langford, Jerome J. 1966. *Galileo, Science and the Church*. Ann Arbor: University of Michigan Press.

Lenoble, Robert. 1956. "L'affaire Galilée: les forces en présence." *Recherches et Débats* 17:154–66.

Lerner, Lawrence S., and Edward A. Gosselin. 1986. "Galileo and the Specter of Bruno." *Scientific American* (November 1986):126–33.

Lindberg, David C., and Ronald L. Numbers, editors. 1986. *God and Nature.* Berkeley: University of California Press.

Maccagni, Carlo, editor. 1972. *Saggi su Galileo Galilei.* Publicazioni del Comitato Nazionale per le Manifestazioni Celebrative del IV Centenario della Nascita di Galileo Galilei, vol. 3, tome 2. Florence: Barbèra.

Mangiagalli, Maurizio. 1984. "Review of Redondi's *Galileo eretico.*" *Rivista di filosofia neo-scolastica* 76:478–87.

Marini, Marino. 1850. *Galileo e l'Inquisizione. Memorie storico-critiche dirette alla Romana Accademia di Archeologia.* Rome.

Martin, R. Niall D. 1987. "Saving Duhem and Galileo: Duhemian Methodology and the Saving of the Phenomena." *History of Science* 25: 301–19.

Martini, Carlo M. 1972. "Galileo e la teologia." In Maccagni (1972), pp. 401–15.

Masini, Eliseo. 1621. *Sacro arsenale overo Prattica dell'officio della Santa Inquisizione.* Genoa: Appresso Giuseppe Pavoni.

McMullin, Ernan. 1967a. "Introduction" to McMullin (1967b).

McMullin, Ernan, editor. 1967b. *Galileo: Man of Science.* New York: Basic Books.

Meagher, Robert E. 1978. *An Introduction to Augustine.* New York: New York University Press.

Mertz, Donald W. 1980. "On Galileo's Method of Causal Proportionality." *Studies in History and Philosophy of Science* 11:229–42.

Millman, Arthur B. 1976. "The Plausibility of Research Programs." In *PSA 1976: Proceedings of the 1976 Biennial Meeting of the Philosophy of Science Association*, edited by Frederick Suppe and P. D. Asquith, vol. 1, pp. 140–48. East Lansing, Mich.: Philosophy of Science Association.

Milton, John. 1927. *Areopagitica and Other Prose Works.* London: J. M. Dent & Sons Ltd.

Montinari, Maddalena, editor. 1983. *Sulla libertà della scienza e l'autorità delle Scritture.* Rome: Edizioni Theoria.

Morpurgo-Tagliabue, Guido. 1981. *I processi di Galileo e l'epistemologia.* Rome: Armando.

Morpurgo-Tagliabue, Guido. 1984. "Galileo: quale eresia?" *Rivista di storia della filosofia* 39:741–50.

Morpurgo-Tagliabue, Guido. 1985. "Sussiste ancora una questione galileiana?" *La nuova civiltà delle macchine*, vol. 3, nos. 1–2, pp. 91–99.

Moss, Jean Dietz. 1983. "Galileo's *Letter to Christina*: Some Rhetorical Considerations." *Renaissance Quarterly* 36:547–76.

Moss, Jean Dietz. 1984. "Galileo's Rhetorical Strategies in Defense of Copernicanism." In Galluzzi (1984), pp. 95–103.

Moss, Jean Dietz. 1986. "The Rhetoric of Proof in Galileo's Writings on the Copernican System." In Wallace (1986), pp. 179–204.

Mourant, John A., editor. 1964. *Introduction to the Philosophy of Saint Augustine: Selected Readings and Commentary.* University Park: Pennsylvania State University Press.

Namer, Emile, editor. 1975. *L'affaire Galilée.* Paris: Editions Gallimard/Julliard.

O'Connor, D. J. 1952. *John Locke.* London: Penguin Books.

O'Meara, John J., editor. 1973. *An Augustinian Reader.* Garden City: Doubleday.

Pagano, Sergio M., editor. 1984. *I documenti del processo di Galileo Galilei.* Vatican City: Pontificia Academia Scientiarum.

Paschini, Pio. 1964. *Vita e opere di Galileo Galilei.* 2 vols. Vatican City: Pontificia Academia Scientiarum.

Petit, Leon. 1964. "L'affaire Galilée vue par Descartes et Pascal." *XVIIIe siècle* 28:231–39.

Pieralisi, Sante. 1875. *Urbano VIII e Galileo Galilei.* Rome.

Pizzorno, Benedetto. 1979. "Indagine sulla logica galileiana." *Physis* 21:71–102.

Portalie, Eugene, S. J. 1960. *A Guide to the Thought of Saint Augustine.* Translated by R. J. Bastian, S. J. Chicago: Henry Regnery.

Poupard, Paul. 1984a. "Introduzione." In Poupard (1984b), pp. 7–20.

Poupard, Paul, editor. 1984b. *Galileo Galilei: 350 anni di storia (1633–1983).* Rome: Edizioni Piemme di Pietro Marietti.

Poupard, Paul, editor. 1987. *Galileo Galilei: Toward a Resolution of 350 Years of Debate—1633–1983.* Translated by I. Campbell. Pittsburgh: Duquesne University Press.

Przywara, Erich, editor. 1958. *An Augustinian Synthesis.* New York: Harper.

Ranke, Leopold. 1841. *The Ecclesiastical and Political History of the Popes of Rome.* 2 vols. Translated by S. Austin. Philadelphia: Lea & Blanchard.

Redondi, Pietro. 1983. *Galileo eretico.* Turin: Einaudi.

Redondi, Pietro. 1985a. *Galilée hérétique.* Translation of Redondi (1983) by Monique Aymard. Paris: Gallimard.

Redondi, Pietro. 1985b. "*Galileo eretico*: Anatema." *Rivista storica italiana* 97:934–56.

Redondi, Pietro. 1985c. "Post-scriptum." In Redondi (1985a), pp. 425–32.

Redondi, Pietro. 1987. *Galileo Heretic.* Translation of Redondi (1983) by Raymond Rosenthal. Princeton: Princeton University Press.

Ricci-Riccardi, Antonio. 1902. *Galileo Galilei e Fra Tommaso Caccini.* Florence: Successori Le Monnier.

Righini Bonelli, Maria Luisa, and William R. Shea, editors. 1975. *Reason, Experiment, and Mysticism in the Scientific Revolution.* New York: Science History Publications.

Rochot, Bernard. 1956. "Remarques sur l'affaire Galilée." *XVIIIe siècle* 30:134–43.

Ronchi, Vasco. 1958. *Il cannochiale di Galilei e la scienza del seicento.* 2nd ed. Turin: Einaudi.

Rosen, Edward. 1958. "Galileo's Misstatements About Copernicus." *Isis* 49:319–30.

Rosen, Edward. 1975. "Was Copernicus's *Revolutions* Approved by the Pope?" *Journal of the History of Ideas* 36:531–42.

Russo, François. 1968. "Lettre de Galilée à Christine de Lorraine, Grande-Duchesse de Toscane (1615)." In *Galilée: Aspects de sa vie et de son oeuvre*, pp. 324–59. Paris: Presses Universitaires de France.

Santillana, Giorgio de. 1955. *The Crime of Galileo*. Chicago: University of Chicago Press.

Santillana, Giorgio de. 1965. "Reply to Stillman Drake." In Geymonat (1965), pp. 221–25.

Santillana, Giorgio de. 1972. "Nuove ipotesi sul processo di Galileo." In Maccagni (1972), pp. 474–86.

Schaff, Philip, and Henry Wace, editors. 1893. *A Select Library of Nicene and Post-Nicene Fathers*. 2nd series, vol. 6: *St. Jerome: Letters and Select Works*. New York: Christian Literature Company.

Schroeder, H. J., translator and editor. 1978. *Canons and Decrees of the Council of Trent*. St. Louis: Herder, 1941. Reprint. Rockford, Ill.: Tan Books and Publishers, 1978.

Shapere, Dudley. 1974. *Galileo: A Philosophical Study*. Chicago: University of Chicago Press.

Shea, William R. 1972. *Galileo's Intellectual Revolution: Middle Period, 1610–1632*. New York: Science History Publications.

Shea, William R. 1986. "Galileo and the Church." In Lindberg and Numbers (1986), pp. 114–35.

Soccorsi, Filippo. 1947. *Il processo di Galileo*. Rome: Edizioni "La Civiltà Cattolica."

Sosio, Libero. 1970. "Galileo e la cosmologia." Introduction to Galileo Galilei, *Dialogo sopra i due massimi sistemi*, edited by L. Sosio, pp. ix–lxxxvii. Turin: Einaudi.

Tertullian, Quintus S. F. 1972. *Adversus Marcionem*. Edited and translated by Ernest Evans. Oxford: Clarendon Press.

Venturi, G. B. 1818–1821. *Memorie e lettere finora inedite o disperse di Galileo Galilei ordinate ed illustrate con annotazioni*. 2 vols. Modena.

Viganò, Mario. 1969. *Il mancato dialogo tra Galileo e i teologi*. Rome: Edizioni "La Civiltà Cattolica."

Wallace, William A., editor and translator. 1977. *Galileo's Early Notebooks: The Physical Questions*. Notre Dame: University of Notre Dame Press.

Wallace, William A. 1981a. *Prelude to Galileo: Essays on Medieval and Sixteenth-Century Sources of Galileo's Thought*. Dordrecht: Reidel.

Wallace, William A. 1981b. "Aristotle and Galileo: The Use of *Hupothesis* (*Suppositio*) in Scientific Reasoning." In *Studies in Aristotle*, edited by D. J. O'Meara. Studies in Philosophy and the History of Philosophy, vol. 9, pp. 47–77. Washington, D.C.: Catholic University of America Press.

Wallace, William A. 1984. *Galileo and His Sources: The Heritage of the Collegio Romano in Galileo's Science*. Princeton: Princeton University Press.

Wallace, William A., editor. 1986. *Reinterpreting Galileo*. Studies in Philosophy and the History of Philosophy, vol. 15. Washington, D.C.: Catholic University of America Press.

Westfall, Richard S. 1984. "Galileo and the *Accademia dei Lincei.*" In Galluzzi (1984), pp. 189–200.

Westfall, Richard S. 1985. "Science and Patronage: Galileo and the Telescope." *Isis* 76: 11–30.

Westman, Robert S. 1986. "The Copernicans and the Churches." In Lindberg and Numbers (1986), pp. 76–113.

Westman, Robert S. 1987. "La préface de Copernic au pape: esthétique humaniste et réforme de l'église. *History and Technology* 4:359–78.

Wisan, Winifred L. 1984. "On Argument *Ex Suppositione* Falsa." *Studies in History and Philosophy of Science* 15:227–36.

Index

Compositor:	G & S Typesetters, Inc.
Text:	10/13 Sabon
Display:	Sabon
Printer:	Maple-Vail Book Mfg. Group
Binder:	Maple-Vail Book Mfg. Group